HIGH ENERGY PHYSICS 1985

Proceedings of the Yale Theoretical
Advanced Study Institute

HIGH ENERGY PHYSICS, 1985

Editors

MARK J BOWICK
FEZA GÜRSEY

VOL. 2

World Scientific

Published by

World Scientific Publishing Co. Pte. Ltd.
P. O. Box 128, Farrer Road, Singapore 9128

Library of Congress Cataloging-in-Publication data is available.

HIGH ENERGY PHYSICS 1985 (VOL. 2)
Copyright © 1986 by World Scientific Publishing Co Pte Ltd.

All rights reserved. This book, or parts thereof, may not be reproduced in any form or by any means, electronic or mechanical, including photocopying, recording or any information storage and retrieval system now known or to be invented, without written permission from the Publisher.

ISBN 9971-50-006-X
     9971-50-007-8 pbk

Printed in Singapore by Fu Loong Lithographer Pte Ltd.

CONTENTS

VOLUME 1

Introduction	v
A. P. BALACHANDRAN: Skyrmions	1
R. JACKIW: Anomalies and Topology	83
A. J. NIEMI: Fractional Fermion Number	125
O. ALVAREZ: Topological Methods in Field Theory	154
M. E. PESKIN: An Introduction to the Theory of Strings	217
P. RAMOND: Group Theory for String States	274
A. H. MUELLER: Lectures on Perturbative QCD	324
M. K. GAILLARD: Aspects of the Physics of Strongly Interacting w's and z's	404
L. M. KRAUSS: Axions . . . The Search Continues	431

VOLUME 2

J. B. HARTLE: Quantum Cosmology	471
P. J. STEINHARDT: Inflationary Cosmology	567
A. CHODOS: Kaluza-Klein Theories	618
S. GEER: Physics Results from the CERN Super Proton Synchrotron Proton-Antiproton Collider	649
E. EICHTEN: Theoretical Expectations at Collider Energies	709
List of Participants	973

QUANTUM COSMOLOGY

James B. Hartle
Department of Physics, University of California
Santa Barbara, CA 93106

1. INTRODUCTION

The traditional enterprise of cosmology has been to construct a model of the universe which agrees with our observations on the largest scales and which, when evolved backwards according to the laws of fundamental physics, gives a consistent historical picture of how the universe came to be the way it is today. Our observations tell us that the universe consists of matter and radiation. The matter that we see in galaxies is distributed roughly homogeneously and isotropically on the largest scales. The cosmic background radiation, in which we see a picture of the universe at an early stage, is remarkably isotropic. As a first approximation, we are thus led naturally to the Friedman-Robertson-Walker cosmological models in which the symmetries of homogeneity and isotropy are enforced exactly. Evolved backward in time using Einstein's gravitational theory and the laws of microscopic physics these models provide a consistent history of the universe. Among other things, they describe the evolution of the background radiation, the origin of the primordial elements, the evolution of the fluctuations which became the galaxies, and perhaps the origin of the baryons. The initial condition implied by the extrapolation is an early state in which the matter is in thermal equilibrium with high temperature and density, distributed homogeneously and isotropically but containing the seeds of condensations later to become galaxies.

The Friedman-Robertson-Walker models are successful and they are simple. Their success and simplicity raise the issue of why does the universe have the properties it does? Can we explain the Friedman-Robertson-Walker models? This is a very different kind of issue from the essentially descriptive questions traditionally asked in cosmology. In effect one is asking for a theory of initial conditions. These lectures are about one approach to this problem: the search for a theory of initial conditions in the application of quantum gravity to cosmology - in two words they are about quantum cosmology.

The lectures are not intended as a review of all models of the universe which involve quantum mechanics or even of those which deal directly with the issue of initial conditions. The subject, although already large, is not yet connected enough to make such a review feasible in the space available.[1] Rather we shall explore a specific proposal for the quantum state of the universe developed by Stephen Hawking and his collaborators.[2] In the process we shall be able to review much about the general issues.

To state the proposal for the quantum state of the universe we shall need some of the framework of quantum gravity. This we describe in Section 3. We shall develop the idea and compare its predictions with observations in Section 6, but, in order to know where we are going, we shall first review the observations we hope to explain in Section 2.

2. THE UNIVERSE TODAY AND THE PROBLEM OF ITS INITIAL CONDITIONS

2.1 Observations

The variety and detail of the observations now available which bear on the structure of our universe in the large is one of the most impressive achievements of contemporary astronomy. The relationships between these observations are complex and deriving an understanding of the universe in the large from them is a complex theoretical story. Emerging from this analysis, however, is a picture of striking simplicity on the largest scales. In this section we shall summarize this picture in a few "observational facts" and briefly indicate the nature of the supporting evidence for each one. These are the facts one seeks to explain in a theory of initial conditions. We can only adumbrate the arguments for these observations here and cannot hope to give a complete list of references to them. For greater detail and references the reader is encouraged to consult the many reviews of this subject.[3]

Fact (1). <u>Spacetime is four dimensional with Euclidean topology</u>

This is so built into our fundamental physics that we usually take it as granted. It is important to remember however that all aspects of geometry have an observational basis.

Fact (2). <u>The universe is large, old and getting bigger</u>

At moderate distances galaxies recede from each other according to Hubble's law

$$\begin{pmatrix} \text{velocity} \\ \text{of recession} \end{pmatrix} = H_0 \text{ (distance apart)} , \quad (2.1)$$

where H_0 is somewhere between 40 and 100 (km/sec)/Mpc.

[A pc is 3.09×10^{18} cm. A Mpc is 10^6 pc.] Inverted, Hubble's law gives us a connection between distance and redshift. Since H_0 is uncertain distances are often quoted as a multiple of h^{-1} where h is $H/[100 (km/sec)/Mpc]$. The background radiation originates at distances of order $c/H_0 \sim 3000\ h^{-1}$ Mpc (the Hubble distance) from us and at times of order $1/H_0 \sim 10^{10}$ yrs. (the Hubble time) ago. These are the largest scales which are directly accessible to observation today. It is perhaps a trivial observation, but these are not the scales of elementary particle physics.

Fact (3). <u>The universe contains matter and radiation distributed homogeneously and isotropically on the largest scales</u>

Direct evidence for the homogeneity of the universe is hard to come by. Ideally one would like to make a three dimensional map showing the distribution of galaxies and this involves measuring distances. Such surveys have been made but only out to limited distances (~100 Mpc). The test that probes homogeneity on the largest scales is the oldest - counts of galaxies vs. limiting flux. One can easily show that if there are several populations of objects distributed <u>uniformly</u> in flat three dimensional space, then the number of objects counted with flux f greater than some limiting flux, f_0, should vary with f_0 as

$$N(f > f_0) \propto f_0^{-3/2} \quad , \qquad (2.2)$$

with calculable corrections for spatial curvature. Modern surveys[4] which probe out to depths comparable to the Hubble distance yield approximate agreement with this law.

If we accept the Copernican principle that we are not at a preferred position in the universe, and there is no evidence that we are, then evidence for isotropy becomes evidence for homogeneity. Evidence for the isotropy of the universe comes from angular surveys of its major constituents.

Most directly there are the galaxies. A plot on the sky of the Shane-Wirtanen catalog of the 10^6 galaxies contained in an effective depth of several hundred Mpc is as close as we can come to a picture of the universe today.[5] It is roughly isotropic on large scales but clearly exhibits structure. A quantitative measure of the isotropy is the galaxy-galaxy angular correlation function. This is the excess probability for finding a second galaxy at some fixed angle from any given one. It is conveniently quoted in terms of a spatial correlation function $\xi(r)$ which would produce the same result assuming homogeneity. $\xi(r)$ is about unity at 7 h^{-1} Mpc and decreases to a few 1/10 ths by 20 h^{-1} Mpc. The galaxy distribution is thus essentially isotropic at large scales.

Radio sources are distributed across the sky in an essentially uniform way. The diffuse X-ray background is isotropic to a few percent at angular scales of $5°$. Since a significant fraction of this radiation comes from distant quasars this becomes a test of isotropy on large scales.

The temperature distribution of the $2.7°K$ cosmic background radiation provides the most accurate test of the isotropy of the universe and the one which probes this feature on the largest scales and therefore the earliest times. The anisotropy in the temperature, $\Delta T/T$, has been well measured on a number of different angular scales.[6] There is a purely dipole anisotropy which is attributable to the motion of our solar system with respect to the rest frame of the radiation. If this is subtracted out no anisotropy has been detected in the residual component on any scale.[7,8] The current best limit on the quadrupole anisotropy, for example, is $(\Delta T/T)_{quadrupole} \lesssim 7 \times 10^{-5}$.

Figure 1 shows this graphically. It is a map of the sky with the dipole anisotropy subtracted out, produced by Lubin and Villela[8] at 3mm. where the background radiation dominates all other sources. It is thus in effect a snapshot of the universe at an age of 10^5 years when the background radiation was emitted. It is essentially featureless.

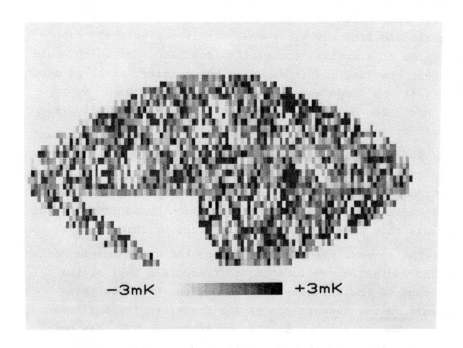

Figure 1. The sky at 3mm. This figure shows the map of the sky at 3mm. observed by Lubin and Villela[8] with the dipole anisotropy removed. The shading in the rectangles, each a few degrees on a side, indicates the temperature deviation from the mean. Since the background radiation is the dominant source of radiation at this wavelength, this is essentially a picture of the universe 300,000 years after the big bang and it is remarkably isotropic.

The observed approximate homogeneity and isotropy on large scales suggest that the Friedman-Robertson-Walker models, in which these symmetries are enforced exactly, should give a good first approximation to the dynamics of the universe. The metric of a spacetime geometry with homogeneous and isotropic spatial sections can, in suitable coordinates, be described by the line element

$$ds^2 = -dt^2 + a^2(t) \left[\frac{dr^2}{1-kr^2} + r^2 d\Omega_2^2 \right] , \qquad (2.3)$$

where $d\Omega_2^2$ is the metric on the unit two sphere. The spatial geometry is open with negative curvature if $k = -1$, open and flat if $k = 0$ and closed with the geometry of a three-sphere if $k = +1$.

All the geometrical information about the dynamics of the universe is contained in the scale factor $a(t)$. Einstein's equation for perfect fluid matter with energy density ρ and cosmological constant Λ implies

$$\left(\frac{\dot{a}}{a}\right)^2 = -\frac{k}{a^2} + \frac{\Lambda}{3} + \frac{8\pi G}{3} \rho . \qquad (2.4)$$

This equation plus the constituative relations of the matter are enough to extrapolate the dynamics of the universe forward and backward in time given the constants k and Λ and the present values of a and ρ or equivalently the present values of ρ and $\dot{a}/a$. The present value of $\dot{a}/a$ is the Hubble constant H_0. It is uncertain because the extragalactic distance scale is uncertain, but most determinations fall in the range $40 - 100$ (km/sec)/Mpc.

Eq. (2.2) shows that, were $\Lambda = 0$, the density today would have to be greater than the critical value

$$\rho_{crit} = \frac{3H_0^2}{8\pi G} , \qquad (2.5)$$

to have a closed ($k = +1$) universe. It has become conventional to quote ρ and even Λ in terms of their dimensionless ratios to this critical density. For example, the present density ρ_0 defines the ratio $\Omega_0 = \rho_0/\rho_{crit}$ and the cosmological constant the ratio $\Omega_\Lambda = (\Lambda/8\pi G)/\rho_{crit}$. We now briefly describe the observational evidence for each of these quantities.

The density in luminous matter, found essentially by counting galaxies whose redshifts and therefore distances are known, corresponds to an Ω of about .01. There is considerable evidence, however, that the universe contains significant amounts of non-luminous matter. The rotational velocity of a galaxy at a given radius from its center can be used to estimate the mass interior to that radius. These velocities do not fall with radius as would be predicted from the density of luminous matter in galaxies. They remain constant as far out as can be measured indicating the presence of a dark component perhaps 10 times more massive than the luminous one. Dynamical analysis of the infall of galaxies towards the center of the Virgo supercluster (of which we are an outlying member) argue for $\Omega_0 \sim .3$ if there is no dumped matter which is non-luminous.[9] Models of the nucleosynthesis of deuterium in the early universe together with its measured abundance today suggest that the Ω corresponding to the density of baryons today is about .1. These arguments suggest a value of Ω_0 of a few tenths. They cannot rule out, however, a larger Ω_0 if there is non-luminous, non-baryonic matter which is not clustered with the galaxies or if there is matter clustered like the galaxies but non-luminous.

It is difficult to measure Ω_Λ from anything other than direct observation of the cosmological expansion. However, it cannot be many orders of magnitude larger than unity or it would imply observable deviations from Newtonian dynamics in clusters of galaxies.[10] Thus there is no direct evidence that $\Lambda = 0$ today. Even an Ω_Λ of 1, however, corresponds to a cosmological constant which is very small on the scale of the Planck mass $m_p = (\hbar c/G)^{1/2}$

$$\Lambda \approx 8.8 \times 10^{-122} m_p^2 \Omega_\Lambda h^2 \quad . \tag{2.6}$$

The available information on the density of energy in the universe is not enough to tell us whether the spatial geometry of the universe is open or closed. It is, however, close to the flat geometry which is the borderline between the two. We might therefore summarize this information in a fourth "observational fact":

Fact (4). <u>The spatial geometry is approximately flat</u>

Fact (5). <u>The spectrum of density fluctuations</u>

The universe is not exactly homogeneous and isotropic. Matter in galaxies is very clumped as measured by the ratio of the difference in their density to the mean density, $\delta\rho/\rho$. The evidence from the background radiation is that earlier the universe was much smoother. The present large scale structure arose from this earlier, smoother distribution through gravitational attraction. At present, direct observations of the background radiation give only upper limits on fluctuations both as to amplitude and spectrum. The amplitude required for those scales where $\delta\rho/\rho \sim 1$ now (superclusters of galaxies) may be found by extrapolating backwards in time using linear perturbation theory and is $(\delta\rho/\rho) \sim 10^{-4}$ at the time the background radiation was emitted. This is consistent with the upper limits. Information on the spectrum can be obtained by assuming appealing candidates at decoupling and extrapolating them forward non-linearly and comparing with the existing large scale structure. The spectrum such that all fluctuations have the same amplitude at the time their scales coincide with the Hubble scale, called the Zel'dovich spectrum, is a popular candidate consistent with all current observations.

Fact (6). <u>The entropy of the universe is low and increasing in the direction of expansion</u>

Today, essentially all of the entropy of matter is in the background radiation. The ratio of the density of entropy s to the density of baryons n_b is

$$s/kn_b \sim 10^9 , \qquad (2.7)$$

so that the total entropy within a Hubble distance is approximately $S/k \sim 10^{87}$ (the word approximately refers to the exponent!). This is a large number but a small fraction of the entropy which could be obtained by clumping all the matter within the Hubble distance into a black hole.[11] A black hole of mass M has entropy $4\pi kGM^2/(\hbar c)$ so that with a reasonable estimate for the mass within the horizon

$$S/k \sim 10^{120} . \qquad (2.8)$$

The 33 orders of magnitude discrepancy between fact and possibility is another way of saying that the universe is still in a reasonably well ordered state. Entropy is increasing and even on the largest scales we seem to see a steady progression from order when the universe is small to disorder when it is large.

We, of course, have more information about the large scale features of the universe than can be summarized in the above six cosmological facts. We observe specific abundances for the elements, a baryon-antibaryon asymmetry, the thermal spectrum of the background radiation and so on. The above list, however, contains those features whose origin is to be found in the earliest stages of our universe.

2.2 Initial Conditions

In most problems in physics we divide the universe up into two parts, the system under consideration and the rest. We use the local laws of physics to solve for the evolution of the system. For example we use Maxwell's equations and Newton's laws of mechanics to predict the evolution of a plasma. The local laws of physics require boundary conditions: sometimes initial conditions, sometimes spatial boundary conditions, sometimes radiative boundary conditions, and often a combination of these. These boundary conditions are set by the physical conditions of those parts of the universe which are not part of the physical system under consideration. There are no particular _laws_ determining these conditions, they are specified by _observations_ of the rest of the universe. The situation is different in cosmology. Boundary conditions are still required to solve the local laws governing the evolution of the universe. They are needed, for example, to solve Einstein's equation (2.2). There is, however, no "rest of the universe" to pass their specification off to. If there is a general specification of these initial conditions it must be part of the laws themselves.

If we extrapolate the Friedman-Robertson-Walker models backward in time we can find initial conditions which give rise to the present universe. What attitude are we to take to these initial conditions? A number of attitudes have been taken. Many of them are summarized in the following four rough categories.

Attitude 1: That's the way it is.

The universe might have been in any one initial state as well as any other. It happens that the one it is in is homogeneous and isotropic on the scales we observe. That's as far as physics can go. Its not the proper subject of physics to explain these initial conditions only to discover what they were.

This is a reasonable but not very adventurous attitude. It certainly has no predictive power concerning what we will see when with increasing time we are able to observe larger and larger regions of the universe. I believe we will only be able to say it is correct when all attempts to explain the initial conditions have failed.

Attitude 2: The conditions which determine the universe are not initial conditions but the fact that we exist.

This attitude is related to the set of ideas called the anthropic principle.[12] The universe must be such as to allow galaxy condensation, star formation, carbon chemistry and life as we know it. This is indeed a restriction on the structure of the universe. Perhaps, if one were given a choice of three or four very different cosmologies one could identify our own using the anthropic principle. As stressed by Penrose,[11] however, the anthropic principle does not seem strong enough to single out the observed universe from among all possibilities. Suppose, for example, the sun had been located in a cloud near the galactic center and we had not been able to make observations of the large scale structure. Would we have been able to predict the large scale homogeneity and isotropy using the anthropic principle?

Attitude 3: Initial conditions are not needed - dynamics does it all.

The idea is that interesting features like the large scale homogeneity and isotropy will arise from any reasonable initial conditions through the action of physical processes over the course of the universe's history. Even if it started in an inhomogeneous and anisotropic state the universe would evolve towards a homogeneous and isotropic one over the scales we can observe it. This is an attractive idea not least because we can achieve determinism with the existing dynamical laws of physics. A variety of physical mechanisms have been proposed to implement this idea beginning with the work of Misner and his co-workers.[13] The most successful mechanism is inflation.[14]

Any dynamical explanation of the large scale structural features has to confront the problem of horizons. We can illustrate this idea for a k = 0 model by rewriting the line element (2.3) in terms of a conformal time coordinate η such that $dt = a d\eta$. Then

$$ds^2 = a^2(\eta)[-d\eta^2 + dr^2 + r^2 d\Omega_2^2] \; , \qquad (2.9)$$

and radial light rays move on 45° lines in an (η, r) plane. Two events are in causal contact if their past light cones intersect each other before they intersect the big bang (Figure 2). Only if two events are in causal contact could they be both influenced by a common event in their past. The horizon size at any time is the largest distance over which two events could have been in causal contact up to that time. As measured in the comoving coordinate r, the horizon radius is

$$r_H = \int_0^\eta d\eta = \int_0^t \frac{dt}{a(t)} \; . \qquad (2.10)$$

Clearly it grows with passing time.

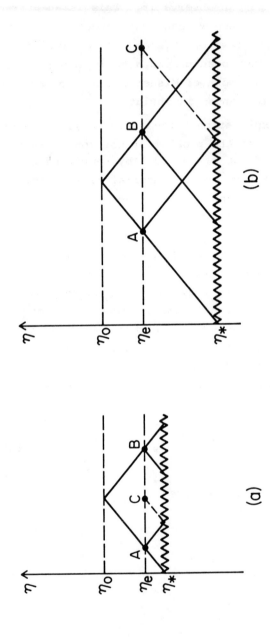

Figure 2. Two possible causal structures. In these diagrams η_0 is the present (conformal) time, η_e is the time of decoupling and η_* the time of the big bang. The history from decoupling to the present is the same in both figures and the times η_0 and η_e have been chosen to correspond. The early histories are different. Were the history like that of (a) two events A and B would not have been in causal contact since the big bang. The size of the observed universe at decoupling AB is larger than the horizon size AC. For a history like that of (b) the horizon size is larger than the size of the observed universe and A and B can be in causal contact.

If we are to have a dynamical explanation of the remarkable isotropy 10^5 years after the big bang, then the different regions from which the radiation was emitted must have been in causal contact. Whether they were able to communicate depends on the history of the universe prior to the time the radiation was emitted i.e. whether there was time enough since the big bang for them to do so. A prerequisite for any dynamical explanation of the observed isotropy of the background radiation is that the horizon size at decoupling be larger than the size of the universe we can observe then.

The most naive extrapolation of the history of the universe to times earlier than decoupling is to assume that the spacetime geometry remains approximately homogeneous and isotropic (the Friedman-Robertson-Walker model), that the matter energy density is dominated by the density of approximately free radiation, and that the evolution is governed by Einstein's equation. With this early history only regions at the time of decoupling now subtending a few degrees on the sky could have been in causal contact since the big bang and no dynamical explanation of the observed isotropy would be possible. This extrapolation, however, is too naive. The horizon can be much larger than the observable universe in models where the geometry is significantly anisotropic before decoupling,[15] as a consequence of quantum effects at the Planck epoch[16] or as a consequence of an inflationary de Sitter-like expansion arising from a matter phase transition at the GUT time. Since this is an important idea it is worthwhile digressing to discuss it briefly.

If the vacuum energy represented by the cosmological constant dominated all other sources of stress energy in Einstein equations the solution would be a geometry called de Sitter space. In particular the solution is

$$ds^2 = -dt^2 + \frac{\cosh^2 Ht}{H^2} d\Omega_3^2 , \qquad (2.11)$$

where $H^2 = \Lambda/3$ and $d\Omega_3^2$ is the metric on the three sphere. de Sitter space is the geometry of the surface of a Lorentz hyperboloid in a 5-dimensional flat, Lorentz signatured spacetime. It contains neither a big bang nor a big crunch. The geometry is non-singular. The spatial three spheres collapse from infinite radius down to a minimum radius

$$a_{min} = H^{-1} = (3/\Lambda)^{1/2} , \qquad (2.12)$$

and then re-expand to infinity. The expansion is exponential - inflationary. During an inflationary expansion the horizon grows at an exponential rate over what we would have guessed from the naive extrapolation. For example, suppose we extrapolate back to a time t_e according to the usual history but before that replace the naive radiation dominated history with an inflationary history back to time t_b. The comoving horizon grows in this epoch according to (2.10) with the revised expansion law. The ratio of the horizon size with inflation to that without is approximately

$$\exp(H(t_e - t_b))/(2Ht_e) , \qquad (2.13)$$

for $t_e \gg t_b$. If $1/H$ is set by a particle physics mass scale it is not difficult to overcome the horizon problem.

In the inflationary history, with only conservative assumptions on the matter physics, the horizon at decoupling could be enormously larger than the observable universe. There is thus opportunity for dynamical processes to act to drive the universe towards isotropy and the inflationary expansion itself provides a mechanism to do this.[17] Further, the region that becomes the observed universe at decoupling is so much smaller than the horizon size at the end of inflation that one might suppose that any reasonable initial conditions which are inhomogeneous on the scale of the horizon will appear smooth on the scale of the observed universe. With an inflationary history no special assumptions on initial conditions are required to explain the observed isotropy. We see the universe as homogeneous and isotropic simply because we do not see a very big part of it.

If the universe is inhomogeneous on a large scale eventually we will find this out. The size of the universe we can observe grows with every second. No dynamical explanation of homogeneity and isotropy can thus ever be an explanation for all time. Eventually we will see outside the horizon and have to face up to the problem of initial conditions. However, if the inflationary history is correct, under even the conservative assumptions mentioned above, we may be able to postpone this discussion for many times the present age of the universe.

Even in the limited region we can observe, no dynamical explanation can ever be a complete explanation for the features of the universe we see. We can always imagine a present state of the observable universe which is highly inhomogeneous. Whatever the laws of geometry and matter, whether they be classical or quantum, whether there are phase transitions or not, these laws can be used to extrapolate this state backward in time and reach some initial condition. No dynamical explanation can, therefore, ever completely exclude a present state of inhomogeneity without some restrictions on the initial conditions. As impressive as they may be in broadening our choice for initial conditions compatible with our present observations, dynamical explanations of these observations can never be complete.

Attitude 4: <u>There is a law of physics specifying the initial condition.</u>

Specification of the boundary conditions is just as much a law of physics as are the dynamical equations governing their evolution. In this view, the question for physics is whether there exists a compelling, simple, and predictive principle which will single out the initial state of our universe. Any search for such a principle is likely to involve the quantum theory of gravity in an essential way.

Classical cosmological spacetimes can either be singular (for example the Friedman model) or non-singular (for example de Sitter space). Depending on the physics of the matter there could either be a big bang or a small bounce. If the past evolution were essentially classical, it would be very difficult to see how to find a principle for the initial conditions. The principle would necessarily be a principle of classical physics and it is difficult to see, for example, what classical principle would single out the spectrum of density fluctuations we are living with.

If there is a singularity in our past then quantum gravity will certainly be important for its structure. Quantum gravity becomes important when the curvature varies significantly over a Planck length, $(\hbar G/c^3)^{1/2} \sim 10^{-33}$ cm. and curvatures of arbitrarily large size are produced in a classical singularity. The big bang singularity gives one a natural place to make a theory of initial conditions, and its quantum fluctuations are a natural starting point for the present spectrum of density fluctuations.

Singularities are not difficult to arrange in general relativity. They are not, for example, artifacts of high symmetry. The singularity theorems of classical general relativity[18] suggest that if we extrapolate the present universe into the past we will <u>generally</u> encounter a singularity provided the matter physics is such that a positive energy condition is satisfied. (This condition is satisfied, for example, in the radiation dominated, pre-phase transition era of the usual inflationary universe.)

It is in the quantum mechanics of the big bang that we shall look for a law of initial conditions. We must therefore now turn to quantum gravity.

3. QUANTUM GRAVITY

3.1 The Problem of Quantum Gravity

We do not possess today a complete, manageable, satisfactorily interpreted, and tested quantum theory of spacetime for application to cosmology. The difficulties with formulating such a theory occur not only at the level of the traditional issues of quantum field theory: What Lagrangian should be used, that of Einstein's well tested classical theory or another with better short distance behavior for which Einstein's theory emerges as a low energy limit? How does one construct a covariant perturbation expansion for the theory? Is this expansion renormalizable and is the resulting scattering theory unitary? If it is not, can the theory still be sensibly implemented through non-perturbative methods? In a quantum theory of spacetime one also encounters difficulties at a more elementary and more fundamental level. What are the physical degrees of freedom of the theory? What variable plays the role of time so central to Hamiltonian quantum mechanics? How does one label the states and what is their Hilbert space? What is the probability interpretation of these states? Is a theory formulated with one time unitarily equivalent to one formulated with another as general convariance would require? These types of "quantum kinematics" issues whose resolution is familiar in flat space quantum field theory become serious problems in the quantum theory of spacetime. Finally, it is clear that in applications to cosmology one will confront the interpretive issues of quantum mechanics in a striking manner.

I cannot present to you a balanced discussion of all these issues for two reasons. First, I believe that a balanced discussion does not exist. Second, I believe that if it did, I would be hard pressed to present it in the compass of a few lectures. What I shall do in this section is the following: I shall assume that we want a quantum theory of spacetime and focus on the quantum kinematics of such a theory. I shall by and large neglect the issues of which theory is correct and those of interpretation. The reasons are part prejudicial and part tactical. It is certainly reasonable to explore at the quantum level Einstein's beautiful idea that gravity is geometry. I shall focus on the quantum kinematics of spacetime theories both because it is possible to say something about these questions and because they may be less familiar to those coming from a particle physics perspective. Where dynamics is needed, I shall for the most part illustrate with the theory constructed on Einstein's action both because this is the simplest illustration and because most of the results in quantum cosmology have been obtained in the semiclassical approximation where the deficiencies of this theory are not immediately present. Further, because it is the correct low energy limit one can hope that some qualitative features of any analysis carried out in Einstein's theory would persist in the correct theory of spacetime. Finally, I shall have little to say about what might be called the "words problem" of quantum mechanics. This is the phenomenon that two scientists can agree on the algorithms in quantum mechanics for predicting the result of every experiment, but disagree passionately on the "words" with which they would like to surround these algorithms. Because of this phenomenon it seems to me likely that interpretative issues in quantum mechanics are in part about the people making the interpretations as well as about the real world. It is not that interpretative issues are uninteresting. Like sex, religion, and politics almost everyone likes to discuss them, and they are sometimes crucial to motivation; it is just that they are difficult to lecture about.

3.2 The Problem of Time

Hamiltonian quantum mechanics is the traditional framework for constructing quantum theories of classical systems. The procedure in this framework is familiar: Construct the Hilbert space of physical states, identify the operators corresponding to observables and the Hamiltonian, then calculate dynamics by solving the Schrödinger equation. There are many good reasons for using this canonical framework, the most important being that it is the most mathematically precise of the several possible frameworks.

The canonical formalism has disadvantages. The most important for the quantization of gravity is the special role played by time. In canonical quantum mechanics time enters as a parameter rather than an operator like other observables. The operators correspond to idealized measurements which take place at one instant of time. The Hilbert space inner product is constructed on a surface of constant time. Indeed, time plays such a distinguished role in the theory that the first problem in constructing a canonical quantum theory of a physical system is to identify the time.

In non-relativistic quantum mechanics there is no difficulty. A time is already singled out in classical physics.

In special relativistic quantum mechanics there is no difficulty but there is an issue. If one reads an elementary book on quantum field theory, one typically finds the canonical quantum mechanics worked out using the time of a particular Lorentz frame. Later, one finds a section "proving the Lorentz covariance of the theory." This means showing that if one had carried out the procedure in a different Lorentz frame one would have obtained physically equivalent results because the corresponding state vectors are connected by a unitary transformation. Thus despite the special role played by time in canonical quantum mechanics it can be made consistent with special relativity.

In general relativity, however, one encounters a crisis. The classical theory does not single out any special set of spacelike surfaces whose labels can play the role of time in canonical quantization. In spacetime physics all spacelike slices are equivalent. There is thus a potential conflict between canonical quantum theory and general covariance unless the canonical quantum theory constructed with one particular set of spacelike surfaces turns out to be equivalent to any other.[19]

Because the problem of choosing a time arises so immediately in canonical quantum theory of spacetime it is useful to examine other frameworks in which this issue is not as central. Feynman's sum over histories formulation, while at present mathematically less precise, is a useful alternative for "the assistance which it gives one's intuition in bringing together physical insight and mathematical analysis."[20]

3.3 Sum Over Histories Quantum Mechanics

The basic elements of a sum over histories formulation of a quantum theory are the following:

(1) <u>The possible histories.</u> A history is a set of observables $\{H\}$ which can describe the results of all possible experiments. The possible histories are the possible histories of all experiments.

(2) <u>The amplitude for a history</u>: This is the joint probability amplitude for the occurrence of a particular history given as

$$\Phi[\{H\}] = \exp(iS[\{H\}]) \quad , \qquad (3.1)$$

where S is the real action functional for the history.

(3) <u>The construction of conditional probability amplitudes by the principle of superposition.</u> In physics we are interested in the results of an experiment. In every experiment the observables constituting a history can be divided into three groups.

    (a) The observables which are fixed by the conditions of the experiment. We call these the conditions $\{C\}$.

    (b) The observables which are measured. We call these observations $\{O\}$.

    (c) The parts of the history which are neither conditioned nor observed $\{U\}$.

The conditional probability of observing $\{O\}$ given $\{C\}$ is

$$\Phi[\{O\}|\{C\}] = \sum_{\{U\}} \Phi[\{H\}] \quad . \qquad (3.2)$$

Typically, the conditions $\{C\}$ define the preparation of the system on which the observations $\{O\}$ are later made.

(4) <u>A probability interpretation.</u> If one can find a complete and exclusive set of observations $\{O_1\},\{O_2\},\ldots$ such that, given the conditions $\{C\}$, one and only one of the observations $\{O_i\}$ is certain to occur, then the probability that it occurs is

$$P[\{O_i\}|\{C\}] = \frac{|\Phi[\{O_i\}|\{C\}]|^2}{\sum_i |\Phi[\{O_i\}|\{C\}]|^2} \quad . \qquad (3.3)$$

(In order for this to make sense the conditions $\{C\}$ must be sufficiently complete to specify the system but this point will become clear through examples.)

The sum over histories formulation of quantum mechanics has been presented somewhat abstractly to emphasize its generality. To apply this framework to a particular theory we must specify (i) the possible histories, (ii) the action functional, (iii) the rules for carrying out the sums in (3.2) and (3.3), and (iv) the complete and exclusive sets of observables. We can make this concrete by considering a few examples:

A non relativistic particle. The histories are the particle paths x(t) which move forward in time i.e. for which $dt/dx \neq 0$. The action is

$$S[x(t)] = \int dt [\tfrac{1}{2} m \dot{x}^2 - V(x)] \quad . \tag{3.4}$$

A typical conditional amplitude is that for the particle to be at x" at time t" given that it was at x' at time t'. In this case the unobserved, unconditioned parts of the history are the parts of the path other than at t' and t". Thus,

$$\Phi[x'',t''|x',t'] = \sum_{\text{paths}} \exp(iS[x(t)]) \quad . \tag{3.5}$$

Where the sum is over all paths which intersect x' at time t' and x" at time t" (Figure 3). (Of course the details of how the sum is carried out - the measure on the space of paths - is also important for the prescription but in this focus on kinematics we are not spelling this out.) An exclusive and complete set of observations are the observations of x at a given t. All particle paths intersect a t = constant surface at at least one x and at no more than one x. Thus the probability (density) that the particle is at x on a constant t surface and nowhere else on that surface, given that it was prepared by passing it at t' through a "slit" characterized[20] by a function f(x') is

$$P[x,t|f,t'] = \frac{|\int dx' \Phi[x,t|x',t'] f(x')|^2}{\int dx |\int dx' \Phi[x,t|x't'] f(x')|^2} \quad . \tag{3.6}$$

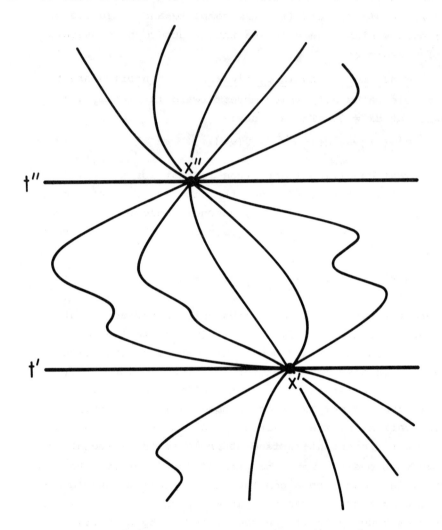

Fig. 3. The conditional probability amplitude for a non-relativistic particle to be at x" at time t", given that it was at x' at time t' is the sum over paths which move forward in time and which intersect the surface t = t' only at x' and the surface t = t" only at x = x".

A relativistic charged particle in an external electromagnetic potential: The histories are again particle paths $x^\alpha(w)$. The paths can't move forward in some time t and be a Lorentz invariant class. We therefore consider paths which move both forward and backward in time and this leads to pair creation. An action is

$$S[x^\alpha(w)] = -m^2 W + \frac{1}{2}\int_0^W dw [\frac{1}{2}\eta_{\alpha\beta}\frac{dx^\alpha}{dw}\frac{dx^\beta}{dw} + e A_\alpha \frac{dx^\alpha}{dw}] \, , \qquad (3.7)$$

where m is the particle's mass, e is its charge and A_α the external vector potential. W is the total parameter time equal on an extremal path to the proper time divided by m. The propagator - the conditional amplitude to find the particle at x''^α given that it was at x'^α - is

$$\Phi[x''^\alpha | x'^\alpha] = \sum_{paths} \exp(iS[x^\alpha]) \, , \qquad (3.8)$$

where, since the parameter time is unobserved, the sum over paths includes not only a sum over the different curves which connect x'^α to x''^α but also the different ways in which the parameter time evolves along these curves. Since the paths move forward and backward in time it is no longer the case that the values of $\vec{x}$ at a given time t constitute an exclusive and exhaustive set of observables. A particle path might intersect a constant time surface many times and one is thus led to a many particle interpretation of this theory. That is, an exclusive and exhaustive set of observations are the number of particles n on a constant t surface and their positions $\vec{x}_1, \ldots, \vec{x}_n$.

A scalar quantum field in flat space: The histories are the possible field configurations in flat spacetime, $\varphi(x)$. An action is

$$S[\varphi(x)] = \frac{1}{2} \int d^4x [- (\nabla_\alpha \varphi)^2 - m^2 \varphi^2 + V(\varphi)] \quad , \quad (3.9)$$

where $V(\varphi)$ is some polynomial interaction in φ. A conditional amplitude is that for the field to take one configuration $\varphi''(\vec{x})$, on a spacelike σ'' surface given that it was $\varphi'(\vec{x})$ on another spacelike surface σ'. This is

$$\Phi[\varphi''(\vec{x}), \sigma'' | \varphi'(\vec{x}), \sigma] = \sum_{\varphi(x)} \exp(iS[\varphi]) \quad , \quad (3.10)$$

where the sum is over all spacetime field configurations which have the prescribed values on σ' and σ''. The different field configurations on such a surface are a set of exhaustive and exclusive observations.

A string: The histories are the world sheet of the string specified by

$$x^A = x^A(\sigma, \tau) \qquad A = 0, 1, 2, \ldots ? \qquad (3.11)$$

The action might be the area of the string. A complete and exhaustive set of variables might be the transverse directions of the string at one instant of time.

Spacetime: Einstein's idea was that gravitational physics is spacetime physics. The histories for gravity are therefore four dimensional geometries, by which one means a four dimensional manifold M with a Lorentz $(-,+,+,+)$ signatured metric $g_{\alpha\beta}(x)$. Two metrics represent the same geometry if they are diffeomorphic i.e. if they can be connected by a coordinate transformation.

To keep our discussion simple I shall assume that we are dealing with cosmologies which are spatially closed and for which the topology is fixed to be $M^3 \times \mathbb{R}$ where M^3 is a closed manifold for space. In the closed Friedman models M^3 is the 3-sphere S^3. This is not the most general class of histories to consider. One might want to consider the possibilities of open cosmologies, for example. One might want to consider summing over manifolds with different topology. Since spacetime is a manifold with metric it seems artificial to sum over metrics but keep the manifold fixed. In particular one might want to sum over histories in which the topology changes, such as that of Fig. 4. Such geometries cannot possess a smooth

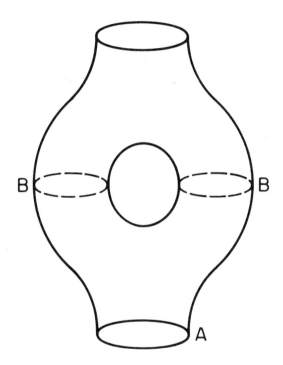

Fig. 4. Topology change in two dimensions. The two-dimensional surface portrayed embedded in three-dimensional flat space corresponds to a history in which the topology of its spatial sections changes from that of a single circle at A to two circles at B. It is not possible to introduce a non-singular vector field corresponding to time on such histories.

decomposition into spacelike slices, there are no non-singular timelike vector fields. I see no particular reason to leave them out, although arguments have been advanced against them.[21] However, we will be able to do all of what we want to do by making the simple assumption described.

The action for Einstein's general relativity is

$$\ell^2 S[g] = 2\int_{\partial M} K h^{1/2} d^3x + \int_M R\sqrt{-g}\, d^4x \quad , \quad (3.12)$$

where ∂M is the spacelike boundary of the manifold M and R is the scalar curvature. h_{ij} is the induced three metric of the spacelike boundary and K is the trace of its extrinsic curvature. That is, if n^α is the unit outward pointing normal to ∂M

$$K = \nabla_\alpha n^\alpha \quad . \quad (3.13)$$

The simplest conditional probability amplitude answers the question: "Given there occurs in the spacetime a spacelike surface with a three geometry described by a three metric h'_{ij} what is the conditional amplitude that there occur a second with a three geometry described by a three metric h''_{ij}?" It is

$$\Phi[h''_{ij}|h'_{ij}] = \Sigma_G \exp(iS[G]) \quad . \quad (3.14)$$

The sum stands for a sum over all physically distinct metrics $g_{\alpha\beta}$ which induce metrics h'_{ij} and h''_{ij} on the two pieces of the boundary. To implement the "physically distinct" part of this prescription requires some "gauge fixing" machinery essentially familiar from gauge theories.

The difficult task in constructing a sum over histories quantum theory of spacetime is the identification of the complete and exclusive sets of observables and the associated probability interpretation. We shall return to these issues in Section 3.5.

3.4 Wave Functions

In quantum mechanics we would like to capture the classical notion of "the state of a system" - a specification of the system at one instant of time. We can then hope to use this information and dynamical laws to evolve the state forward in time. When spacetime is a fixed, flat background, it is straightforward to start with the sum over histories formulation of quantum mechanics and identify the correct notion of state. Suppose the system is prepared by conditions which lie to the past of some spacelike surface. Probabilities for future observations are determined by sums of histories which cross this surface. We may reasonably regard the state of the system on the spacelike surface as specified by the collection of sums over histories proceeding from the given conditions in the past to a fixed value on the surface. This defines the wave function on the surface.

For example, fixing the history of a non-relativistic particle on a constant time surface means fixing its position on the surface. Thus we write for the wave function

$$\Psi_C(x,t) = \int_C \delta x(t) \exp(iS[x(t)]) \quad , \quad (3.15)$$

where the sum is over all paths to the past of t which meet the conditions C. There is a wave function for each set of conditions, that is, for each way of preparing the system.

The probability interpretation for sums over histories immediately assigns a probability interpretation for the wave function. The values of x at a given t are a set of complete and exclusive observations. The conditional amplitude $\Phi[x,t|C]$ factors into a part from the sum over paths to the past of t (the wave function) and a part from the sum over the paths to the future where there are no conditions. The part from the paths to the future is not very well defined mathematically and has to be assigned a constant value for consistency. In any event it cancels in the formation of probabilities and we have

$$P[x,t|C] = \frac{|\psi_C(x,t)|^2}{\int dx |\psi_C(x,t)|^2} \quad . \tag{3.16}$$

Fixing a field configuration on a spacelike surface corresponds to fixing the value of the field on the surface. Thus we write

$$\psi_C[\varphi(\vec{x}),\sigma] = \int_C \delta\varphi \exp(iS[\varphi]) \quad , \tag{3.17}$$

where the sum is over field configurations which match the configuration $\varphi(\vec{x})$ on the surface σ and satisfy the conditions C to the past. The wave function can be assigned a similar probability interpretation because the values of $\varphi(\vec{x})$ on the surface are a complete and exclusive set of possibilities.

The wave function is the quantum analog of the classical motion of "state of the system at one time." We would like to derive a dynamical equation for it. Solving the equation would then be an alternative way of calculating the wave function and a useful one because we are more used to solving differential equations than evaluating functional integrals. In theories with a flat background spacetime we are familiar with how to do this.[20] In the quantum mechanics of a single particle for example, we calculate $\psi(x,t+\epsilon)$ from $\psi(x,t)$ by doing the sum over histories to calculate the propagator between infinitesimally separated slices. For small ϵ the integrals can be done by steepest descents and only the values of

$\psi(x,t)$ for nearby values of x contribute to $\psi(x,t+\epsilon)$. In this way we recover the Schrödinger equation

$$i \frac{\partial \psi}{\partial t} = [- \frac{1}{2m} \frac{d^2}{dx^2} + V(x)]\psi \quad . \tag{3.18}$$

The derivation of quantum dynamics from the sum over histories formulation and the assignment of a probability interpretation to the wave function is straightforward when spacetime is a fixed background. It is less straightforward when spacetime is part of the dynamical variables as it is in gravitational theories. To understand the issues involved we begin by considering a simple model.

3.5 A Parametrized Model

Suppose we are presented with a classical theory involving n+2 variables q^α, $\alpha = 0,1,\cdots n$ and L described by an action

$$S[q^\alpha, L] = \int_M dt [(L - \frac{1}{2} L^2 \dot{q}^0) \delta_{ab} \dot{q}^a \dot{q}^b - V(q^a) \dot{q}^0] \quad . \tag{3.19}$$

Here, M is a finite interval in t, V is a given function, a,b range from 1 to n, and a dot denotes differentiation with respect to time. We shall now describe both the classical and quantum dynamics of this theory.[23]

The theory described by (3.19) has a symmetry. If, for arbitrary f equal to unity on the endpoints of M, we make the transformation

$$L \to \dot{f}(t) L(f(t)) \quad , \tag{3.20a}$$

$$q^\alpha \to q^\alpha(f(t)) \quad , \tag{3.20b}$$

the action remains unchanged. One can easily see this by simultaneously changing the variable of integration

$$t \to f(t) \quad , \tag{3.21}$$

and for this reason the transformation (3.20) is called a reparametrization of the time.

The classical equations of motion found by varying q^a, q^0 and L are

$$-\frac{d}{dt}[(L - \frac{1}{2}L^2\dot{q}^0)\dot{q}^a] - \frac{\partial V}{\partial q^a}\dot{q}^0 = 0 \quad , \quad (3.22a)$$

$$\frac{d}{dt}[\frac{1}{2}L^2\delta_{ab}\dot{q}^a\dot{q}^b + V(q^a)] = 0 \quad , \quad (3.22b)$$

$$\dot{q}^0 = L^{-1} \quad . \quad (3.22c)$$

From (3.22c) we learn that the theory has a constraint. We are not free to specify all the q^a, q^0, L, $\dot{q}^a$, $\dot{q}^0$, $\dot{L}$ on some initial surface and integrate forward in time. First, there is not even an evolution equation for L. With an appropriate transformation of the form (3.20) we could, in fact, pick L = 1 for all times. Second, $\dot{q}^0$ is not freely specifiable but must satisfy the condition (3.22). This is the constraint. Only the q^a are the true degrees of freedom whose value and first time derivative are freely specifiable at an initial time.

The constraint takes an interesting form if we re-express it in terms of the momenta conjugate to q^α. (There is no momentum corresponding to L because it is undifferentiated in (3.19).) These momenta are

$$p_0 = -\frac{1}{2}L^2\delta_{ab}\dot{q}^a\dot{q}^b - V(q^a) \quad , \quad (3.23a)$$

$$p_a = L(2-L\dot{q}^0)\dot{q}^a \quad . \quad (3.23b)$$

Then from (3.22c)

$$p_0 + [\frac{1}{2}\delta^{ab}p_a p_b + V(q^a)] = 0 \quad . \quad (3.24)$$

The left hand side of (3.24) is the total Hamiltonian

$$H = p_\alpha \dot{q}^\alpha - \mathcal{L} \quad , \quad (3.25)$$

multiplied by L (when (3.22c) is satisfied). Thus

$$H = 0 \quad . \tag{3.26}$$

This is characteristic of theories invariant under reparametrizations of the time.

Working in the gauge where $L = 1$ and eliminating q^0 using the constraint we can see what this theory really is. Then $q^0 = t$ and the equation of motion (3.22a) is

$$\ddot{q}^a + \frac{\partial V}{\partial q^a} = 0 \quad . \tag{3.27}$$

This is the equation of motion of a particle in a potential V. Eq. (3.22b) is the associated conservation of energy. Classically, the theory described by the action (3.19) is the same as the theory of a non-relativistic particle in a potential V, but written in a form where the time appears as one of the dynamical variables q^0.

It is instructive to construct the sum over histories quantum mechanics of the theory with the action (3.19) as if we did not know that it was the theory of a non-relativistic particle. Recall that one needs the action, the histories, the measure and the sets of complete and exclusive observations. The action is (3.19). The histories can be described by the functions $q^\alpha(t)$, $L(t)$. Two sets of $\{q^\alpha, L\}$ which are equivalent under reparametrization of the time [eq. (3.20)] describe the same history and are to be counted only once in the sum. If we restrict the paths so that the q^a move forward in q^0 the quantum theory will correspond to non-relativistic quantum mechanics. Other classes of paths could be investigated but lead to different theories with pair creation and annihilation.

Wave functions of states in the quantum theory should depend on the variables which describe the restriction of a history to a spacelike surface. To see what this means imagine a particle path crossing such a surface. We could describe this path by many different $\{q^\alpha(t), L(t)\}$ each related to the other by a reparametrization transformation. The values of q^α by themselves describe a history restricted to a spacelike surface. To specify L in addition would be to specify too much because we can always choose a reparametrization gauge where L has an arbitrary value. Thus we write

$$\psi = \psi(q^\alpha) \qquad (3.28)$$

We do not include an additional label "t" for two reasons: (1) t is not a measurable physical variable but only a parameter label. It can be changed by a reparametrization transformation. (2) Even in a particular reparametrization gauge to include t as a label would be redundant. The value of q^0 already locates the particle along its trajectory in time.

The path integral for the wave function determined by some previous conditions C is

$$\psi_C(q^\alpha) = \int_C \delta q^\alpha \delta L \det(\dot{T}) \delta(q^0 - T(t)) \exp(iS[q^\alpha, L]) . \qquad (3.29)$$

Here, we have introduced a "gauge-fixing" δ-function to enforce the gauge condition $q^0 = T(t)$ with T a monotomically increasing function matching q^0 on boundary. This is the simplest way to fix f in (3.20). $\det(\dot{T})$ is the corresponding "Faddeev-Popov" determinant.

The action (3.19) is quadratic in L. With an appropriate choice of measure, the integral over L is gaussian and can be carried out explicitly. The integral over q^0 can be carried out using the δ-function. The result is familiar

$$\psi_C(q^\alpha) = \int_C \delta q^a \exp(i \int dt [\tfrac{1}{2} \delta_{ab} \dot{q}^a \dot{q}^b - V]) . \qquad (3.30)$$

Thus with the choice of paths and measure described above

the wave functions of the quantum theory built on the action (3.19) are those of the quantum mechanics of the free non-relativistic particle.

It follows from (3.30) that the evolution equation for $\psi_C(q^\alpha)$ is just the Schrödinger equation

$$[-i\frac{\partial}{\partial q^0} - \frac{1}{2}\delta^{ab}\frac{\partial^2}{\partial q^a \partial q^b} + V(q^a)]\psi_C(q^\alpha) = 0 \ . \qquad (3.31)$$

This is the operator form of the classical constraint equation (3.24). We could write

$$H\psi(q^\alpha) = 0 \ . \qquad (3.32)$$

Thus in quantum mechanics as in classical mechanics the vanishing of the Hamiltonian does not imply the absence of dynamics. In fact this condition becomes the dynamical equation of the theory when expressed in terms of its true degrees of freedom.

3.5 General Relativity

3.5.1 The classical theory in 3+1 form

The structure of the general theory of relativity is similar in many ways to the model of non-relativistic particle mechanics with parametrized time that was discussed in the preceding section. Like this model, general relativity is invariant under reparametrizations of the time. If one singles out a family of spacelike surfaces, labels them by a time coordinate, t, then the action for general relativity [eq. (3.12)] is unchanged by a relabeling of these surfaces with a different t-coordinate. Of course, general relativity is also invariant under the larger group of diffeomorphisms corresponding to general coordinate transformations.

Because of its invariance under diffeomorphisms general relativity is a theory with constraints - restrictions on the values of the metric and its first time derivative that can be specified on an initial value surface. To spell these constraints out we begin by rewriting the classical action (3.12) in a form which emphasizes spacelike surfaces. This is the 3+1 formulation of Arnowitt, Deser and Misner (ADM).[24]

Suppose that one has a family of spacelike surfaces labeled by a time coordinate t. The metric can generally be written as

$$ds^2 = -N^2 dt^2 + h_{ij}(dx^i + N^i dt)(dx^j + N^j dt) \quad . \qquad (3.33)$$

The tensor h_{ij} is the induced metric of the spacelike surface. N and N^i can be described as follows: Given two neighboring spacelike surfaces, labeled by t and t+dt, the displacement between a point in the first surface and a point in the second be decomposed into a displacement in the first surface and another displacement normal to it. If the points are labeled by x^i and $x^i + dx^i$ respectively then $dx^i + N^i dt$ is the displacement vector in the surface and Ndt is the length of the normal displacement. N^i is therefore called the "shift vector" and N is called the "lapse function."

The 3-metric h_{ij} specifies the intrinsic geometry of a surface of constant t. With it we can associate a spatial covariant derivative D_i and compute the spatial curvatures, $^3R_{ijk\ell}$. The lapse and shift can be thought of as scalar and vector fields on this surface and tensor operations carried out accordingly. The differential change in the unit normal projected into the surface

$$K_{ij} = -\nabla_i n_j \quad , \qquad (3.34)$$

is a measure of how the spacelike surface is curved in

4-dimensional space. It is called the extrinsic curvature. Explicitly, in terms of the decomposition (3.33) it is

$$K_{ij} = \frac{1}{N}\left[-\frac{1}{2}\frac{\partial h_{ij}}{\partial t} + D_{(i}N_{j)}\right] \quad . \tag{3.35}$$

The action for general relativity, eq. (3.12) can be rewritten in terms of the decomposition (3.33) as

$$\ell^2 S_E = \int d^4x\, h^{1/2} N(K_{ij}K^{ij} - K^2 + {}^3R - 2\Lambda) \quad . \tag{3.36}$$

Here, $K = K^i{}_i$, ${}^3R$ is the scalar curvature of the surface and tensor operations are carried out in the geometry of the surface. To this must be added the action of the matter fields, but for simplicity we shall consider pure gravity until Section 5. Written in this ADM form it is clear that general relativity is a theory with constraints because the lapse N and the shift N^i occur in S_E undifferentiated with respect to t. The corresponding equations of motion may therefore be expressed entirely in terms of the metric h_{ij} and its conjugate momentum π_{ij}. They are thus constraints on initial data.

The momentum conjugate to h_{ij} is easily found from (3.36) and (3.35) and is

$$\ell^2 \pi_{ij} = -h^{1/2}(K_{ij} - h_{ij}K) \quad . \tag{3.37}$$

The three constraints following from varying (3.36) with respect to the N^i are

$$D_i \pi^{ij} = 0 \quad . \tag{3.38}$$

The one following from varying N may be written

$$\ell^2 G_{ijk\ell}\pi^{ij}\pi^{k\ell} + \ell^{-2}h^{1/2}(-{}^3R + 2\Lambda) = 0 \quad , \tag{3.39}$$

where

$$G_{ijk\ell} = \frac{1}{2}h^{-1/2}(h_{ik}h_{j\ell} + h_{i\ell}h_{jk} - h_{ij}h_{k\ell}) \quad . \tag{3.40}$$

This constraint will be important for us. It is the constraint associated with the invariance of the theory under a reparametrization of the spacelike surfaces. The model of parametrized time particle quantum mechanics suggests, and one can check, that this constraint implies the vanishing of the total Hamiltonian density

$$\mathcal{H} = \ell^2 G_{ijk\ell} \pi^{ij} \pi^{k\ell} + \ell^{-2} h^{1/2} (-{}^3R + 2\Lambda) = 0 \quad . \quad (3.41)$$

As in the model, this equation summarizes the dynamics of classical general relativity without matter.

3.5.2 Quantum mechanics of closed cosmologies

To investigate the quantum mechanics of a closed cosmology we must first describe correctly a quantum state. Recall from the discussion of Section 3.3 that a state is described by a wave function whose arguments are the variables describing a history fixed on a spacelike surface. The histories are the 4-geometries on $M^3 \times \mathbb{R}$. Each may be described by a metric $g_{\alpha\beta} = \{N, N^i, h_{ij}\}$ but there will be many metrics corresponding to the same geometry. This complicates the identification of the arguments of the wave function, as it did in the case of parametrized time particle quantum mechanics, but we may proceed in the analogous way. Consider the 4-geometries in which a given spacelike surface with definite 3-geometry occurs, but which are otherwise free to vary off this surface. By a suitable choice of coordinates, say $N = 1$ and $N^i = 0$, the metrics for all these geometries could be brought to a standard form where h_{ij} is the only variable. Thus, we may take the 3-metric h_{ij} as describing a history fixed on a spacelike surface and write for the wave function of a closed cosmology

$$\Psi = \Psi[h_{ij}] \quad . \quad (3.42)$$

There is no additional dependence on the coordinate t.
First, t is not a physical label but may be prescribed
at will. Second, the three geometry already carries
information about the location of the surface in time.
A generic 3-geometry will fit into a generic 4-geometry
in a locally unique way (and in particular at a unique
"time") if it fits in at all. The counting of variables
implied by this labeling of the wave function is correct.
There are 6 components of h_{ij} at each space point. Three
of these "are pure gauge," that is, could be chosen
arbitrarily by a suitable choice of spatial coordinates.
If one component corresponds to time there remain two.
This is the correct number of degrees of freedom of a
massless spin-2 field.

The sum over histories for the wave function corresponding to a set of conditions C is

$$\Psi_C[h_{ij}] = \int_C \delta g \, \exp(iS_E[g]) \quad . \quad (3.43)$$

The integration is over a class of 4-geometries defined on
the manifold which is that part of $M = M^3 \times \mathbb{R}$ to the past
of a bounding $M^3 = \partial M$. (See Figure 5.) The class consists
of those 4-geometries which induce the metric h_{ij} on ∂M
and which satisfy the conditions C to its past.

Wave functions constructed as sums over histories
should automatically satisfy the operator form of the
constraints of the theory as did the wave function for
the model of parametrized time particle quantum mechanics.
An operator form of the classical constraints may be obtained by replacing $\pi^{ij}(x)$ by $-i\delta/\delta h_{ij}(x)$ in eqs. (3.39)
and (3.40). In the case of (3.39) we have

$$D_i \left(\frac{\delta \Psi}{\delta h_{ij}(x)} \right) = 0 \quad . \quad (3.43)$$

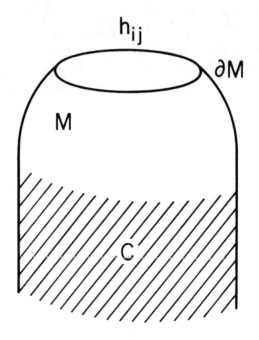

Fig. 5. The sum over histories for a wave function $\Psi_C[h_{ij}]$ is a sum over geometries on the manifold $M^3 \times (-\infty, 0]$ which induce the metric h_{ij} on the boundary $\partial M = M^3$ and which satisfy the conditions C to the past of this surface.

For the Hamiltonian constraint (3.40) there are many possible operator equations depending on the choice of factor ordering. They all have the form

$$[-\ell^2 \nabla_x^2 + \ell^{-2} h^{1/2}(x)(-{}^3R(x) + 2\Lambda)] = 0, \qquad (3.44)$$

where

$$\nabla_x^2 = G_{ijk\ell} \frac{\delta^2}{\delta h_{ij}(x)\, \delta h_{k\ell}(x)} + \begin{pmatrix} \text{linear derivative} \\ \text{term depending on} \\ \text{factor ordering} \end{pmatrix}.$$

(3.45)

Eqs. (3.43) and (3.44) are an infinite set of equations – one for each point on the spacelike surface at which Ψ is defined. Eq. (3.44) is called the Wheeler-DeWitt equation. Like its analog (3.31) in parametrized time particle quantum mechanics, the Wheeler-DeWitt equation determines the quantum dynamics of general relativity.

We shall not derive the Wheeler-DeWitt equation from the functional integral (3.43) although it would be possible to do so[*], at least formally.[25,26] The formal derivation of (3.43), however, is simple enough that we can give it here to indicate the methods involved: Consider an infinitesimal coordinate transformation on M which relabels the spatial coordinates of each constant t surface in an

---

[*]There should be a connection between the measure used for the sum over histories and the factor ordering in the Wheeler-DeWitt equation. This connection bears on the long standing problem in the canonical theory[27] of whether there exists a factor ordering such that the algebra of the constraints closes. Either the construction of wave functions by sums over histories resolves this problem or there is a restriction on the permissible measures in the sum arising from it. (See also Ref. 28.) The absence of a clear understanding of the connection between measure and the form of the Wheeler-DeWitt equation is the reason we have not presented a detailed derivation. As we intend to solve the equation for the most part in the semiclassical approximation these ambiguities will not affect our limited conclusions.

identical fashion. Such a transformation has the form

$$t \to t, \qquad x^i \to x^i + \xi^i(x^k) \quad , \qquad (3.46)$$

for infinitesimal ξ^i. The spatial metric transforms as

$$h_{ij} \to h_{ij} + 2D_{(i}\xi_{j)} \quad , \qquad (3.47)$$

and the 4-metric as

$$g_{\alpha\beta} \to g_{\alpha\beta} + 2\nabla_{(\alpha}\xi_{\beta)} \quad , \qquad (3.48)$$

with $\xi^0 = 0$. In eq. (3.43), shift simultaneously the argument of the wave function on the left by the amount in (3.47) and the integration variables on the right by the amount in (3.48). There remains an identity. The integral on the right hand side of this relation is identical to what it was before because both action and measure are invariant under coordinate transformations. Thus we conclude

$$\Psi[h_{ij} + 2D_{(i}\xi_{j)}] = \Psi[h_{ij}] \quad , \qquad (3.49)$$

or equivalently

$$\int_{\partial M} d^3x \, D_{(i}\xi_{j)} \frac{\delta\Psi}{\delta h_{ij}(x)} = 0 \quad . \qquad (3.50)$$

Integrating this relation by parts and noting that ξ^i is arbitrary, one recovers the operator form of the constraints (3.43). This derivation shows that the physical content of these three constraints is that the wave function does not depend on the choice of coordinates in the spacelike surface as it should not.

3.5.3 Superspace

As a consequence of the constraint (3.43) the wave function may be thought of as a function on the space of three geometries - the space of gauge inequivalent metrics on M^3. This is called "superspace." The quantity

$G_{ijk\ell}$ may be used to form a metric on superspace. If δh_{ij} and $\delta h'_{ij}$ are the changes in 3-metric corresponding to an infinitesimal displacement in superspace we may define the inner product of these two displacments to be

$$(\delta h, \delta h') = \int d^3x G^{ijk\ell} \delta h_{ij} \delta h'_{k\ell} \qquad (3.51)$$

Where $G^{ijk\ell}$ is the inverse of $G_{ijk\ell}$ considered as a 6 x 6 matrix in the space of symmetric index pairs (ij). Other metrics on superspace may be defined by inserting an arbitrary function in (3.51). The metric (3.51) suggests that an elegant way for choosing the factor ordering in the Wheeler-DeWitt equation is to take ∇_x^2 to be the "covariant Laplacian" in this metric. (For the issues raised by such a choice see Ref. 28.)

At each space point the signature of the 6 x 6 metric $G_{ijk\ell}$ is $(-,+,+,+,+,+)$. The Wheeler-DeWitt equation is thus a kind of "hyperbolic" equation in superspace. We recall that one of the six degrees of freedom in the 3-metric represents time in the sense of locating the spacelike surface locally in a 4-geometry. We may think therefore, of a fixed choice for this time as defining a family of hypersurfaces in superspace and the Wheeler-DeWitt equation as specifying the propagation of the wave function from one hypersurface to another. It is in this sense that the Wheeler-DeWitt equation implements quantum dynamics.

3.5.4 Use of the wave function

In non-relativistic particle quantum mechanics (Section 3.3) we were able to assign a probability interpretation to the wave function because the values of the particle's position at a given time constituted a complete and exclusive set of observations. The identification of a complete and exclusive set of observables in the quantum theory of spacetime is a much more difficult problem. We lack, by and large, the simple analyses of thought experiments which guide our intuition in particle quantum mechanics. Even formally, the question is complicated by the problem of the choice of time (Section 3.2). By analogy with particle quantum mechanics one would expect a complete and exclusive set of observables to be different values of some part of the 3-metric "at one time" - but which "time" should be used? The sum over histories formulation of quantum mechanics may guide us to a resolution of such questions but in the meantime we can proceed qualitatively through an analysis of the correlations in the wave function. We expect variables to be correlated on those regions of superspace where the wave function is large and anticorrelated where it is small. This minimal interpretation will be enough to take the first steps in quantum cosmology.

Of the four ingredients of a sum over histories formulation of quantum cosmology - histories, action, measure, observables - we have discussed the first two. Discussions of the measure, not given here because they are complex and will not bear directly on the semiclassical approximations to the wave function we shall mostly consider, may be found in Refs. 29. The interpretation of Ψ at the level of correlations will be sufficient for a first analysis of the proposal for the quantum state of the universe that we shall now discuss.

4. THE QUANTUM STATE OF THE UNIVERSE

4.1 A State of Minimum Excitation

The universe is in a state of low excitation. The large scale distribution of matter and metric are nearly homogeneous and isotropic. The entropy of the matter is low compared to that of the highly clumped and irregular configurations it might have had. Whatever the state of the universe is, it is close in some sense to a state of minimum excitation. We, therefore, begin a discussion of the state of the universe with a discussion of the state of minimum excitation for closed cosmologies.

In the quantum mechanics of a particle in a potential there are two ways of calculating the wave function of the state of minimum excitation. We can calculate it as the lowest energy eigenfunction of the Hamiltonian (the ground state)

$$H\psi_0(x_0) = E_0\psi_0(x_0) \quad . \tag{4.1}$$

Completely equivalently, the wave function of the ground state may be calculated as a Euclidean functional integral.

$$\psi_0(x_0) = \int \delta x(\tau) \exp(-I[x(\tau)]) \quad . \tag{4.2}$$

Here, $I[x(\tau)]$ is the Euclidean action functional

$$I[x(\tau)] = \int d\tau \left[\tfrac{1}{2}m\dot{x}^2 + V(x)\right] \quad . \tag{4.3}$$

The sum is over all paths which start at x_0 at time $\tau = 0$ and proceed in the infinite past to a configuration of minimum action.

It is not difficult to sketch the demonstration of the equivalence of (4.1) and (4.3). One begins with the path integral for the propagator

$$\langle x'',t''|x't'\rangle = \int \delta x(t)\exp(iS[x(t)]) \quad . \tag{4.4}$$

Here, S is the usual action (i.e. (4.3) with the opposite sign for V(x)) and the sum is over paths which start at x' at the time t' and wind up at x" at time t". Consider the particular propagator $\langle x_0, 0 | 0, t \rangle$ and expand it in a complete set of energy eigenstates as follows:

$$\langle x_0, 0 | 0t \rangle = \Sigma_n \langle x_0, 0 | n \rangle \langle n | 0, t \rangle$$
$$= \Sigma_n e^{iE_n t} \psi_n(x_0) \psi_n^*(0) , \quad (4.5)$$

where ψ_n is the wave function of the energy eigenstate with eigenvalue E_n. Equate the last line of (4.5) to the right hand side of (4.4) and rotate the time to imaginary values, $t \to -i\tau$, on both sides of the equation. One has

$$\Sigma_n e^{E_n \tau} \psi_n(x_0) \psi_n^*(0) = \int \delta x(\tau) \exp(-I[x(\tau)]) . \quad (4.6)$$

Then take the limit $\tau \to -\infty$. If one normalizes the energy so that the lowest eigenvalue is zero, only the ground state term survives in the sum on the left hand side of (4.6). The sum over paths on the right hand side becomes the sum described above and one recovers after a normalization eq. (4.2).

In the quantum mechanics of closed cosmologies one does not expect to recover the wave function of the state of minimum excitation as the lowest eigenvalue of a Hamiltonian. This is because there is no natural notion of energy for closed cosmologies.

In classical gravity, the principle of equivalence shows us that there can be no definition of local energy for the gravitational field. All gravitational effects vanish locally in a freely falling frame. Alternatively note that energy is the conserved quantity which arises from time translational symmetries of spacetime and in a general spacetime there will be no time translation invariance. For spacetimes with special symmetries one can define an energy. For example, asymptotically flat spacetimes have a time translation symmetry "at infinity." Correspondingly, we can define a total energy which is conserved. For closed cosmological models, however, there is no such symmetry.

One might pick arbitrarily a family of spacelike slices, identify the generator which takes us from one slice to the next, call that the Hamiltonian, and find the lowest eigenstate. As the above argument suggests, however, and has been shown by Kuchař[30] there is no slicing for which the resulting Hamiltonian is time independent and thus none for which one could construct a unique ground state.

While the construction of the wave function of the state of minimum excitation as the lowest eigenstate of the Hamiltonian fails for closed cosmologies the construction using a Euclidean functional integral can be generalized.[25] Schematically, including a generic matter field φ, one would write

$$\Psi_0[h_{ij},\varphi_0] = \int \delta g \, \delta\varphi \exp(-I[g,\varphi]) \ . \tag{4.7}$$

The sum is over a class of Euclidean four geometries which have a boundary on which the induced 3-metric is h_{ij} and matter configurations which match the values $\varphi_0(\vec{x})$ on the boundary. (These are the analog of the paths starting at x_0 in (4.2)). The action is the sum of the Euclidean action for the matter and the Euclidean action for general relativity. On a manifold M with boundary ∂M the latter is

$$\ell^2 I_E[g] = -2\int_{\partial M} K h^{1/2} d^3x - \int_M (R-2\Lambda) g^{1/2} d^4x \ . \tag{4.8}$$

where, as in eq. (3.12), K is the trace of the extrinsic curvature of the boundary. To completely specify the wave function Ψ_0 it remains to complete the specification of the class of geometries and field configurations summed over (the analog of the paths going to minimum action at $\tau \to -\infty$ in (4.2)). The proposal is that one should sum over compact Euclidean four geometries and field configurations which are regular on them. The remaining boundary condition for geometries contributing to the state of minimum excitation is that there is no other boundary. In Section 5 we shall examine evidence for the appropriateness of this choice.

The Euclidean construction of the wave function of the state of minimum excitation does not change the form of the constraints it satisfies. Ψ_0 continues to satisfy the generalizations of (3.43) and (3.44) to include matter. These are

$$iD_j\left(\frac{\delta \Psi}{\delta h_{ij}}\right) = \ell^2 T_n^i(\varphi, -i\frac{\delta}{\delta\varphi})\Psi , \qquad (4.9a)$$

$$\frac{1}{2}[\ell^2 \nabla_x^2 + \ell^{-2} h^{1/2}(^3R - 2\Lambda)]\Psi = h^{1/2} T_{nn}(\varphi, -i\frac{\delta}{\delta\varphi})\Psi . \qquad (4.9b)$$

Here $T_{\alpha\beta}(\varphi, \pi)$ is the stress energy of the matter expressed in terms of the field and its canonical momentum. This becomes the operators in (4.9) when projected appropriately onto the direction n^α normal to the constant t surfaces and when π is replaced by $-i\delta/\delta\varphi$. One can derive (4.9a) simply by following the derivitation sketched in Section 3.5.2. The Wheeler-DeWitt eqn can be derived formally in a similar fashion. (See, e.g. Ref. 2.)

The Wheeler-DeWitt equation and the associated constraints (4.9) presumably have many solutions. The sum over histories (4.7) singles out one of them. The Euclidean functional integral prescription may therefore be thought of as supplying boundary conditions for the Wheeler-DeWitt equation and this will prove a useful approach to take when actually solving for Ψ_0.

4.2 The Conformal Factor

Some attention must be given to the meaning of the sum in (4.7). The Euclidean Einstein action is not positive definite. One can see this[31] by considering the family of metrics generated from a given one, $\tilde{g}$, by conformal transformations

$$g_{\alpha\beta} = \Omega^2 \tilde{g}_{\alpha\beta} \quad . \tag{4.10}$$

In terms of Ω and $\tilde{g}$ the Euclidean action becomes

$$\ell^2 I_E[\Omega,\tilde{g}] = -2\int_{\partial M} d^3x \tilde{h}^{1/2}\Omega^2 \tilde{K}$$
$$- \int d^4x \tilde{g}^{1/2}[\Omega^2 \tilde{R} + 6(\tilde{\nabla}\Omega)^2 - 2\Lambda\Omega^4] \quad . \tag{4.11}$$

By making Ω rapidly varying, the action can be made as negative as desired. A sum over real geometries of the form (4.7) will therefore not converge.

One might think that the indefiniteness of the Euclidean Einstein action was an indication of some instability in the quantum theory. There are such situations in particle quantum mechanics.[32] Consider, for example, a particle moving in a potential V(x) of the form shown in Figure 6 and its Euclidean propagator $\langle 0,0|0,\tau\rangle$. From (4.6) it follows that the large negative time behavior of this propagator is proportional to $\exp(E_0\tau)$ where E_0 is the energy of the ground state. Also from (4.6) it follows that we could calculate this energy by evaluating the path integral on the right hand side. Let us do this by the method of steepest descents. There are two stationary paths which satisfy the Euclidean equations of motion (the usual equations with the sign of V reversed) and the boundary conditions. There is first the solution $\bar{x}(\tau) = 0$. If V had risen monotonically with increasing x this would be the only stationary path, the action would be always positive, and quantum state associated with the classical minimum would be stable. Since, however, V(x) turns over and again intersects the x-axis there is another "tunneling solution" $\bar{x}_t(\tau)$ which proceeds from x = 0 to the turning point and back again (Fig. 6). The tunneling solution is not a true minimum of the action as one can see from the expression for the action of small fluctuations about it.

$$I_2[\delta x(\tau)] = \frac{1}{2}\int_0^T d\tau \left[m\left(\frac{d\delta x(\tau)}{d\tau}\right)^2 + V''(\bar{x}_t(\tau))(\delta x(\tau))^2 \right]. \quad (4.12)$$

Here, V" is the second derivative of V(x). By choosing $\delta x(\tau)$ concentrated where V" < 0 the action can be made negative and thus less than its value for tunneling

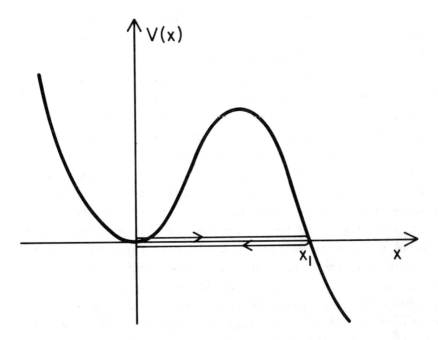

Fig. 6. The minimum of this potential at x = 0 is classically stable but quantum mechanically unstable via tunneling. This is reflected in the indefiniteness of the action for fluctuations about the tunneling motion from x = 0 to x = x_1 and back again.

solution.* Since the action is negative the path integral
over the fluctuations will not converge. To keep it convergent the contour of path integration in x must be
distorted into the complex plane in a way which can be
justified by starting with a V(x) for which x = 0 is a
global minimum and smoothly distorting it to the shape
of Figure 6. In the process of this distortion the energy
E_0 acquires an imaginary part. The former ground state
thus becomes unstable when the action becomes negative.

Unlike the above example, the negative definiteness
of the gravitational action does not signal a tunneling
instability. One can see this clearly in the case of pure
gravity and asymptotically flat spacetimes where the total
energy is defined. Classically there is certainly a ground
state. This is flat space and it has zero energy. The
positive energy theorems guarantee that all other classical
configurations have higher energy.[33] Because flat space
is a global minimum of the energy we do not expect the
associated quantum state to be unstable and indeed Witten[34]
has shown that there are no tunneling solutions. Yet, the
action is not positive definite.

The quantum mechanics of the fluctuations about flat
space provides simple model in which the significance of
the indefiniteness of the gravitational action can be
understood. If we write

---

*One might worry that the positive kinetic energy term
in (4.12) would defeat this argument. This is not the
case.[32] An infinitesimal time translation of the
tunneling solution is a zero mode with zero action of
the quadratic operator defined by (4.12). It has one
mode at $x = x_1$ so there must be a mode with lower,
negative action.

$$g_{\alpha\beta} = \delta_{\alpha\beta} + h_{\alpha\beta} , \qquad (4.13)$$

then the Euclidean action (4.8) to quadratic order in the $h_{\alpha\beta}$ is

$$\ell^2 I_2[h] = \frac{1}{2} \int d^4x \left[\frac{1}{2} \nabla_\alpha \bar{h}_{\beta\gamma} \nabla^\alpha h^{\beta\gamma} - (\nabla_\alpha \bar{h}^{\alpha\beta})^2 \right] , \qquad (4.14)$$

+ (surface terms)

where

$$\bar{h}_{\alpha\beta} = h_{\alpha\beta} - (1/2) \delta_{\alpha\beta} h . \qquad (4.15)$$

The action (4.14) is just that for a free spin-2 field in a flat background. Its quantum mechanics is equivalent to an assembly of independent harmonic oscillators and the ground state is, therefore, certainly stable. The action (4.14), however, is no more positive definite than that for the full theory (4.8). (Try $h_{\alpha\beta} = 2\delta_{\alpha\beta}\chi$ which is a linearized conformal transformation of flat space.) What's going on?

The point is that gravity formulated as a field theory in the metric is a theory expressed in terms of redundant variables. Part of the metric is arbitrary corresponding to the choice of coordinates (gauge) and the remainder is connected by the constraints. In fact, there are only two physical degrees of freedom. Issues concerning stability are best discussed in terms of these physical degrees of freedom, if they can be identified.

In linearized gravity one can fix a gauge, solve the constraints and identify the physical degrees of freedom. They are the two transverse-traceless "TT" parts of the metric fluctuations satisfying

$$\nabla^\alpha h^{TT}_{\alpha\beta} = 0 , \qquad n^\alpha h^{TT}_{\alpha\beta} = 0 , \qquad h^{TT\alpha}_{\alpha} = 0 , \qquad (4.16)$$

where n^α is the unit normal to the constant t surfaces.

It is not difficult to check that the two $h_{\alpha\beta}^{TT}$ are invariant under infinitesimal coordinate transformations and satisfy the linear versions of the constraints (3.38) and (3.39).

On the physical degrees of freedom the Euclidean action is

$$\ell^2 I_2[h_{\alpha\beta}^{TT}] = \frac{1}{4}\int d^4x[(\dot{h}_{ij}^{TT})^2 + (\nabla_i h_{jk}^{TT})^2] \quad , \qquad (4.17)$$

where $(a_{ij...})^2 = a_{ij...}a^{ij...}$. This is the action of two free fields. It is positive. Sums over Euclidean histories expressed in terms of the physical degrees of freedom will converge. Flat space is stable in linearized gravity.

The moral of the example of linearized gravity is that the quantum mechanics of a theory with redundant variables is best analyzed in terms of its physical degrees of freedom. Sums over histories, for example, can be carried out in the physical configuration space without gauge fixing and without ghosts. But it is often convenient to have the quantum mechanics expressed in terms of the redundant variables, for example, to display the invariances of the theory. In a theory like general relativity such a formulation is essential because it appears not to be possible to solve the constraints to exhibit the physical degrees of freedom explicitly. To pass from a sum over histories in the physical degrees of freedom to one in terms of redundant variables one simply adds back in the extra integrations in such a way as to not affect the value of the integral.[35] Typically, one might make use of identities like

$$1 = \int_{-\infty}^{+\infty} dx \left|\frac{\partial \Phi}{\partial x}\right| \delta(\Phi(x)) \quad , \qquad (4.18)$$

for adding back in gauge variables and

$$1 = \int_{-\infty}^{+\infty} \frac{dx}{\sqrt{\pi M}} e^{-Mx^2} \quad , \qquad (4.19)$$

for adding back in the gauge invariant ones. Identity (4.18) gives rise to "gauge fixing δ-functions" and "Faddeev-Popov determinants." Identity (4.19) modifies the action. One is free to use any identities one wants as long as they converge and the value of the sum over histories is left unaffected.

In the theory of linearized gravity integrations cannot be added to a sum over histories with the action (4.17) to obtain one with the action (4.14). The starting integral is convergent the resulting integral is divergent. One can, however, arrive at a coordinate invariant result by the following procedure[36]: Decompose the fluctuations as

$$h_{\alpha\beta} = \varphi_{\alpha\beta} + 2\delta_{\alpha\beta}\chi \quad , \qquad (4.20)$$

and fix the decomposition by requiring one condition on $\varphi_{\alpha\beta}$. It is convenient to take

$$R_1(\varphi_{\alpha\beta}) = 0 \quad , \qquad (4.21)$$

where R_1 is the linearized scalar curvature. The function χ is thus a gauge invariant scalar generating linearized conformal transformations. Then,

$$\ell^2 I_2[h] = \ell^2 I_2[\varphi_{\alpha\beta}] - 6\int d^4x (\nabla\chi)^2 \quad , \qquad (4.22)$$

where the action I_2 on $\varphi_{\alpha\beta}$ is positive definite. A physical sum over histories cannot be manipulated into a form involving the action (4.22) but one can arrive at one with the form

$$\ell^2 I_2[h] = \ell^2 I_2[\varphi_{\alpha\beta}] + 6\int d^4x (\nabla\chi)^2 \quad . \qquad (4.23)$$

This action is gauge invariant, O(4) invariant, positive definite, and it is physically equivalent to $I_2[h]$. It is the action to use in constructing convergent Euclidean functional integrals for fluctuations about flat space and use of this action gives the correct ground state wave function for linearized gravity.[37] A sum over histories based on the action (4.23) may be thought of as a sum based on the action (4.22) but carried out along a functional contour in which χ is purely imaginary. It is in this form that we shall find it most convenient to summarize the result.

For the full general theory of relativity, while we cannot explicitly identify the physical degrees of freedom, we can carry out a procedure analogous to that of linearized gravity. Consider first the case of the sums over histories which determine vacuum expectation values in asymptotically flat spacetimes with $\Lambda = 0$. These are integrals over asymptotically Euclidean spacetimes with the action (4.8). Split the integral over all metrics into an integration over a conformal factor and an integration over metrics in a conformal equivalence class. That is, write

$$g_{\alpha\beta} = \Omega^2 \tilde{g}_{\alpha\beta} \quad , \qquad (4.24)$$

with $\Omega = 1$ at infinity and require

$$R(\tilde{g}) = 0 \quad . \qquad (4.25)$$

If we write $\Omega = 1+Y$ and carry out the formal rotation $Y \to iY$, the action becomes [cf. (4.11)]

$$\ell^2 I_E[g] = \ell^2 I_E[\tilde{g}] + 6 \int_M d^4x \sqrt{\tilde{g}} (\tilde{\nabla} Y)^2 \quad . \qquad (4.26)$$

The last term is positive definite so the integral over the conformal factor converges. There remains the integral over metrics $\tilde{g}$ satisfying (4.25). The positive action theorem[38] shows that the action on such metrics is positive. These integrals thus converge.

The analysis of the asymptotically flat case and the case of small fluctuations about flat space suggests that a similar procedure should be used to define the Euclidean functional integral giving the state of minimum excitation for closed cosmologies. One divides the integral into conformal equivalence classes in (4.24) using perhaps a condition of constant curvature rather that (4.25). One writes the conformal factor as $1+Y$ where Y vanishes on the boundary where the argument of the wave function is given. One rotates Y into iY making the second term in (4.10) positive. The resulting action, however, is no longer manifestly positive. In fact it is complex. We do not yet have a demonstration that the resulting integrals converge but the preceding two examples give some hope that they may. We will write the prescription for the state of minimum excitation as

$$\Psi_0[h_{ij},\varphi] = \int_{C_0} \delta g\, \delta\varphi \exp(-I[g,\varphi]) \quad , \quad (4.27)$$

where the C_0 indicates that an appropriate complex contour must be taken. This contour will ensure that Ψ_0 is real. However, since C_0 is complex, we cannot conclude that Ψ_0 is positive. In general it will oscillate and this will be important for its interpretation.

4.3 The Wave Function of the Universe

The wave function so naturally identified by the Euclidean functional integral prescription (4.27) displays many properties one would associate with a state of minimum excitation when analyzed in simple models as we shall show. As our own universe is not in a state of very high excitation, and as this state emerges so simply in the theory, it is a natural conjecture[2] that this wave function is the wave function of our universe and that the law (4.27) is the law specifying the initial conditions. We shall examine this conjecture in what follows.

5. MINISUPERSPACE MODELS

5.1 Minisuperspace

To test the conjecture that the wave function for our universe is that constructed by a sum over compact Euclidean four geometries, we want to calculate it and compare its predictions with the observations summarized in Section 2. The sum cannot be done exactly, we can only make approximations to it. Approximations may be constructed by singling out a family of geometries described by only a few parameters or functions and carrying out the sum only over geometries in this restricted class. Such a restriction on the 4-geometries which occur in the sum implies a restriction on the 3-geometries which can occur as arguments of the resulting wave function. This restriction reduces the configuration space on which the wave function is defined from the superspace of all 3-geometries to a smaller class - a minisuperspace. For this reason such approximations are called minisuperspace approximations.

One way of constructing a minisuperspace approximation is to restrict geometries and field configurations to have a certain symmetry. This type of approach has had a long and useful history in quantum cosmology.[39] Minisuperspace models based on symmetry are easy to implement and generally easy to interpret. They do not, however, offer the possibility of systematic improvement. That can be achieved in the lattice approximation to general relativity called the Regge calculus. There, curved geometries are built out of flat 4-simplices in much the same way that a geodesic dome is built out of triangles. The lengths of the edges of the simplices making up a 3-geometry become the parameters of a minisuperspace. Such simplicial minisuperspace approximations offer the hope of systematic improvement and are well adapted for the study of topological questions.[40]

Minisuperspace methods have already been extensively applied to construct approximations to the wave function of the universe.[2,25,28,41-61] We shall discuss only three of these models here and these can only be treated briefly. In two of these models the essential minisuperspace restriction is that geometries and field configurations be homogeneous and isotropic. The models differ in their assumptions about the matter. A conformally invariant scalar field gives a model which is not very realistic but easy to analyze. A massive scalar field provides a model which possesses many of the features of our universe.[2] Finally we discuss the model of Halliwell and Hawking[49] in which the origin of deviations from exact homogeneity and isotropy can be predicted.

5.2 Homogeneous, Isotropic Geometries with Scalar Field

5.2.1 Framework

The simplest class of minisuperspace models are those obtained by restricting the geometry to be homogeneous and isotropic and thus close to that of the present universe. The line element is then

$$ds^2 = \sigma^2[-N^2(t)dt^2 + a^2(t)d\Omega_3^2] \quad , \tag{5.1}$$

where $N(t)$ is an arbitrary lapse function and $\sigma^2 = \ell^2/24\pi^2$ is a normalizing factor chosen for later convenience. $d\Omega_3^2$ is the metric on the unit three sphere. The Euclidean histories with the same symmetries which enter into the sum defining the wave function have the metric

$$ds^2 = \sigma^2[N^2(\tau)d\tau^2 + a^2(\tau)d\Omega_3^2] \quad . \tag{5.2}$$

For the matter, we take a single scalar field with mass M, coupling to curvature ξ, and potential $V(\Phi)$ whose action generally is

$$I_\Phi = \frac{1}{2}\xi \int_{\partial M} d^3x h^{1/2} K\Phi^2$$
$$+ \frac{1}{2} \int_M d^4x g^{1/2} [(\nabla\Phi)^2 + \xi R\Phi^2 + M^2\Phi^2 + V(\Phi)] \quad (5.3)$$

We restrict the matter field to be homogeneous following the symmetries of the geometry.

With these restrictions, the three geometry of a constant t spacelike surface is characterized by a single number a_0 and the field by its homogeneous value on the surface Φ_0. The minisuperspace is thus two dimensional and we write

$$\Psi_0 = \Psi_0(a_0, \Phi_0) \quad . \quad (5.4)$$

The gravitational action on this minisuperspace is

$$I_E = \frac{1}{2} \int d\tau [\frac{N}{a}] [-(\frac{a\dot{a}}{N})^2 - a^2 + H^2 a^4] \quad , \quad (5.5)$$

where

$$H^2 = \sigma^2 \Lambda/3 \quad . \quad (5.6)$$

(The H used in this section thus differs by the normalizing factor σ^2 from that defined previously.) The action for the matter may be conveniently written in terms of the rescaled variables

$$\varphi = (2\pi^2\sigma^2)^{1/2}\Phi \quad , \quad m = \sigma M \quad ,$$
$$v(\varphi) = \sigma^2 V(\Phi) \quad . \quad (5.7)$$

It is

$$I_\varphi = \frac{1}{2} \int d\tau (\frac{N}{a}) [\frac{a^4}{N^2}(\dot{\varphi} + 6\xi \frac{\dot{a}}{a}\varphi)^2 + 6\xi a^2 \varphi^2$$
$$+ a^4 (m^2\varphi^2 + v(\varphi))] \quad . \quad (5.8)$$

The kinetic energy part of (5.8) may be diagonalized by a further rescaling

$$\varphi = \chi/a^{6\xi} \quad . \quad (5.9)$$

For the total action we then have

$$I = \frac{1}{2} \int d\tau (\frac{N}{a}) [-(\frac{a}{N}\dot{a})^2 + (\frac{a^{2-6\xi}}{N}\dot{\chi})^2 + U(a,\chi)] \quad , \quad (5.10a)$$

where

$$U(a,\chi) = -a^2 + H^2 a^4 + 6\xi a^2 \varphi^2 + a^4 (m^2 \varphi^2 + v(\varphi)) \quad . \quad (5.10b)$$

and φ is understood to be a function of a and χ through (5.9).

The Hamiltonian constraint follows from the action by varying it with respect to the lapse N and expressing the resulting classical equation in terms of the variables and conjugate momenta. Recalling that, for example, $\pi_\chi = -i\partial L_{Euclidean}/\partial \dot{\chi}$ one finds

$$\frac{1}{2a}[-\pi_a^2 + a^{12\xi-2}\pi_\chi^2 + U(a,\chi)] = 0 \quad . \quad (5.11)$$

We can write this as

$$\frac{1}{2} G^{AB} \pi_A \pi_B + \frac{1}{2a} U(a,\chi) = 0 \quad , \quad (5.12)$$

where G_{AB} is a minisupermetric on our minisuperspace. In (a,χ) coordinates

$$G_{AB} = \begin{pmatrix} -a & 0 \\ 0 & a^{3-12\xi} \end{pmatrix} \quad . \quad (5.13)$$

The Wheeler-DeWitt equation is the operator form of (5.12). In constructing it there are ambiguities of factor ordering which can only be resolved through a careful analysis of the measure of the sum over histories. As the precise form will not be very important for us we shall simply write

$$\frac{1}{2} [\nabla^2 - \frac{1}{a} U(a,\chi)] \Psi(a,\chi) = 0 \quad . \quad (5.14)$$

Here, ∇^2 is the "covariant" Laplacian constructed from G_{AB},

$$\nabla^2 = \frac{1}{\sqrt{-G}} \frac{\partial}{\partial x^A} (\sqrt{-G}\, G^{AB} \frac{\partial}{\partial x^B}) \quad , \quad (5.15)$$

in "general coordinates" on the two dimensional minisuperspace or

$$\nabla^2 = -\frac{1}{a^{2-6\xi}} \frac{\partial}{\partial a} (a^{1-6\xi} \frac{\partial}{\partial a}) + \frac{1}{a^{3-12\xi}} \frac{\partial^2}{\partial \chi^2} \quad , \quad (5.16)$$

in the coordinates (a,χ). From either (5.11) or (5.16) we recognize that the Wheeler-DeWitt equation is hyperbolic with $\underline{a}$ being a "timelike direction" in the minisuperspace.

The wave function of minimum excitation, $\Psi_0(a_0,\chi_0)$, in these minisuperspace models is the sum of $\exp(-I)$ over all compact geometries of the form (5.2) with a single three sphere boundary of radius a_0 and over all regular configurations of scalar field which match χ_0 on the boundary. If we fix the last gauge freedom explicitly by taking $N = a$, and denote the time coordinate in this special gauge by η, then

$$\Psi_0(a_0,\chi_0) = \int_{C_0} \delta a\, \delta\chi \exp(-I[a,\chi]) \quad , \quad (5.17)$$

where

$$I[a,\chi] = \frac{1}{2} \int_{-\infty}^{0} d\tau [-a'^2 + (a^{1-6\xi}\chi')^2 + U] \quad . \quad (5.18)$$

and a prime denotes an η-time derivative. With this choice of gauge, the "south pole" of the geometry is located at $\eta = -\infty$ (see Fig. 7). We have used the residual η-time translation invariance to locate the boundary by $\eta = 0$. This fixes the limits of the η-coordinate range. We integrate over those $a(\eta)$ which vanish at $\eta = -\infty$ so the geometry is regular at its "south pole" and over field configurations $\chi(\eta)$

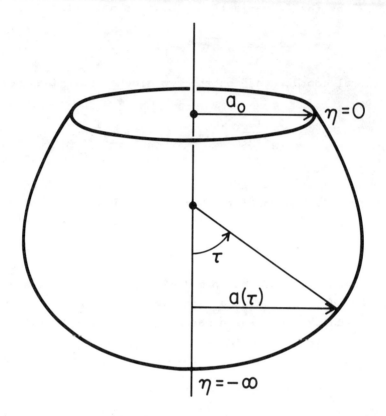

Fig. 7. A two dimensional representation of a homogeneous and isotropic 4-geometry contributing to the sum for the state of minimum excitation $\Psi_0(a_0,\chi_0)$. Shown embedded in a flat 3-dimensional space is a 2-dimensional slice of such a geometry whose intrinsic geometry is

$$d\Sigma^2 = d\tau^2 + a^2(\tau)d\varphi^2$$

τ is thus a "polar angle" and a the "radius from the axis." The geometry is compact and has only one boundary at which the radius is a_0, the argument of Ψ_0. In the time coordinate η such that $ad\eta = d\tau$ the "south pole" is located at $\eta = -\infty$ and the boundary at $\eta = 0$. The field configurations $\varphi(\tau)$ which contribute to the sum are those which are regular on this surface and which match the argument of the wave function Ψ_0 on the boundary.

which are regular on this geometry. Thus $\chi = 0$ at $\eta = -\infty$ when $\xi > 0$. Ψ_0 satisfies the Wheeler-DeWitt equation for the minisuperspace, (5.14). The functional integral supplies the boundary conditions for singling out the state of minimum excitation from among all other solutions of the Wheeler-DeWitt equation.

We shall now calculate Ψ_0 approximately for two different model actions for the scalar field. To do this we shall move back and forth between evaluating the integral and solving the equation.

5.2.2 A conformally invariant field

If $m = 0$ and $v = 0$ and $\xi = 1/6$ the scalar field is conformally invariant. This is not a very realistic model of matter - it lacks any particle physics scale, for example - but it does lead to an easily analyzable example. The reason is that geometries of the form (5.2) are conformally static. (To see this just put $N = a$.) Since the field is conformally invariant its dynamics are thus essentially trivial.

The action (5.18) for this special case reads

$$I = \frac{1}{2} \int_{-\infty}^{0} d\tau [-a'^2 - a^2 + H^2 a^2 + \chi'^2 + \chi^2] \quad . \quad (5.19)$$

The scalar field action decouples from the gravitational one. Indeed, since the action for χ is just the Euclidean action for a harmonic oscillator the integral over χ in (5.17) is purely gaussian and easily evaluated to find

$$\Psi_0(a_0, \chi_0) = \exp(-\frac{1}{2}\chi_0^2) \Phi(a_0) \quad . \quad (5.20)$$

The one homogeneous mode of the scalar field is in its ground state as one would expect for the state of minimum excitation.

Ψ_0 must satisfy the Wheeler-DeWitt equation and this gives a differential equation for Φ. Choosing the operator ordering as in (5.16) we have

$$-\frac{d^2\Phi}{da^2} + (a^2 - H^2 a^4)\Phi = \Phi \quad . \tag{5.21}$$

This is just a "Schrödinger equation" for a particle in a potential

$$V(a) = a^2 - H^2 a^4 \quad . \tag{5.22}$$

The boundary conditions for Φ are to be extracted from the integral (5.17). Under the simplest interpretation of the measure which is consistent with (5.16) we have $d\Phi/da = 0$ at $a = 0$. The overall normalization is as yet arbitrary. Some typical solutions are shown in Figure 8.

It will be of later interest to see how solutions to (5.21) arise semiclassically from (5.17). The integral defining $\Phi(a_0)$ is

$$\Phi(a_0) = \int_{C_0} \delta a \exp(-I_E[a]) \quad , \tag{5.23}$$

where I_E is (5.5) with $N = a$. Evaluation of (5.23) by the method of steepest descents gives the semiclassical approximation. For this we must find the extrema of I_E through which the contour of integration can be distorted. We begin with values of a_0 less than H^{-1}. The possible extrema of I_E are just the solutions of

$$a'' - a - 2H^2 a^3 = 0 \quad . \tag{5.24}$$

The equation has an "energy integral" whose value may be found from the regular vanishing of a at $\eta = -\infty$. Expressing this integral in terms of τ gives

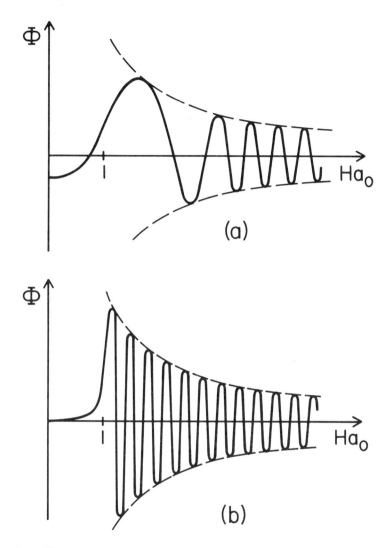

Fig. 8. The wave function Φ for the homogeneous, isotropic minisuperspace model with conformally invariant scalar field. Figure 8a shows a sketch of Φ for $H \approx 1$, Figure 8b for a much larger value of H. As H decreases the amplitude to find a 3-sphere of radius $a_0 < 1/H$ becomes very small. This is the classically forbidden region for de Sitter evolution (Figure 11). For $a_0 > 1/H$ the envelope approaches the distribution of 3-spheres in de Sitter space.

$$\left(\frac{\dot{a}}{a}\right)^2 = \frac{1}{a^2} - H^2 \quad . \tag{5.25}$$

This is the Euclidean Einstein equation for a metric with the symmetries of the model as it must be. The solution is illustrated in Figures 9 and 10 and is just the 4-sphere of radius $1/H$. For $a_0 < 1/H$ there are thus two possible extrema which are compact 4-geometries with a 3-sphere boundary of radius a_0. One for which the boundary bounds less than a hemisphere of the 4-sphere and another for which it bounds more. The action for the 4-sphere is negative and therefore one might think that the extremum encompassing more 4-sphere should dominate. One must remember, however, that because of the conformal rotation the contour of $\underline{a}$ integration is in the imaginary direction in the immediate vicinity of the extremum. Extrema of analytic functions are saddle points so that a maximum in a real direction is a minimum in an imaginary direction. The stationary configuration which contributes to the steepest descent evaluation of (5.23) is the one which is a maximum of the action in real directions and a least action configuration in imaginary directions. The extremum corresponding to the smaller part of the 4-sphere, therefore, provides the steepest descent approximation to the wave function. In fact, the contour cannot be distorted to pass through the other extremum. We thus have for $a_0 < 1/H$

$$\Phi(a_0) \approx N[-1+a_0^2-H^2 a_0^4]^{-1/4}$$
$$\times \exp[-\frac{1}{3H^2}(1-H^2 a_0^2)^{3/2}] \quad , \tag{5.26}$$

where N is an arbitrary normalizing factor.

If a_0 is increased to a value larger than $1/H$ there are no longer any real extrema because a 3-sphere of radius

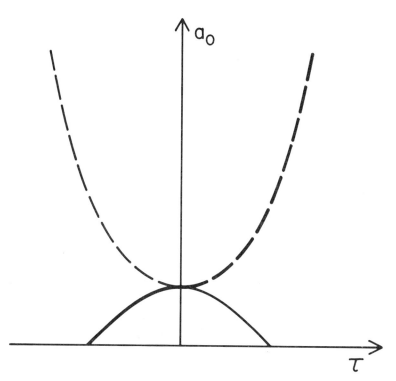

Fig. 9. The extremizing scale factor for the homogeneous, isotropic minisuperspace model with conformally invariant scalar field. The solid line is the solution of (5.25) for real Euclidean extrema of the action. The complete range of a from zero to maximum and back again describes the geometry of the 4-sphere (Figure 10). The dashed curve is the solution of (5.27) for complex Euclidean (Lorentzian) extrema. It describes the geometry of de Sitter space (Figure 11). For each value of a_0 there are thus two possible extremizing solutions. Choosing the trajectory to start on the left at $a_0 = 0$ the Euclidean prescription for the state of minimum excitation singles out the heavy curve shown. This gives the semiclassical approximation to the wave function Ψ_0.

$a_0 > 1/H$ cannot fit into a 4-sphere of radius $1/H$. There are, however, complex extrema. These can be obtained by changing $\tau \to \pm it$ in Eq. (5.25) so they solve

$$\left(\frac{\dot{a}}{a}\right)^2 = H^2 - \frac{1}{a^2} \quad . \tag{5.27}$$

They are thus the solutions of the Lorentzian Einstein equations with positive cosmological constant. This solution is called de Sitter space (Figure 11). These complex extrema must contribute in complex conjugate pairs so that the wave function is real. By a standard WKB matching analysis we can establish the form of the wave function for $a_0 > H^{-1}$

$$\Phi(a_0) \approx 2N [H^2 a_0^4 - a_0^2 + 1]^{1/4}$$
$$\times \cos\left[\frac{(H^2 a_0^2 - 1)^{3/2}}{3H^2} - \frac{\pi}{4}\right] \quad . \tag{5.28}$$

This form could be derived by carefully following the extremum configuration as a_0 is increased along the heavy curve shown in Fig. 9.

The complete wave function Ψ_0 on the minisuperspace of homogeneous isotropic geometries with conformally invariant scalar field is given by

$$\Psi_0(a_0, \varphi_0) = \exp[-\varphi_0^2/(2a_0^2)] \Phi(a_0) \quad , \tag{5.29}$$

where Φ is given approximately by (5.26) and (5.28). From this correlations between field and geometry can be extracted. The exponential factor gives an inverse correlation between φ_0 and a_0. Large φ_0 occurs at small a_0 and vice versa. This is the type of correlation that occurs in classical evolution.

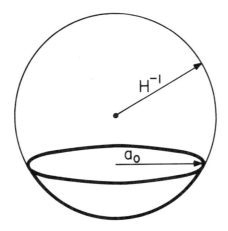

Fig. 10. The real Euclidean extrema of the homogeneous, isotropic minisuperspace model with conformally invariant scalar field have the geometry of a 4-sphere of radius H^{-1}. The extremizing configuration which gives the semiclassical approximation to Φ at $a_0 < 1/H$ is a part of the 4-sphere with a single 3-sphere boundary of radius a_0. There are two possibilities corresponding to more than a hemisphere or less. The Euclidean functional integral prescription for Φ identifies the smaller part of the 4-sphere as the contributing extremum. For $a_0 > H^{-1}$ there are no real extrema.

The factor Φ suppresses correlations for $a_0 < H^{-1}$ but affects them only weakly for $a_0 > H^{-1}$. The classical solution of Einstein's equation restricted by the minisuperspace assumptions is de Sitter space. This homogeneous, isotropic, empty geometry may be thought of (Figure 11) as the evolution of a 3-sphere which contracts to a minimum radius H^{-1} and reexpands. The region of minisuperspace with $a_0 < H^{-1}$ is thus classically forbidden and Φ suppresses correlations there. The region $a_0 > H^{-1}$ is classically allowed and there Φ varies only weakly.

The analysis leading to (5.28) gives another way of summarizing the information contained in Ψ_0 in the semiclassical limit. In the classically allowed region, $a_0 > H^{-1}$, the semiclassical approximation to Ψ_0 is given by a complex Euclidean but real Lorentzian extremum of the action. The wave function oscillates proportionally to cos(S) where S satisfies the Lorentzian Hamilton-Jacobi equation. This action specifies a solution to the equations of motion up to initial conditions. In this semiclassical approximation the wave function thus corresponds to an ensemble of Lorentzian de Sitter spaces which differ from one another only in the time assigned the minimum radius. It is in this way we recover the classical limit.

5.2.3 A massive scalar field

The conformally invariant scalar field is not a realistic model of the matter in the universe. It contains no scale. Within the general framework of the minisuperspace models discussed in Section 5.2.1, a more realistic model is provided by a free, massive, minimally coupled scalar field. This was discussed by Hawking[2] in the case $\Lambda = 0$. The parameters of this model are thus $\xi = 0$, $\Lambda = 0$, $v = 0$, and $m \neq 0$.

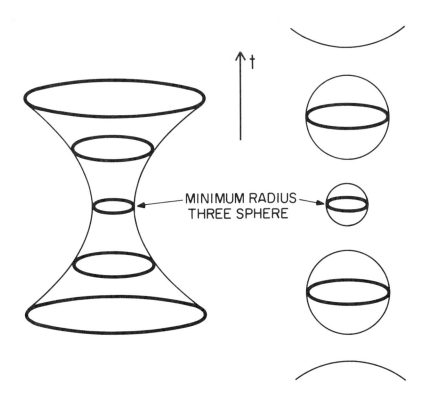

Fig. 11. Lorentzian de Sitter Space. In the classically allowed region of the minisuperspace of homogeneous isotropic geometries with conformally invariant scalar field, the wave function corresponds semiclassically to Lorentzian de Sitter space. This is the most symmetric solution of the source free Einstein's equation with positive cosmological constant. It is the geometry of a Lorentz hyperboloid in a 5-dimensional Lorentz signatured spacetime. It may be thought of as a three sphere which collapses to a minimum radius $(3/\Lambda)^{1/2}$ and then re-expands.

The metric on minisuperspace in this model is [(5.13)]

$$G_{AB} = a \begin{pmatrix} -1 & 0 \\ 0 & a^2 \end{pmatrix} . \qquad (5.30)$$

This is conformal to the metric on the interior of the forward light cone in a two dimensional Minkowski space. To see this, introduce new coordinates

$$x = a \sinh\varphi$$
$$y = a \cosh\varphi \qquad (5.31)$$

($\chi = \varphi$ in this case when $\xi = 0$.) The metric in x, y coordinates is

$$G_{AB} = (y^2-x^2)^{1/2} \begin{pmatrix} -1 & 0 \\ 0 & 1 \end{pmatrix} . \qquad (5.32)$$

The new and old coordinates can be conveniently plotted on an x-y diagram as in Figure 12a. The Wheeler-DeWitt equation becomes

$$[-\frac{\partial^2}{\partial y^2} + \frac{\partial^2}{\partial x^2} - U(x,y)]\Psi(x,y) = 0 , \qquad (5.33)$$

where $U(x,y)$ is the "potential" of eq. (5.10b). Expressed in terms of x and y it is

$$U(x,y) = x^2-y^2+m^2(y^2-x^2)^2[\tanh^{-1}(\frac{x}{y})]^2 . \qquad (5.34)$$

The first term is from the spatial curvature in the Wheeler-DeWitt equation. The second is the contribution of the scalar field's mass to its energy.

Eq. (5.33) is a wave equation with potential in one space and one time dimension. It could be integrated numerically if boundary conditions could be found for Ψ_0. These boundary conditions are supplied by sum over compact Euclidean histories (5.17) which defines the quantum state

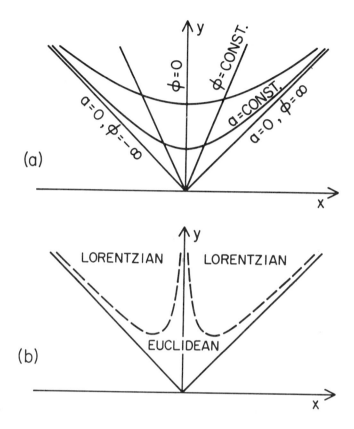

Fig. 12. (a) Two sets of coordinates on the minisuperspace of homogeneous isotropic geometries with minimally coupled massive scalar field. The minisuperspace is conformal to the forward light cone of a 2-dimensional Minkowski space (y,x). Curves of constant spatial volume are the hyperbolae a = constant. Curves of constant scalar field are straight lines through the origin.

(b) The classically allowed and classically forbidden regions of minisuperspace. In the classically allowed region the semiclassical approximation is given by a Lorentzian extremum of the action and the wave function oscillates. In the classically forbidden regions the semiclassical approximation is given by a Euclidean extremum and the wave function is nonoscillatory.

of the universe and distinguishes it from all other solutions of the Wheeler-DeWitt equation. A convenient place to evaluate these boundary conditions is the characteristic surface $y = |x|$ corresponding to $a = 0$, $\varphi = \pm \infty$. The boundary condition needed to integrate the Wheeler-DeWitt equation is the value of Ψ_0 on this surface. To evaluate this and also to obtain a qualitative understanding of its behavior we consider the semiclassical approximation to (5.17).

Semiclassicaly, we expect to find

$$\Psi_0(a_0,\varphi_0) \approx A(a_0,\varphi_0)\exp[-I(a_0,\varphi_0)] \quad , \qquad (5.35)$$

for those values of a_0 and χ_0 for which there is a real Euclidean extremum of the action. For those values for which the extremum is complex Euclidean (i.e. real Lorentzian) we expect

$$\Psi_0(a_0,\varphi_0) \approx A(a_0,\varphi_0)\cos[S(a_0,\varphi_0)] \quad , \qquad (5.36)$$

where S is the Lorentzian action. In this case the wave function will oscillate. The equations of motion which a Euclidean extremum must satisfy follow from varying (5.10a) with respect to N and φ. In the gauge where $N = 1$ they are

$$\ddot{\varphi} + \frac{3\dot{a}}{a}\dot{\varphi} - m^2\varphi^2 = 0 \quad , \qquad (5.37a)$$

$$\left(\frac{\dot{a}}{a}\right)^2 = \frac{1}{a^2} + \dot{\varphi}^2 - m^2\varphi^2 \quad . \qquad (5.37b)$$

A solution which is a compact geometry starts at some τ (say $\tau = 0$) where $a = 0$ (Figure 7). There, $\dot{\varphi} = 0$ in order for the field to be regular. The value $\varphi(0)$ is arbitrary. There are thus a one parameter family of solutions to (5.37) which correspond to compact geometries with regular field configurations. A typical one looks schematically like Figure 13. The solution whose action gives the wave function

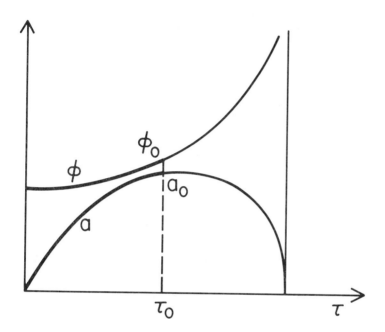

Fig. 13. A typical Euclidean extremizing configuration. The geometry has a "south pole" where $a = 0$, $\dot{\varphi} = 0$ and φ takes some value $\varphi(0)$. (cf. Figure 7.) If eqs. (5.37) are integrated forward in the "polar angle" τ a "north pole" is eventually reached where $a = 0$ and both geometry and field are singular. A compact, non-singular geometry with 3-sphere boundary on which $a = a_0$ and $\varphi = \varphi_0$ is obtained by locating the τ_0 for which $a = a_0$ and then varying $\varphi(0)$ until $\varphi = \varphi_0$ at that value of τ_0. This extremizing configuration is indicated by the heavier curves.

at a given value of a_0 and φ_0 is determined by adjusting $\varphi(0)$ until φ assumes the value φ_0 at the value of τ for which $a = a_0$. There may be several solutions for which this is possible. This phenomenon is familiar from the example of the conformally invariant scalar field where there were solutions to the equation of motion (5.25) corresponding to less than a hemisphere of the 4-sphere and also to more than a hemisphere. The solution which gives the semiclassical approximation, in that case as in this, is the one for which the action is a maximum for real variations. Explicit calculation[2,54] shows that such solutions are confined to a region of minisuperspace for which approximately

$$a_0 < (m|\varphi_0|)^{-1} \text{ or } |\varphi_0| \lesssim 1 \quad . \tag{5.38}$$

Inside this region the wave function varies without oscillation. Outside this region it oscillates. In the semiclassical approximation this behavior emerges because, outside the region (5.38), the extrema are complex Euclidean and the wave function behaves as in (5.36). These complex extrema are real Lorentzian geometries and field histories which obey

$$\ddot{\varphi} + \frac{3\dot{a}}{a}\dot{\varphi} + m^2\varphi = 0 \quad , \tag{5.39a}$$

$$\left(\frac{\dot{a}}{a}\right)^2 = -\frac{1}{a^2} + \dot{\varphi}^2 + m^2\varphi^2 \quad , \tag{5.39b}$$

a dot denoting differentiation with respect to Lorentzian time. Indeed, semiclassically, the quantum state may be thought of as corresponding to an ensemble of classical histories which obey these equations.

The values $a_0 \to 0$ and $\varphi_0 \to \pm\infty$, of interest for evaluating the characteristic initial value data for the Wheeler-DeWitt equation, are within the region of minisuperspace defined by (5.38). There is therefore always a real solution of (5.37) giving the semiclassical approximation to Ψ_0. The action approaches zero as $a_0 \to 0$ since φ is regular. Thus, to the extent the variation in the prefactor A can be neglected* in (5.35) we have

$$\Psi_0 = \text{constant} . \qquad (5.40)$$

on the characteristic initial value surface $y = |x|$.

A numerical integration of the Wheeler-DeWitt equation with the boundary condition (5.40) is shown in Figure 14. (For more see Refs. 41, 51, 63.) In the classically forbidden region (Figure 12b, Eq. (5.38)) the wave function varies smoothly. In the classically allowed region it oscillates.

Figure 15 shows a schematic representation of a typical solution to (5.39). For sufficiently large φ the explicit computation[2,54] shows that the transition from Euclidean to Lorentzian extremum occurs when $a \approx (m|\varphi|)^{-1}$ [cf.(5.38)] and when $\dot\varphi$ and $\dot a$ are approximately zero. From (5.39), a classical trajectory thus starts from a field value φ_1 which is a local maximum and a scale factor which is a local minimum, $a_1 \approx (m|\varphi_1|)^{-1}$. In the subsequent evolution, while φ is nearly constant the term $m^2\varphi^2$ in (5.39b) behaves as an effective cosmological constant. The universe thus inflates with a time scale $m|\varphi_1|$. Eventually φ decreases, begins to oscillate and the matter field acquires kinetic energy of its own.

In the classically allowed region of minisuperspace the wave function constructed as the sum over compact, regular histories corresponds semiclassically to an ensemble of classical histories each characterized by a value of φ_1. The histories have an initial inflationary

---

*The prefactor in the case of pure gravity has been evaluated by K. Schleich.[62]

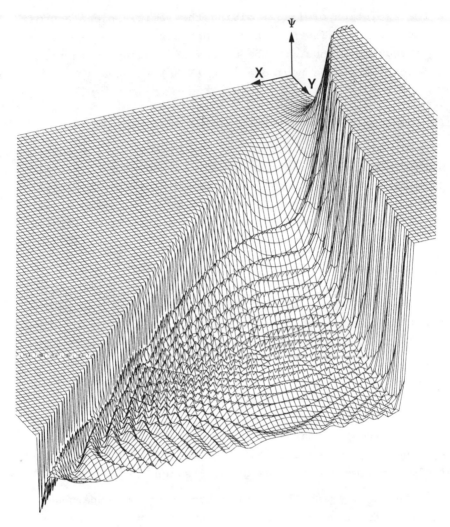

Fig. 14. A numerical integration of the Wheeler-DeWitt equation in the homogeneous, isotropic, massive scalar field minisuperspace model. The figure shows the integration of E. P. Shellard[63] of the Wheeler-DeWitt equation (5.33) with the boundary conditions (5.39). Ψ_0 is plotted as a function of the coordinates x and y. Oscillatory and non-oscillatory regions can clearly be distinguished and these correspond to those predicted semiclassically and shown in Figure 12b.

epoch and, later, an epoch in which the matter field has kinetic energy and the expansion behaves as though matter dominated. Thus through the inflationary mechanism, the universe, although in the analog of the ground state, can become large, approximately flat and contain matter.

5.3 Linearized Fluctuations about Homogeneous Isotropic Models

Minisuperspace models which assume homogeneity and isotropy can neither provide an explanation of this large scale feature nor of the observed spectrum of fluctuations away from it. On the one hand an explanation of homogeneity and isotropy can only come by comparing the wave function on configurations which have these symmetries with those which do not, and on the other fluctuations cannot be studied in geometries which do not have them. To progress with either question one needs to enlarge the minisuperspace to include geometries which are not homogeneous and isotropic. Of the several models of this type[44,47,48] perhaps the most complete is that of Halliwell and Hawking.[49] They considered linear fluctuations away from the homogeneous and isotropic models with massive scalar field discussed in the preceding subsection. They discuss the most general fluctuations and thus explore completely a small domain of superspace about exact homogeneity and isotropy. Their model is thus not strictly a minisuperspace model but contains an infinite number of degrees of freedom.

In the following we shall sketch the assumptions and method of Halliwell and Hawking's calculation and quote some of their results.

The model considers Euclidean histories which deviate only slightly from exact homogeneity and isotropy. Metric and field can thus be written

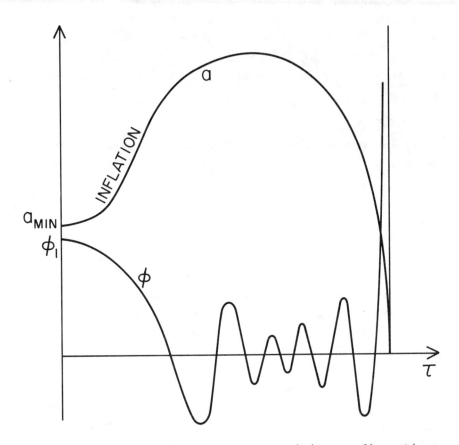

Fig. 15. A typical Lorentzian extremizing configuration. The solution shown schematically here starts at a minimum radius with $\dot\varphi = 0$. In the domain where φ varies slowly the universe follows a de Sitter like inflationary expansion with $H = m\varphi_1$. Later the scalar field begins to oscillate and the universe evolves approximately as though matter dominated. Eventually a maximum expansion is reached, the universe recollapses and matter and geometry become singular. A sufficiently large $m\varphi_1$ would provide a long enough inflationary period to explain the present large size of the universe and its approximate spatial flatness. The oscillation of the scalar field models the creation of matter.

$$ds^2 = d\hat{s}^2 + \epsilon_{\alpha\beta}(x) dx^\alpha dx^\beta \quad , \tag{5.41a}$$

$$\varphi = \hat{\varphi}(\tau) + f(x) \quad . \tag{5.41b}$$

Here $d\hat{s}^2$ is the homogeneous, isotropic line element (5.2), $\hat{\varphi}$ a homogeneous field configuration and $\epsilon_{\alpha\beta}$ and f the inhomogeneous deviations from these quantities. Both $\epsilon_{\alpha\beta}$ and f can be expanded in the harmonics of the homogeneous, isotropic 3-sphere. Schematically, these expansions have the form

$$\epsilon_{\alpha\beta}(\tau, x^i) = \Sigma_{(n)} \epsilon^{(n)}(\tau) Q^{(n)}_{\alpha\beta}(x^i) \quad , \tag{5.42a}$$

$$f(\tau, x^i) = \Sigma_{(n)} f^{(n)}(\tau) Q^{(n)}(x^i) \quad , \tag{5.42b}$$

where $Q^{(n)}_{\alpha\beta}$ and $Q^{(n)}$ are a complete set of tensor and scalar harmonics. Near the homogeneous, isotropic configurations, the $\epsilon^{(n)}$ and $f^{(n)}$ may be regarded as "coordinates" on superspace and we write

$$\Psi_0 = \Psi_0(\hat{a}, \hat{\varphi}, \epsilon^{(1)}, f^{(1)}, \epsilon^{(2)}, f^{(2)}, \ldots) \quad . \tag{5.43}$$

Classically the linearized field equations for the fluctuations $\epsilon^{(n)}$, $f^{(n)}$ decouple into a set for each mode. Quantum mechanically, the Wheeler-DeWitt equation and the associated constraints (4.9a) separate when written to quadratic order in the variables describing the fluctuations. That is, the wave function is a product

$$\Psi_0 = \hat{\Psi}(\hat{a}, \hat{\varphi}) \prod_{(n)} \psi^{(n)}(\hat{a}, \hat{\varphi}, \epsilon^{(n)}, f^{(n)}) \quad . \tag{5.44}$$

The Wheeler-DeWitt equation implies for $\hat{\Psi}$ an equation of the form

$$\left[\nabla^2 - \frac{1}{2a} U(a, \varphi) + \begin{pmatrix} \text{source term} \\ \text{quadratic} \\ \text{in the } \psi^{(n)} \end{pmatrix}\right] \hat{\Psi}(\hat{a}, \hat{\varphi}) = 0 \quad . \tag{5.45}$$

The first two terms in this equation are the Wheeler-DeWitt equation of the homogeneous, isotropic massive scalar field model [eq. (5.14)]. The additional term represents the expected energy of the fluctuations. In the semiclassical approximation one expects solutions for $\hat{\Psi}$ of the form (5.35) or (5.36).

In the classically allowed region, the semiclassical approximation to $\hat{\Psi}$ is of the form (5.36). The action $S(a_0, \varphi_0)$ is the action of classical histories $a(t)$, $\varphi(t)$ which satisfy the classical equations of motion. When $\hat{\Psi}$ is approximated in this way, the Wheeler-DeWitt equation implies a "Schrödinger" equation for $\psi^{(n)}$

$$i \frac{\partial \psi^{(n)}}{\partial t} = H^{(n)} \psi^{(n)} . \qquad (5.46)$$

$\psi^{(n)}$ and $H^{(n)}$ become functions of time through the connection between a and t provided by the classical trajectory $a(t)$. This is generally the way that the notion of time is recovered in the semiclassical approximation to the quantum dynamics of spacetime.[64] The dynamics of the fluctuations, in effect, becomes the ordinary quantum dynamics of fields moving in the background spacetime which provides the semiclassical approximation to $\hat{\Psi}$.

Halliwell and Hawking solve (5.45) and (5.46) with boundary conditions extracted from the Euclidean sum over histories specification of the wave function. They argue that for this wave function the additional source term in (5.44) is small after appropriate renormalization. The wave function $\hat{\Psi}$ and the classical trajectories which give its semiclassical approximation are thus those of the minisuperspace model discussed in the previous subsection. These display an early inflationary phase followed by a transition to a matter dominated evolution. Eq. (5.46) shows that the fluctuations will evolve as quantum fields in this background spacetime. Semiclassically the evolution will therefore be much like the evolution of fluctuations in the standard inflationary universe history.[65] What the wave function of the universe supplies is the boundary conditions to begin this evolution.

Halliwell and Hawking find first that at all stages in which their approximation is valid the wave function is peaked about isotropy and homogeneity. The fluctuations away from this symmetry begin in their ground state. They remain in their ground state until they expand outside the Hubble radius. Their amplitude then remains frozen until they reenter the Hubble radius in the matter dominated era. There results a scale free (Zel'dovich) spectrum of fluctuations which, for the correct choice of the scalar field mass m, can have the amplitude to correctly reproduce the spectrum of density fluctuations we observe.

6. OBSERVATIONS AND PREDICTIONS

The accompanying table offers a comparison between the large scale observations of our universe reviewed in Section 2, and the predictions of the proposal for the quantum state of universe worked out in minisuperspace models. The predictions are consistent with the observations. Essential to this consistency is the action of an inflationary mechanism when the universe is small. Without such action there would be no natural way for the cosmological analog of the ground state to explain the large size of the universe, its approximate spatial flatness and its matter content. It is encouraging that inflation appears to occur naturally in a wide range of matter models.[2,43,50] Inflation, however, makes it difficult to test the proposal for the quantum state in a definitive way. Inflation, it will be recalled from Section 2, was the dynamical mechanism which successfully explained several large scale features for a wide range of "reasonable" initial conditions. It is thus difficult to test any theory which makes specific predictions about initial conditions and which involves an inflationary mechanism in an essential way.

Views on what are "reasonable" initial conditions and what are not are inevitably subjective. A more fundamental question is whether the states which are consistent with our present observations are a large subset of the set of all possible states or a small one. Arguments of Penrose[11] suggest that it is small by a very large factor. In this sense, present observations already lead to a strong test of the proposal.

OBSERVED PROPERTY	MINISUPERSPACE MODELS	RESULTS	SELECTED REFERENCES
Spacetime is 4-dimensional with Euclidean topology	Local properties unstudied but some Kaluza-Klein models	There are preferred compactifications in Kaluza-Klein models	40,45,46, 53,60
The universe is large and old	Homogeneous and isotropic geometry with either massive scalar field or in (curvature)^2 theories	There are trajectories along which the universe "inflates" to a large size	2,42,43, 50,52,56, 57
Matter and geometry are nearly homogeneous and isotropic	Homogeneous but anisotropic models	The wave function is sharply peaked about zero anisotropy	44,47,48
Space is nearly flat	Homogeneous and isotropic geometry with either massive scalar field or in (curvature)^2 theories	The distribution of Ω_0 is sharply peaked about $\Omega_0 = 1$	2,28
The spectrum of density fluctuations	Homogeneous and isotropic geometry with massive scalar field plus linear inhomogeneous perturbations in matter and geometry	The wave function predicts initial quantum fluctuations which evolve classically during an inflationary epoch to give a scale free spectrum with a plausible amplitude	49
The entropy of the universe is low and increasing in the direction of expansion	Homogeneous and isotropic geometry with massive scalar field plus linear inhomogeneous perturbations in matter and geometry	Order when the universe is small evolves to disorder when the universe is big	49,61

557

Evaluating the wave function on some of the more exotic regions of superspace is important for this kind of test. As yet, the quantum state of the universe has only been calculated in minisuperspace models built on symmetries which closely resemble those of the present universe or which deviate slightly from them. These explore but a small part of the whole of superspace. The observed universe is located in this part but it is important to demonstrate that the wave function of the universe has support on no other.

The proposal developed by Stephen Hawking and his collaborators for the quantum state of the universe is one of compelling simplicity and beauty. In the models tested to date it agrees remarkably well with observations. In the process of exploring this idea on ever larger domains of superspace with ever better theories of gravity and matter we shall certainly learn more about quantum gravity. We may well be exploring the state of the universe in which we live.

7. CONCLUDING REMARKS

The point of view developed in these lectures might be summarized in a minimal way in the following three statements. If the reader can carry only three ideas away from these lectures I would hope that they would be these:

a) Cosmology requires a law for initial conditions. This will involve quantum gravity.

559

b) There are basic issues in the kinematics and interpretation of a quantum gravitational theory which are not those of standard flat space field theory. The sum over histories formulation of quantum mechanics may guide their resolution.

c) As conjectured by Stephen Hawking and his collaborators, the universe may be in its ground state and all the features of the universe we see about us may have their origin in the special properties of this state and its quantum fluctuations.

Acknowledgments

The author has benefited from discussions with many colleagues in the preparation of these lectures. He would like to thank C. Hogan, M. Turner, G. Steigman and D. Wilkinson for help reviewing the current status of the observations. He is grateful to P. Lubin for supplying Figure 1 and E.P. Shellard for Figure 14. The preparation of these lectures was supported in part by the National Science Foundation under grant PHY 85-06686.

A. Notational Appendix

For the most part we follow the conventions of Ref. 10 with respect to signature, curvature and indices. In particular:

Signature: $(-,+,+,+)$ for Lorentzian spacetimes.

$(+,+,+,+)$ for Euclidean spacetimes.

Indices: Greek indices range over spacetime from 0 to 3. Latin indices range over space from 1 to 3.

Units: We use units in which $\hbar = c = 1$. The Planck length is $\ell = (16\pi G)^{1/2} = 1.15 \times 10^{-32}$ cm.

Minkowski metric: $\eta_{\alpha\beta} = \text{diag}(-1,1,1,1)$.

Covariant Derivatives: ∇_α denotes a spacetime covariant derivative and D_i a spatial one.

Traces and Determinants: Traces of second rank tensors $K_{\alpha\beta}$ are written as $K = K^\alpha{}_\alpha$ except when the tensor is the metric in which case g is the determinant of $g_{\alpha\beta}$ and h the determinant of h_{ij}.

Squares: If $A_{\alpha\beta\gamma\ldots}$ is a tensor, $(A_{\alpha\beta\gamma\ldots})^2$ means $A_{\alpha\beta\gamma\ldots} A^{\alpha\beta\gamma\ldots}$.

Symmetrization: $A_{(\alpha\beta)} = \frac{1}{2}(A_{\alpha\beta} + A_{\beta\alpha})$

Extrinsic Curvatures: If n_α is the unit normal to a spacelike hypersurface in a Lorentzian spacetime we define its extrinsic curvature to be

$$K_{ij} = -\nabla_i n_j \ .$$

If the surface is embedded in a Euclidean spacetime we define the extrinsic curvature to be

$$K_{ij} = \nabla_i n_j \ .$$

Intrinsic Curvatures: Intrinsic curvatures are defined so that the scalar curvature of a sphere is positive.

Metric on the unit n-sphere: This is denoted by $d\Omega_n^2$ and in standard polar angles is

$$d\Omega_2^2 = d\theta^2 + \sin^2\theta d\varphi^2 \qquad n = 2$$

$$d\Omega_3^2 = d\chi^2 + \sin^2\chi d\Omega_2^2 \qquad n = 3$$

References

1. For some recent articles which bear on the question of initial conditions in quantum cosmology see Atkatz, D. and Pagels, H., Phys. Rev. D25, 2065 (1982); Vilenkin, A., Phys. Lett. B117, 25 (1983), Phys. Rev. D27, 2848 (1983), Phys. Rev. D30, 509 (1984); Narlikar, J.V. and Padmanabhan, T., Phys. Reports 100, 151 (1983), Banks, T., Nucl. Phys. B249, 332 (1985); Banks, T., Fischler, W. and Susskind, L. (preprint), Fischler, W, Ratra, B, Susskind, L., Nucl. Phys. B259, 730 (1985).

2. Hawking, S.W. Nucl. Phys. B239, 257 (1984), and other references cited below.

3. See, e.g. Balian, R., Audouze, J. and Schramm, D., Physical Cosmology: Les Houches 1979 (North Holland, Amsterdam, 1980) and Proceedings of the Inner Space/Outer Space Conference, Fermilab, May 1984 (to be published).

4. Kron, R., Physica Scripta 21, 652 (1980).

5. Peebles, P.J.E., The Large Scale Structure of the Universe (Princeton University Press, Princeton, 1980).

6. See, e.g., the review of Wilkinson, D.T. in Proceedings of the Inner Space/Outer Space Conference, May 1984 (to be published).

7. Fixen, D.J., Cheng, E.S. and Wilkinson, D.T., Phys. Rev. Lett. 50, 620 (1983).

8. Lubin, P. and Villela, T. in Proceedings of the Third Rome Conference on Astrophysics, December 1984 (to be published).

9. See, e.g. Davis, M. and Peebles, P.J.E., Ann. Rev. Astron. Ap. 21, 109 (1983).

10. Misner, C., Thorne, K. and Wheeler, J.A. Gravitation, (W.H. Freeman, San Francisco, 1970) p. 411.

11. Penrose, R., in General Relativity: An Einstein Centenary Survey, Hawking, S.W. and Israel, W. (eds.), (Cambridge University Press, Cambridge 1979).

12. See, for example, Dicke, R.H., Nature 192, 440 (1961), Carter, B, in Confrontation of Cosmological Theories with Observational Data (IAU Symp. 63), Longair, M.S. (ed.), (D. Reidel, Dordrecht, 1974) and Carr, B.J. and Rees, M.J., Nature 278, 605 (1979).

13. Misner, C.W., Ap. J. 151, 431 (1968), Matzner, R.A. and Misner, C.W., Ap. J. 171, 415 (1972), Matzner, R.A., Ap. J. 171, 433 (1972).

14. See, e.g., Guth, A. in Proceedings of the Inner Space/Outer Space Conference, Fermilab, May 1984 (to be published).
15. Misner, C.W., Phys. Rev. Lett. $\underline{22}$, 1071 (1969), Chitre, D.M.,"Investigations of the Vanishing of the Horizon for a Bianchi IX (Mixmaster) Universe," Ph.D. dissertation, University of Maryland (1972).
16. See, e.g., Starobinsky, A.A., Phys. Lett. $\underline{91B}$, 99 (1980) and Anderson, P., Phys. Rev. $\underline{D28}$, 271 (1983), Phys. Rev. $\underline{D29}$, 615 (1984).
17. See, e.g. Boucher, W. and Gibbons, G.W., in The Very Early Universe, Gibbons, G.W., Hawking, S.W., and Siklos, S.T.C. (eds.), (Cambridge University Press, Cambridge, 1983).
18. See, e.g., Hawking, S.W. and Ellis, G.F.R., Ap. J. $\underline{152}$, 25 (1968) and Geroch, R. and Horowitz, G.T., in General Relativity: An Einstein Centenary Survey, Hawking, S.W. and Israel, W. (eds.), Cambridge University Press, Cambridge (1979).
19. For more on the canonical quantum mechanics of gravity see Kuchař, K. in Relativity, Astrophysics and Cosmology, ed. by Israel, W. (D. Reidel, Dordrecht, 1973) and in Quantum Gravity 2 ed. by Isham, C., Penrose, R. and Sciama, D.W., (Clarendon Press, Oxford, 1981).
20. See, e.g., Feynman, R. and Hibbs, A., Quantum Mechanics and Path Integrals, (McGraw Hill, New York, 1965).
21. See, e.g. DeWitt, B. in Relativity, Groups and Topology II, ed. by DeWitt, B.S. and Stora, R. (Elsevier, Amsterdam, 1984) and Anderson, A. and DeWitt, B. in Proceedings of the Third Moscow Quantum Gravity Seminar (to be published).
22. For a discussion in depth of the sum over histories quantization of gravity, see Teitelboim, C., Phys. Rev. $\underline{D25}$, 3159 (1983), Phys. Rev. $\underline{D28}$, 297 (1983), Phys. Rev. $\underline{D28}$, 310 (1983).
23. Hartle, J.B. and Kuchař, K., J. Math. Phys. $\underline{25}$, 117 (1984).
24. See, e.g., Arnowitt, A., Deser, S. and Misner, C. in Gravitation ed. by Witten, L. (Wiley, New York, 1962).

25. Hartle, J.B. and Hawking, S.W., Phys. Rev. $\underline{D28}$, 2960 (1983).
26. Ponomarev, V.N., Barvinsky, A.O. and Obukhov, Yu. N., Geometrodynamical Methods and the Gauge Approach to Gravity Theory, (Enerogatomizdat, Moscow, 1985) Chapter 7. (In Russian.)
27. See, e.g., Ashtekar, A. (to be published).
28. Hawking, S.W. and Page, D. "Operator Ordering and the Flatness of the Universe" (preprint).
29. Leutwyler, H., Phys. Rev. $\underline{134}$, B1155 (1964), DeWitt, B.S. in Magic Without Magic: John Archibald Wheeler, ed. by Klauder, J. (Freeman, San Francisco, 1972), Faddeev, L. and Popov, V., Usp. Fiz. Nauk. $\underline{111}$, 427 (1973) [Sov. Phys.-Usp. $\underline{16}$, 777 (1974)], Fradkin, E. and Vilkovisky, G., Phys. Rev. $\underline{D8}$, 4241 (1973), Kaku, M., Phys. Rev. $\underline{D15}$, 1019 (1977).
30. Kuchař, K., J. Math. Phys. $\underline{22}$, 2640 (1981).
31. Gibbons, G., Hawking, S.W. and Perry, M., Nucl. Phys. $\underline{B138}$, 141 (1978).
32. Callan, C. and Coleman, S., Phys. Rev. $\underline{D16}$, 1762 (1977).
33. Schoen, R. and Yau, S.-T., Comm. Math. Phys. $\underline{65}$, 45 (1979), Phys. Rev. Lett. $\underline{43}$, 1457 (1979).
34. Witten, E., Comm. Math. Phys. $\underline{80}$, 381 (1981).
35. Faddeev, L., Teor. Mat. Fiz. $\underline{1}$, 3 (1969) [Theor. Math. Phys. $\underline{1}$, 1 (1970)]; Faddeev, L. and Popov, V., Usp. Fiz. Nauk $\underline{111}$, 427 (1973) [Sov. Phys. Usp. $\underline{16}$, 777 (1974)]; Fradkin, E. and Vilkovisky, G. "Quantization of Relativistic Systems with Constraints" CERN Report TH-2322 (1977).
36. Hartle, J.B. and Schleich, K. "The Conformal Rotation for Linearized Gravity" in the festschrift for E. Fradkin (to be published).
37. Hartle, J.B., Phys. Rev. $\underline{D29}$, 2730 (1984).
38. Schoen, R. and Yau, S.-T., Phys. Rev. Lett. $\underline{42}$, 547 (1979).
39. For reviews of earlier work see, e.g. Misner, C.W., in Magic Without Magic: John Archibald Wheeler, ed. by Klauder, J.R. (Freeman, San Francisco, 1972); Ryan, M. Hamiltonian Cosmology (Springer, New York, 1972) MacCallum, M.A.H., in Quantum Gravity ed. by Isham, C.J., Penrose, R., Sciama, D. (Clarendon, Oxford, 1975).

40. See, e.g. Hartle, J.B., J. Math Phys. 26, 804 (1985), J. Math. Phys. (to appear), Class. Quant. Grav. 2, 707 (1985).
41. Moss, I. and Wright, W., Phys. Rev. D29, 1067 (1983).
42. Hawking, S.W., in Relativity Groups and Topology II, ed. by DeWitt, B.S. and Stora, R. (North Holland, Amsterdam, 1984).
43. Hawking, S.W. and Luttrell, J.C., Nucl. Phys. B247, 250 (1984).
44. Hawking, S.W. and Luttrell, J.C., Phys. Lett. 143B, 83 (1984).
45. Wu, Z.C., Phys. Lett. B146, 307 (1984).
46. Hu, X.M. and Wu, Z.C., Phys. Lett. B149, 87 (1984).
47. Wright, W. and Moss, I., Phys. Lett. 154B (1985).
48. Amsterdamski, P., Phys. Rev. D31, 3073 (1985).
49. Halliwell, J.J. and Hawking, S.W., Phys. Rev. D31, 1777 (1985).
50. Horowitz, G.T., Phys. Rev. D31, 1169 (1985).
51. Hawking, S.W. and Wu, Z.C., Phys. Lett. 151B, 15 (1985).
52. González-Díaz, P.F., Phys. Lett. 159B, 19 (1985).
53. Wu, Z.C., "Dimension of the Universe" (preprint).
54. Page, D.N., "Hawking's Wave Function of the Universe" (preprint).
55. Wada, S., "Quantum-Classical Correspondence in Wave Functions of the Universe" (preprint).
56. Kazama, Y. and Nakayama, R. "Wave Packet in Quantum Cosmology" (preprint).
57. Carow, U. and Watamura, S., "Quantum Cosmological Model of the Inflationary Universe" (preprint).
58. Wada, S., "Quantum Cosmology and Classical Solutions in the Two Dimensional Higher Derivative Theory" (preprint).
59. Wu, Z.C., "Primordial Black Holes" (preprint).
60. Halliwell, J.J., "Quantum Cosmology of the Einstein-Maxwell Theory in Six Dimensions" (preprint).
61. Hawking, S.W., "The Arrow of Time in Cosmology" (preprint).

62. Schleich, K., Phys. Rev. D (to be published).
63. Shellard, E.P., unpublished Ph.D. dissertation, Cambridge University 1985.
64. Banks, T., Nucl. Phys. $\underline{B249}$, 332 (1985).
65. Guth, A. and Pi, S.-Y., Phys. Rev. Lett. $\underline{49}$, 1110 (1982), Hawking, S.W., Phys. Lett. $\underline{B115}$, 295 (1982), Bardeen, J., Steinhardt, P. and Turner, M., Phys. Rev. $\underline{D28}$, 679 (1983). For a lucid review see Brandenberger, R.H., Rev. Mod. Phys. $\underline{57}$, 1 (1985).

INFLATIONARY COSMOLOGY

Paul J. Steinhardt
Department of Physics, University of Pennsylvania
Philadelphia, PA 19104-6396

ABSTRACT

The present status of the inflationary universe model is reviewed. The requirements for a particle field theory necessary to implement the model are emphasized.

1 INTRODUCTION

Twenty years ago the cosmic microwave background was discovered and the hot big bang model became established as the standard model for describing the evolution of our universe. In addition to the microwave background, the model successfully explains the Hubble expansion of the universe and the primordial nucleosynthesis of the elements. However, as the model has become better understood and with the advent of grand unified theories (GUTs), a number of problems with the hot big bang picture have become apparent.

In this series of lectures, we discuss a radically new model of the very early evolution of our universe — the "inflationary universe model" — which is designed to overcome all of the problems of the hot big bang model while maintaining all of its successful predictions. According to the inflationary universe model, the universe underwent a burst of tremendous exponential expansion during the first 10^{-32} seconds of its existence triggered by a strongly first order phase transition. The transition was associated with the spontaneous breaking of some symmetry among elementary particle forces as described by a unified field theory.

Thus, by its very nature, the inflationary universe model derives from numerous subfields of physics: astrophysics, cosmology, particle physics, and statistical mechanics. These lectures have been written for readers with a background in elementary particle theory, especially unified field theories. The inflationary model, if correct, provides important hints about the behavior of elementary particles at very high energies (near Planck scale) and may be an important guide in model building. The emphasis, then, will be on equipping the particle theorist with the necessary concepts from astrophysics, cosmology and statistical mechanics in order to understand the detailed constraints the inflationary model imposes on particle physics.

The organization of the lectures is as follows: In Section 2, we discuss the development of the standard hot big bang model and describe its successes and failures. In Section 3, we describe the "old" inflationary model proposed by Guth[1] and a fundamental flaw in the scenario that makes it untenable. We then discuss the "new" inflationary model developed independently by A. Linde[2] and by A. Albrecht and P. Steinhardt,[3] which is designed to overcome the flaw and to obtain a consistent scenario. In Section 4, we discuss how the inflationary model also leads to a prediction of the primordial spectrum of density fluctuations that eventually accounts for the formation of galaxies. We show how the model leads to a spectrum that is qualitatively consistent with the present astrophysical understanding of galaxy formation. In order to be quantitatively correct, the inflationary model requires new constraints on the particle physics model which generates the inflationary phase transition. In Section 5, we present a prescription for designing a field theory satisfying all the constraints imposed by the inflationary model. In Section 6, we briefly describe some of the efforts that have been made to design yet "newer" models of inflation.

2 THE STANDARD HOT BIG BANG MODEL

2.1 HISTORICAL PERSPECTIVE

[4] Historically, cosmology has been much more a theoretical rather than an empirical science. As a result, the rise and fall in acceptance of various theories depends on issues of aesthetics and simplicity to a much greater degree than in most other subfields of physics. In order to understand the place of the inflationary and hot big band models with respect to other cosmological theories, it is useful to trace the development of cosmological ideas and the change in aesthetic views that has occurred in the last seventy years since Einstein introduced his Theory of General Relativity.

When Einstein published his Theory of General Relativity in 1915, it was recognized as a major intellectual achievement but it was generally felt to be of little practical significance. It was believed that gravitational fields could never be so strong that there would be much difference between its predictions and those of the classical Newtonian theory of gravity. This attitude changed as the Theory came to be applied to cosmology. In 1917, Einstein introduced his Theory of Static Cosmology. The model is infinite in time, finite and unbounded in space. The theory was based on the prejudice that the universe is not evolving with time. In order to balance the gravitational attraction of the "static" matter, Einstein introduced the cosmological constant, Λ. He believed his to be the only positive Λ non-singular cosmological solution. (Later, when it was discovered that the universe is evolving, Einstein regarded the introduction of the cosmological constant as the "biggest blunder" of his life. Today, we would say that the possibility of a cosmological constant is unavoidable and the problem is to explain why its value happens to be zero today.)

In 1919, de Sitter discovered a non-singular solution to the Einstein field equations with energy density, ρ, equal to zero. Technically, the solution is static in the absence of matter; however, if test particles are introduced, the solution evolves homogeneously and isotropically. The solution corresponds to an exponentially expanding cosmology closely related to the inflationary epoch of the inflationary universe model.

In the 1920's, Hubble and Slipher discovered the apparent recession of the galaxies. They found that the recession velocity of distant galaxies, v, is proportional to their distance, ℓ:

$$v = H\ell \qquad (1)$$

where the proportionality constant, H, is now known as the Hubble constant. Slipher and Hubble reported a value of $H = 500$ km/sec/Mpc, although today the accepted value is $H = 50 - 100$ km/sec/Mpc (1 Mpc $= 3.26 \times 10^6$ ly $= 3.1 \times 10^{24}$ cm).

Friedmann in 1922 and Le Maitre in 1927 discovered another homogeneous and isotropic solution to Einstein's equations in which $\rho \neq 0$. Unlike the Einstein and de Sitter solutions, the new solution could not be viewed as static; it required that the universe expand with time. Le Maitre noted that the solution might be relevant to the recent results of Hubble and Slipher. The solution used the "Robertson-Walker" line element:

$$d\ell^2 = dt^2 - R^2(t) \left(\frac{dr^2}{1 - kr^2} + r^2(d\theta^2 + \sin^2\theta d\phi^2) \right); \qquad (2)$$

where $d\ell$ measures the proper (physical) distance between two two observers; $R(t)$, called the *scale factor*, describes the evolution of the universe with time and k is the *3-space curvature constant*. If k is positive, the universe is *closed*; if k is negative, the universe is *open*; if k is equal to zero, the universe is *flat*. In 1935, Robertson and Walker showed that the line element in Eq. (2) is the most general possible for a homogeneous and isotropic spacetime with Riemannian geometry.

Thus, there is a close connection between the Hubble expansion law and homogeneity and isotropy. If the universe is homogeneous, isotropic and Riemannian, then it must be described by Eq. (2). In such a homogeneous and isotropic universe, there is only one parameter, $R(t)$, which describes the evolution of the universe. Consider two observers with $dt = d\theta = d\phi = 0$, $dr = r > 0$ and $k = 0$. What is the velocity of expansion between the observers? According to Eq. (2), the physical distance between the two observers is $\ell = R(t)r$, and therefore

$$v \equiv \frac{d\ell}{dt} = \frac{dR}{dt}r = \frac{\dot{R}}{R}\ell. \qquad (3)$$

If we define $H \equiv \frac{\dot{R}}{R}$, we obtain directly the Hubble expansion law. Thus, homogeneity and isotropy implies directly the Hubble law. It should be noted in passing, that the Hubble constant is independent of space but is *not* independent of time, in general; in this sense, the term is a misnomer.

As a further illustration of the properties of an expanding universe, consider a particle travelling from observer 1 to observer 2 with velocity v, where dr for the observers is fixed (they are said to be at fixed "comoving" coordinates). (The following two exercises are due to Peebles.[4]) If the universe were not expanding, it would appear to both observers that the particle is travelling with speed v. However, if the universe is expanding, the velocity will appear to decrease as the universe expands. Suppose that the particle passes observer 1 at time $t = 0$ with velocity $v(0)$; then at time dt, the particle travels a distance $v(t)dt$ to observer 2. In the meantime, the second observer is moving away from the first at a rate

$$dv = [v(t)dt]\dot{R}/R, \qquad (4)$$

so that observer 2 sees the particle pass with velocity

$$v(t + dt) = v - v\frac{\dot{R}}{R}dt. \qquad (5)$$

Thus, we obtain a differential equation for v whose solution is:

$$v(t) = \frac{v(0)}{R(t)} \qquad (6)$$

This measured velocity is known as the "peculiar velocity" of the particle. The expression can be integrated to obtain the "coordinate" distance (the physical distance is $R(t)$ times the coordinate distance) travelled by the particle in time t:

$$x = \int_0^t \frac{v(0)dt}{R(t)}. \qquad (7)$$

Two observers expanding with the universe maintain the same coordinate distance. Although this analysis is quite straightforward, it leads to a result that many physicists who have not considered the argument previously find very counterintuitive: Because there is no constraint on $R(t)$, it may increase so rapidly (e.g. exponentially) that x is bounded! That is, even a signal travelling at the speed of light may travel only a finite coordinate distance in an infinite time. Two observers beyond that distance can never communicate! The result appears at first to contradict principles of special relativity, but it does not. It is still true that the fastest particles are photons, but there is no constraint on the expansion rate of the universe which involves the stretching of space and time rather than the transmission of particles. Thus, two observers may *expand* away from one another at a peculiar velocity which is greater than the speed of light. Another consequence of the notion of peculiar velocity is that light travelling in a expanding universe is redshifted.

In 1948, Bondi, Gold and Hoyle published their Steady State Model of the universe. In their model, the separation of matter increases exponentially and continuous matter creation is required to explain the density of matter near us. (In many ways, their description reads very much like the description of the inflationary epoch, although in the latter case the expansion occurs for only a short time and instead of matter creation there is vacuum energy creation.) By the 1960's, studies of radio sources showed that they were older than the Steady State Model would predict. Later, the model was never able to simply explain the observed microwave background.

In the same year, Alpher and Gamow used the Le Maitre model to develop a theory of the primordial formation of the nuclear elements. Their discussion forms the the basis of the hot big bang model. In 1965, Penzias and Wilson, attempting to develop a satellite communication system, discovered the "cosmological noise" we know today as the $3K$ microwave background, a clear prediction of the hot big bang model. The discovery essentially eliminated all competing cosmological models.

In the late sixties, Hawking and Penrose developed the famous singularity theorems for the Friedmann-(Le Maitre)-Robertson-Walker (FRW) big bang model. The existence of the microwave background substantiated the fact that the universe is expanding and, therefore, must also have a finite lifetime. Assuming the equation of state is such that the energy density, ρ, and the pressure, p, have the property that

$$\rho + 3p > 0 \quad (the\ positivity\ condition) \tag{8}$$

they showed that the universe must begin from a singularity. Since all forms of matter considered at the time obeyed the positivity condition, this appeared to make the existence of the cosmological singularity inevitable. (Interestingly enough, during an epoch of exponential expansion, as occurs in the inflationary universe model, the positivity condition mentioned above is violated. Thus, it is at least possible that in an inflationary model, the universe may have had a nonsingular beginning. To some this would be regarded as aesthetically pleasing.)

In 1980, Guth[1] introduced his inflationary universe model and discussed its fatal flaw. The model was designed so that the universe underwent a burst of exponential (de Sitter-like) expansion during the first few instants of its existence during the course of a first order phase transition. Once the transition was completed, the universe returned to a FRW universe and the usual big bang picture is followed. The period of exponential expansion could resolve many problems associated with the big bang picture, Guth argued. (Other physicists had earlier proposed models bearing certain similarities with Guth's.) However, as Guth himself noted, it seemed almost impossible to find a mechanism to ever end the phase transition. Thus, the model appeared to be useless since it led to a universe that expanded exponentially forever, in clear contradiction with the universe we observe. In 1982, Linde[2] and Albrecht and Steinhardt[3] introduced a variation of the model in which the transition can complete itself. These events marked the beginning of great interest in the inflationary picture as a model for our universe.

2.2 Basic Assumptions of the Big Bang Model

The hot big bang picture is based upon Einstein's Theory of General Relativity, conventional nuclear and particle physics, and three basic assumptions:

A1 - **Homogeneity and Isotropy:** As has been described, the assumption that the universe is homogeneous and isotropic necessarily implies that it is described by the Robertson-Walker line element. Einstein's equations then reduce to simple differential equations for the scale factor, $R(t)$. The two most useful equations for our purposes are:

$$H^2 \equiv \left(\frac{\dot{R}}{R}\right)^2 = \frac{8\pi}{3M_p^2}\rho - \frac{k}{R^2} + \Lambda, \qquad (9)$$

and

$$\ddot{R} = -\frac{4\pi}{3M_p^2}(\rho + 3p)R, \qquad (10)$$

where k is the 3-space curvature constant, ρ is the energy density, p is the pressure. Λ is the cosmological constant and $M_p = 1.2 \times 10^{19}$ GeV is the Planck mass. From astrophysical measurements, the cosmological constant is known to be very small (consistent with zero) in the present universe. Note that $R(t)$ is always decelerating if the positivity condition, $\rho + 3p > 0$, is obeyed but can accelerate if the condition is violated, as is the case in the inflationary epoch. The energy density and the pressure are related by the conservation of energy-momentum equation:

$$\frac{d}{dt}(\rho R^3) = -p\frac{dR^3}{dt}. \tag{11}$$

A2 - **Adiabaticity:** The second assumption of the big bang model is that the universe has been expanding adiabatically; that is, there has been no major departure from thermal equilibrium. Specifically, this means that the total entropy, S, is conserved, or, if σ is the entropy density:

$$\frac{dS}{dt} = \frac{d(\sigma R^3)}{dt} = 0 \tag{12}$$

In reality, even before the inflationary model, it was well known that there were some departures from thermal equilibrium in the early history of the universe as the temperature dropped below the masses of heavy particles and they fell out of equilibrium. However, such small departures from thermal equilibrium have a negligible effect on the evolution of the universe. On the other hand, the assumption of adiabaticity is strongly violated in the inflationary model. During a strongly first order phase transition, as occurs in the inflationary picture, large amounts of entropy can be generated with the release of the latent heat in the transition. In the inflationary picture, the violation of the adiabaticity is for only a very short period of time ($< 10^{-32}$ sec); this period of violation, though, is sufficient to dramatically alter the history of the universe.

A3 - **Radiation Dominated:** The big bang model is referred to as "hot" because it is assumed that the universe was radiation dominated in its early history. Radiation dominated means that the dominant contribution to the energy density of the universe came from the kinetic energy of highly relativistic (hot) particles. This assumption provides an equation of state relating the energy density to the temperature and, via the adiabaticity assumption, to the scale factor R. According to standard results of thermodynamics, the energy and entropy density in a radiation dominated universe are given by:

$$\begin{array}{rcl} \rho & = & \frac{\pi^2}{30}(N_b + \frac{7}{8}N_f)T^4 \\ \sigma & = & \frac{2\pi^2}{45}(N_b + \frac{7}{8}N_f)T^3 \end{array} \tag{13}$$

where T is the temperature, and N_b and N_f are the numbers of "nearly massless" bosonic and fermionic degrees of freedom, respectively. By nearly massless one refers only to highly relativistic particles. The quantities N_b and N_f are only weakly temperature dependent, generally changing when the temperature drops below the masses of some particles. For the purposes of discussion, we will treat them as if they are temperature independent.

Given the equation relating σ to the temperature and the adiabaticity assumption which states that σR^3 is constant, we obtain the result

$$RT \approx constant \qquad (14)$$

Then, Eq. (13) implies that $\rho \approx R^{-4}$ and this can be substituted into Eq. (9) to solve for the evolution of $R(t)$ with time. It is an astrophysical observation that throughout the history of the universe the 3-space curvature term has been negligible compared with the energy density term and the cosmological constant today is zero. (These facts form the basis of a number of cosmological puzzles to be discussed later.) Thus, it is a good approximation to consider only the reduced equation:

$$\left(\frac{\dot{R}}{R}\right)^2 \approx \frac{8\pi}{3M_p^2} R^{-4} \qquad (15)$$

The solution is $R(t) \approx t^{\frac{1}{2}}$ — the evolutionary behavior in a "radiation dominated" FRW universe. Later, when the universe cools enough that the dominant contribution to the energy density is from the rest mass of non-relativistic particles, $\rho \approx R^{-3}$ and from the same equations we obtain $R(t) \approx t^{\frac{2}{3}}$ — the evolutionary behavior in a "matter dominated" FRW universe. Combining these simple relations with a knowledge of the behavior of elementary particles and nuclei, the history of the universe can be derived (see Fig. 1).

2.3 SUCCESSES OF THE HOT BIG BANG PICTURE

The acceptance of the hot big bang model as the standard model of cosmology is based upon three important experimentally testable "predictions" derived from it. (The term "prediction" is a little strange in this circumstance since we are predicting events that occurred earlier in the history of the universe.) First, the model predicts that the universe was and is expanding in such a way that two distant objects recede from one another with a velocity proportional to their separation. As we have emphasized, this derives directly from the assumption of homogeneity and isotropy and corresponds to the Hubble expansion law that we have already described. From astrophysical measurement, we believe that

$$H_{today} = 50 - 100 \text{ km/sec/Mpc} \qquad (16)$$

or $H^{-1} = 10 - 20 \times 10^9$ yrs. The measurement of H has had a long and strange history (it's accepted value has decreased by a factor of 5 to 10 from Hubble's original value.) Today, the uncertainty of a factor of two in the measurement is

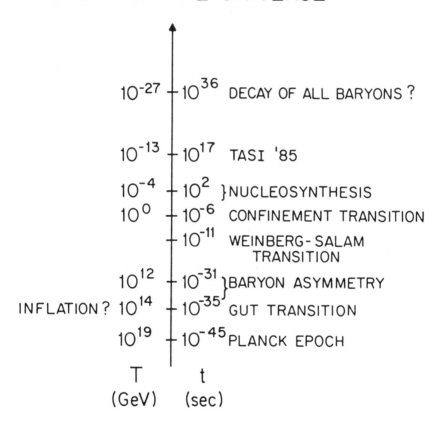

Figure 1: History of the Universe according to the standard big bang model.

much greater than the experimental uncertainty of any one measurement. The situation is that there are two groups who consistently report values of 50 or 100 km/sec/Mpc each with small error bars. The difference is due to systematically different choices of standard "candles" (astronomical objects whose distance from us is believed by one group or the other to be well determined) to determine the separation of more distant objects. Over the past few years, a third group is gearing up to make a third independent measurement, and their result at the moment lies somewhere between the two previous values.

The situation is a very serious one with especially serious implications for the inflationary model. The inflationary model, as we shall see, predicts that Ω, the ratio of the energy density to the critical density, is equal to unity (the critical density, defined as $\rho_c = \frac{3M_p^2 H^2}{8\pi}$, is the energy density in a flat $k=0$ universe). For a flat universe, the age of the universe is $\frac{2}{3}H^{-1}$. Astrophysical observations indicate that the oldest globular clusters are at least 10^{10} years old, which implies that $H^{-1} > 15 \times 10^9$ yrs or $H < 66$ km/sec/Mpc. Obviously, the inflationary model depends upon the correct value of H being closer to the lower of the two presently accepted ones.

The hot big bang model also successfully explains the 3 K cosmic microwave background radiation first observed by Penzias and Wilson. According to the model, the radiation is a remnant of photons that were in thermal equilibrium with matter during the first 100,000 years or so; as the universe continued to expand the photons decoupled and have been expanding and redshifting with the universe ever since. Today they appear to be thermally distributed with an effective blackbody temperature of a few degrees Kelvin.

It may seem peculiar at first that the photons have a thermal distribution when they are *not* in equilibrium today. However, it is a simple consequence of the expansion. Let $I(t,\nu(t))$ be the intensity of light at time t with frequency $\nu(t)$. Once the photons decouple from the matter, we have a conservation law:

$$\frac{d}{dt}\left[R^3 I(t,\nu(t))\right] = 0. \tag{17}$$

Using the same argument as we applied to computing the peculiar velocity in Section 1, we can compute that the frequency of light in an expanding universe redshifts according to $\nu(t) = R^{-1}(t)$. If there is a blackbody distribution of photons when they are coupled to matter in the early universe, then at decoupling, $t = t_d$,

$$I(t_d,\nu(t_d)) = \frac{2h\nu^3}{c^2}\left(\exp^{h\nu/kT_d} - 1\right)^{-1}. \tag{18}$$

Then, at a later time t *after decoupling*, the frequency is redshifted, but according to Eq. (17),

$$I(t, \nu_0) = \left(\frac{R(t_d)}{R(t)}\right)^3 I\left(t_d, \nu_0 \frac{R(t)}{R(t_d)}\right)$$
$$= \frac{2h\nu_0^3}{c^2} \left(\exp^{h\nu_0/kT(t)} - 1\right)^{-1} \quad (19)$$

where $T(t) = T_d(\frac{R(t_d)}{R(t)})$. Remarkably enough, the spectrum looks just like a blackbody spectrum but with a reduced (redshifted) temperature. By correlating this result with the predictions of primordial nucleosynthesis to be discussed below, one derives a predicted "temperature" for the radiation of today in the range of a few degrees K, in agreement with observation. The experimental confirmation of this prediction has pushed the big bang model into pre-eminence over all competitors.

The big bang model also leads to successful predictions of primordial nucleosynthesis. Nucleosynthesis is the formation of heavier nuclei through the fusing of light nuclei and it is an important process that occurs in stellar interiors. However, stellar nucleosynthesis is not sufficient to explain the abundances of the nuclear elements; in particular, it fails to explain the abundances of the lightest elements: helium, deuterium, and, to a certain degree, lithium. The hot big bang model offers an alternative source of nucleosynthesis deriving from the epoch when the universe as a whole was hot enough to form the lighter nuclei but cool enough that they did not get destroyed by subsequent interactions, a period ranging from the first few seconds to minutes after the big bang. A massive calculation using all the current data on nuclear interactions has been performed by Yang, et al.,[5] and to compute the production of light nuclei predicted by the big bang model. We shall give a brief description of the results here.

The abundances of elements with $A > 11$ can be successfully explained by stellar nucleosynthesis alone. For Li, Be, and B, cosmic ray spallation, the bombarding of C, N, and O by cosmic rays to form other nuclei, makes an important contribution. It appears to be enough to explain the abundances of Be and B, but probably not enough to explain the abundance of ^7Li. Deuterium is easily destroyed, so it is hard to find a production site that does not overproduce other elements; it is believed that all deuterium must be primordial. ^3He abundance is poorly known and may be accounted for by either stellar or primordial nucleosynthesis. ^4He, the second must abundant element, cannot be totally accounted for by stellar nucleosynthesis — most is primordial. In general, any production mechanism can overproduce D and ^7Li since they are easily destroyed, but ^3He and ^4He are more stable and their production is more constrained.

In Fig. 2 is shown the calculation of primordial nucleosynthesis by Yang, et al., where Y_P is the mass in ^4He, $\tau^{1/2}$ is the neutron half-life in minutes, N_ν is number of (nearly) massless neutrino degrees of freedom and η is the ratio of baryons to photons. The horizontal dashed lines represent the present experimental limits on the various abundances (in some cases, error bars are

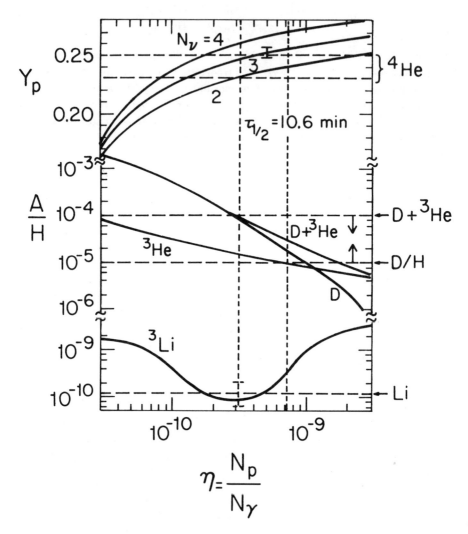

Figure 2: The computed abundances of light nuclear elements generated by primordial nucleosynthesis (solid lines) according to Yang, Turner, Steigman, Schramm and Olive and the present experimental limits on the primordial abundances (dashed lines).

also indicated). The latest addition is the improved theoretical and experimental information about ^7Li abundance. Taking the limits from all the data, we find a window, indicated by the vertical dashed lines, where all the theory is consistent with all the data. The window corresponds to $\eta = 3 - 7 \times 10^{-10}$ and $N_\nu \leq 4$. The value of η, so specifically determined by the computation, is especially important to particle physicists. It measures the excess of baryons over antibaryons in the early universe. It is believed that the value of η can be computed assuming a GUT theory with baryon number violating interactions and some out-of-equilibrium decay of GUT particles at high energies, as we shall discuss shortly. It should also be emphasized that the window of agreement is in Fig. 2 extremely narrow. This may be taken as encouraging as long as there is a window left, but there is a possibility that more accurate determination of nuclear abundances will eliminate the window altogether. Until then, we will take these results as astounding support for the big bang model.

2.4 Failures of the Big Bang Model

In spite of the successes of the big bang model, a number of problems become apparent as the model is extrapolated back further before the epoch of nucleosynthesis, that is, before the first seconds after the big bang. These problems should be examined closely since many involve aesthetic issues rather than outright contradictions with experiment. Views about what is and is not aesthetic have changed over the years and will continue to do so. Since the inflationary model is directed towards resolving these problems, different aesthetic viewpoints will result in different viewpoints about what is or is not supporting evidence for the model.

P1 - **Entropy Problem:** A lower bound on the entropy of the universe can be obtained by computing the contribution of the photons in the cosmic microwave background. If the universe is $R = 10^{10}$ lyrs across and the microwave background is at temperature $T = 2.7$ K, then the entropy, $S_p = R^3 T^3 > 10^{85}$ in dimensionless units. If the universe expanded adiabatically, then the entropy was at least this great throughout its history, even at the Planck epoch when one might have expected the entropy to be of order unity. The entropy problem is the puzzle of explaining this extraordinarily large value. (Interestingly, some physicists view the problem quite differently. They consider the entropy of the universe to be anomalously low considering the fact that the entropy of a black hole with the mass of the universe would be many orders of magnitude greater. Here is an example where there are two very different aesthetic viewpoints.)

P2 - **Horizon Problem:**[6] Any cosmological model according to which the universe has a finite lifetime has the property that there is a maximum distance that light (or any other particles or information) can have travelled up to any finite time t — the causal horizon distance, $\ell(t)$. According to the big bang model, the observable universe has always been much bigger than a single causal horizon distance, and yet the model also assumes (and it is verified by astrophysical experiments) that the universe is uniform on large scales. The horizon problem is the puzzle of explaining how the universe could become so homogeneous and isotropic on large scales when distant regions were never causally connected in the past. A light pulse beginning at the time of the big bang ($t = 0$) travels (recall $R(t) \propto t^{1/2}$):

$$\ell(t) = R(t) \int_0^t dt' \frac{c}{R(t)} = 2ct \qquad (20)$$

which, up to a constant factor of order unity, corresponds to the Hubble radius (H^{-1} normalized so that it is expressed in units of distance.) We can estimate the radius of our observable universe, $L(t)$, from the assumption that the universe has been adiabatically expanding:

$$s(t)L(t) = S_p \qquad (21)$$

where S_p is the total entropy today (for which a lower bound was estimated above). Comparing these two relations we find:

$$\frac{\ell^3}{L^3} = S_p^{-1} \frac{t^{3/2}}{t_{Pl}^{3/2}} \qquad (22)$$

where $t_{Pl} \approx 10^{-45}$ sec is the Planck time. According to this computation and given the entropy bound computed above, the universe contained more than 10^{86} distinct causal horizon volumes at the Planck epoch!

One of the best measures of the uniformity of the universe on large scales is the observed isotropy of the microwave background. The photons in the microwave background decoupled from matter several 100,000 years after the big bang and have been expanding with the universe ever since. We find that the temperature of the background coming from opposite directions in the sky is uniform at least to within one part in 10^4 and yet, according to Eq. (22), these photons last scattered from regions more than 90 causal horizon lengths (ℓ) apart. There is no natural explanation within the big bang model for the uniformity of causally disconnected regions.

P3 - **Flatness/Oldness:**[7] The Einstein equation, Eq. (9), can be recast in the form:

$$\Omega \equiv \frac{\rho}{\rho_c} = 1 + \Delta(t) \quad \text{where} \quad \Delta(t) \equiv \frac{k}{R^2 H^2} \qquad (23)$$

where $\rho_c \equiv 3M_p^2 H^2/8\pi$, and Ω is a dimensionless number that measures the ratio of the energy density to the critical value. A value of $k < 0$ corresponds to an open universe and implies $\rho < \rho_c$; a value of $k > 0$ corresponds to a closed universe and implies $\rho > \rho_c$; $k = 0$ corresponds to the flat universe and implies $\rho = \rho_c$. From various astrophysical measurements, we know today that $.1 < \Omega < 2$ (and many astrophysicists would claim to know the value even more precisely) which implies that $\Delta(t) < 1$ today. At first, this does not seem to be particularly puzzling, but it must be noted that Ω in Eq. (23) evolves with time. If $R(t) \propto t^{1/2}$ and $H \propto t^{-1}$, then $\Delta(t) \propto t$; that is, the deviation of Ω from unity is linear with time. This means that the flat condition, $\Omega = 0$, is a highly unstable condition. If $\Delta(t) < 1$ today, then at the time of nucleosynthesis (≈ 1 sec) $\Delta(t) < 10^{-17}$ and the universe was extraordinarily close to the flat condition $\Omega = 1$. The flatness problem is the puzzle of explaining why the universe has remained so close to this flat condition for so long (compared to a Planck time, say) when the flat condition is unstable. (That is, although the universe approaches spatial flatness when extrapolated back to $t = 0$, the deviation from flatness should have been O(1) in a typical evolution time, which is of order the Planck time. Yet, we find a universe which is 10^{62} Planck times old in which the deviation from flatness is still O(1) or less.)

These first three problems with the big bang model relate to the initial conditions for the universe and, in this sense, are "softer" than the problems that follow. One could eliminate them (as was done prior to the inflationary model) by assuming that the universe began with extraordinary entropy, uniformity on scales large compared to the causal horizon lengths, and spatial flatness. However, the aesthetic viewpoint at present is these are rather unnatural assumptions and it would be much better to replace them with some dynamical mechanism that would produce such conditions. The inflationary model is designed to accomplish this goal.

P4 - **Primordial Fluctuations Problem:**[4] In addition to requiring rather special initial conditions, the big bang model is incomplete. A basic assumption of the model is that the universe is homogeneous and isotropic. Although this is a good approximation on large scales (on the scale of superclusters or larger), it is obviously a poor assumption on smaller scales where matter clumps into galaxies, stars, planets, etc. The big bang model offers no information about the density fluctuations in the early universe that evolved to form the observed small scale inhomogeneity. To make matters worse, we know the primordial density fluctuations had to be very special. First, in the context of the big bang model, the fluctuations must have been generated by acausal processes (so they cannot correspond to thermal fluctuations, for example), since the galactic scale, for example, was much greater than a causal horizon size in the early universe. In this sense, the primordial fluctuation problem is closely related to the horizon problem. Also, we know that the fluctuation spectrum (the amplitude of the fluctuations as a function of distance scale) had to be quite special: large enough on the galactic scales to account for galaxy formation and yet small enough on large scales so that there are not large fluctuations in the temperature of the microwave background (the microwave background senses distance scales much greater than the galactic scale). A thermal spectrum, for example, does not have this property. The primordial fluctuation problem is the puzzle of understanding how such a special fluctuation spectrum could be generated on scales larger than a causal horizon volume. This problem is not placed in the same category as the previous three problems because, in the author's opinion at least, it is unreasonable to accept such a special spectrum as a natural initial condition.

P5 - **Monopole and Domain Wall Problem:**[8] The big bang model encounters further difficulty when combined with our present understanding of particle physics. One of the key notions in particle physics today is grand unification: the idea that the strong, weak and electromagnetic forces, and probably the gravitational force as well, are really manifestations of a simple, symmetric unified gauge field theory. All the forces are related through a "grand unified symmetry" (in the simplest model the symmetry is $SU(5)$). The forces have different couplings and characteristics today because the symmetry is spontaneously broken when a scalar Higgs field, Φ, obtains a non-zero expectation value (the field actually has some 24 or more components only some of which may obtain non-zero expectation values). In most models with spontaneous symmetry breaking, the symmetry is restored at high temperatures. Thus, it is natural to expect that, as the universe expanded and cooled, it underwent a "phase transition" from a symmetric state (where $\Phi = 0$) to a symmetry breaking state (where $\Phi \neq 0$). The transition is expected to occur when the universe had a temperature of about the grand unification scale, $M_G \approx 10^{14}$ GeV, some 10^{-35} seconds after the big bang.

The monopole and domain wall problems result from the fact that there are many equivalent symmetry breaking states (depending upon which components of Φ obtain non-zero expectation values). When the phase transition occurs, the universe breaks into many domains with a radius of order a causal horizon length, $O(H^{-1})$; inside each domain is a different symmetry breaking state. Where domains come together, topological defects or knots in the Higgs field will form, analogous to defects that form when liquids crystallize. Where domains come together in which the values of Φ are related by a continuous subgroup of the GUT, point-like defects called "magnetic monopoles" form; where domains come together that are related by a discrete subgroup of the grand unified symmetry, surface-like defects called "domain walls" form. (Monopole defects occur in any GUT provided that the unification symmetry corresponds to some simple group G which is spontaneously broken to $G' \times U(1)$ (where G' need not be simple), as in any realistic unified theory; domain wall defects occur only in theories that have spontaneously broken discrete symmetries.) Both of these defects have numerous fascinating properties but for the purpose of this discussion all one need know is that they are stable and extremely massive (e.g. mass of a monopole $> 10^{16}$ mass of a proton). The number of such defects after the phase transition is roughly equal to the number of domains: since each domain has a radius of order a horizon length and the universe is much larger than a horizon length in the big bang model, many monopoles and domain walls are expected to be produced in the transition (roughly one for every 100 baryons). Since the mass of the defects is so much greater than the mass of a baryon, the mass of the defects produced would dominate the energy density of the universe. The subsequent behavior of the universe would be as if the universe were matter dominated rather than radiation dominated as in the usual hot big bang picture. From our discussion above, we see that this means that the expansion of the universe speeds up from $R(t) \propto t^{1/2}$ to $R(t) \propto t^{2/3}$ at a much earlier time, $t = O(10^{-35}$ sec) than occurs in the usual big bang scenario, $t = O(10^{30}$ sec). As a result, the universe cools much more rapidly and reaches the presently observed microwave background temperature, $T \approx 3$ K, only 30,000 years after the big bang! As a result, all the successful predictions of the hot big bang theory are lost.

2.5 More About Phase Transitions

Since phase transitions play such an essential role in the inflationary model, they shall be discussed in a little more detail. To understand why phase transitions occur in theories with spontaneous symmetry breaking, it is useful to consider the scalar effective potential for the Higgs field that drives the symmetry breaking phase transition in a GUT.

The effective scalar potential is altered by quantum corrections and finite temperature contributions. Let us consider the famous example first studied by

Coleman and E. Weinberg[9] for which the uncorrected tree-level potential is given by:

$$V(\phi) = \frac{1}{2}\phi^2 + \frac{\lambda}{4!}\phi^4 \qquad (24)$$

where $\phi = |\Phi|$. As Coleman and Weinberg showed, quantum corrections at zero temperature lead to a correction of the form:

$$\gamma\phi^4 \left[\ln\frac{\phi^2}{M^2} - \frac{1}{2}\right] \qquad (25)$$

where M is a renormalization mass and $\gamma = \frac{3}{64}(g^2/4\pi)^2$. The potential can be reparameterized in terms of two constants, A and B:

$$V_{eff}^{T=0} = (2A - B)\sigma^2\phi^2 - A\phi^4 + B\phi^4 \ln\frac{\phi^2}{\sigma^2} \qquad (26)$$

where the σ corresponds to the minimum of the potential, and the Higgs mass is given by $M_H^2 = 4\sigma^2(3B - 2A)$. (A case of special interest for the inflationary model will be the "Coleman-Weinberg limit", $2A = B$, in which the mass term is absent. For the $SU(5)$ model, $\sigma = 4.5 \times 10^{14}$ GeV.) The finite temperature corrections to lowest order are given by:[10]

$$\Delta V_T = \sum_n \frac{T^4}{2\pi^2} \int d^3p \ln[1 - \exp(-\beta\sqrt{p^2 + M_n^2(\phi)})] \qquad (27)$$

where the sum over n is over the gauge meson degrees of freedom with mass M_n (which depends on ϕ), and $\beta = 1/kT$ where k is Boltzmann's constant. This correction is essentially what is expected from a hot gas of classical particles with mass M_n. This expression can be expanded for $\phi \ll T$:

$$\Delta V_T \approx aT^4 + bT^2\phi^2 + ... \qquad (28)$$

for some constants a and b. For our purposes, the first term is uninteresting because it is ϕ-independent, but the second term is very interesting because it distorts the *shape* of the effective potential, as shown in Fig. 3. In particular, at high enough temperatures, the ground state corresponds to $\phi = 0$ and the symmetry is "restored." This behavior is typical of a theory with spontaneous symmetry breaking: *As the temperature is decreased, the theory undergoes a "phase transition" (or series of phase transitions) from the unbroken symmetry phase at high temperature to the broken symmetry phase at low temperatures.*

The *critical temperature*, T_c, for a phase transition is the temperature at which the energy density of the unbroken symmetry phase is equal to that of the broken symmetry phase (see Fig. 3). At $T > T_c$ the unbroken phase is energetically favored; at $T < T_c$ the broken phase is favored. There are several types of

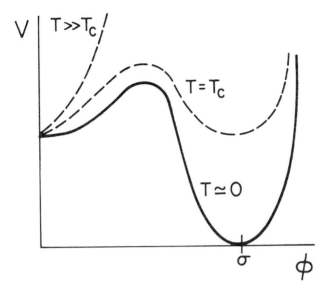

Figure 3: Free energy density versus ϕ for $T \gg T_c$, $T = T_c$ and $T \approx 0$ for a typical first order phase transition in a GUT. The energy density curves have been shifted so that $V(0)$ is the same for the three.

phase transitions depending upon the detailed form of the effective potential as a function of ϕ and T.

A *second order or continuous* phase transition is one in which the energy barrier separating the two phases is zero at $T = T_c$. Thus, as the temperature drops infinitesimally below the critical temperature, the state of the system shifts continuously from the unbroken to the broken symmetry phase. No latent heat is released in the transition. The second order phase transition is the one which exhibits the well-known critical behavior near $T = T_c$.

A *first order or discontinuous* phase transition is one in which there is a barrier separating the unbroken and broken phases at $T = T_c$. Thus, if a system begins at high temperature ($T > T_c$) in the unbroken phase (because it is energetically favored) it may be trapped by the energy barrier in the unbroken phase even after T drops below T_c. Such a phenomenon is known as *supercooling* and is characteristic of a first order phase transition. There are three mechanisms possible for completing a first order phase transition after supercooling. The first method, *homogeneous nucleation*, is the method most familiar to particle physicists and condensed matter physicists. Due to quantum or thermal fluctuations, a small finite region undergoes a transition from the unbroken phase to the energetically favored broken phase. If the region is greater than a certain critical size, the region grows, converting a system from unbroken to broken phase. The region tends to be spherical and is known as a "bubble." The value of ϕ is roughly uniform in the bubble and corresponds to one of the broken symmetry ground states. The energy gained as the bubble grows is stored almost entirely in the bubble wall and is used to propel the wall outwards. The bubble wall accelerates outward rapidly approaching the speed of light. Once many bubbles nucleate (the rate depends upon the activation energy), they grow and coalesce to convert the system totally from unbroken to broken phase. Where the bubbles come together, the orientation of the order parameter ϕ in group space (i.e. which compponents of Φ are non-zero) will vary and topological defects or knots in the field will form. For the case of unified field theories, these knots correspond to point defects that we call magnetic monopoles. (Roughly one monopole forms for every 10 bubbles.)

A second method of completing a first order phase transition is known as *inhomogeneous nucleation*. In most realistic systems in condensed matter physics, there are impurities which act as preferred nucleation sites for bubbles. Whereas the nucleation probability is uniform in space for homogeneous nucleation, the probability is non-uniform for inhomogeneous nucleation.

The third transition mechanism, *spinodal decomposition*, is the least known. Actually, J. Willard Gibbs described how all three mechanisms are possible (depending on the details of the transition) in his original treatise, but this aspect of his work was not appreciated. In fact, there grew to be a debate for many years between those who believed the first order transition mechanism was homogeneous nucleation and those who thought it was spinodal decomposition. It was almost 100 years before it was realized (again) that both mechanisms are possible, depending upon the details of the transition.[11]

In spinodal decomposition, the nucleation probability remains low as the system supercools until the barrier separating the unbroken and broken symmetry phases disappears altogether. (This is quite different from a second order phase transition because $T \ll T_c$ and the energy density of the two phases is quite different.) At this point, small fluctuations drive the system away from symmetric phase. Each region is driven nearly simultaneously towards one symmetry breaking phase or another. The result is a system divided into different domains of symmetry breaking phase. In the end, the latent heat of transition (associated with the energy difference) is released. The spinodal transition will play an important role in the new inflationary universe picture.

2.6 BARYON ASYMMETRY

Any attempt to revise the big bang description of the very early evolution of the universe must take into account a significant success obtained by combining

GUTs with cosmology — the successful prediction of the baryon asymmetry of the universe.[12] One of the surprising features of our universe is that it is filled predominantly with matter rather than with an equal distribution of matter and anti-matter. The primordial nucleosynthesis computations result in an accurate measure of this matter asymmetry (or, since the matter is chiefly baryonic, the baryon asymmetry): there is roughly one excess baryon for every 10^9 photons (see the parameter η discussed in Fig. 2 above). Before grand unification, it was believed that the baryon number is conserved in any interaction; thus, the only explanation for the baryon asymmetry was that the universe began with the peculiar asymmetry as an initial condition and has maintained the value ever since. This seems highly unnatural, especially since, as we have seen, the successful prediction of nucleosynthesis depends sensitively on the quantitative value of the asymmetry.

One of the essential elements of grand unification is that, because there is a hidden symmetry between baryons and leptons, baryons can decay via the exchange of certain Higgs fields and gluons. During the symmetry breaking phase transition at $T_G \approx 10^{14}$ GeV, the Higgs fields and gluons gain a mass of order the symmetry breaking scale, $M_G \approx T_G$. At present temperatures, $T \ll T_G$, baryon decay occurs very rarely since the particles exchanged in the decay interaction are so massive. However, as the universe cooled from high temperatures, baryon number violating interactions occurred frequently and it is widely believed that this epoch may account for the observed baryon asymmetry of the universe.

The typical scenario is as follows. Three features are necessary to produce baryon asymmetry dynamically. Of course, the first is non-negligible baryon number violating interactions. Second, both C and CP must be violated or else compensating and symmetrical interactions would exist that maintain baryon asymmetry. Finally, a departure from thermal equilibrium is necessary: otherwise CPT plus thermal equilibrium would insure compensating interactions that maintain baryon asymmetry. All three conditions can be met by a combination of GUTs and big bang cosmology. The baryon number, C and CP violating interactions can be naturally incorporated into a realistic GUT. The departure from thermal equilibrium occurs as the universe cools and undergoes the symmetry breaking phase transition. The Higgs and gluon fields responsible for the baryon number violation obtain a large mass $O(M_G)$ and, as the temperature continues to drop, they fall out of thermal equilibrium. If the temperature at which they fall from equilibrium is greater than their mass (which depends upon their coupling to lighter fields) no asymmetry is produced because equilibrium interactions compensate. In the simplest $SU(5)$ model, though, the color triplet Higgs field falls out of thermal equilibrium at a temperature below its mass and its decay can then lead to a baryon asymmetry. The baryon asymmetry is calculable for any given theory and is found to be $O(10^{-(10-14)})$ for typical GUTs depending upon the details. Most calculations lead to values that are too low by a few orders of magnitude, but this is generally believed to be resolvable. The dynamical generation of baryon asymmetry is generally believed to be a successful consequence of grand unification plus big bang.

Since the baryon generation occurs at temperature scales $O(M_G)$, any model that revises the early history of the universe must return to the big bang scenario at this temperature or above in order that, during subsequent cooling, the baryon asymmetry can be generated. This represents one of the most stringent constraints on the inflationary model. Recently, proposals have been discussed for generating baryon asymmetry at much lower temperatures.[13] These approaches could prove to be quite exciting and highly relevant for inflationary cosmology, but at the moment they depend upon unproven physics or unknown couplings.

3   OLD AND NEW INFLATIONARY UNIVERSE MODELS

3.1   Old Inflationary Universe Scenario

The inflationary universe model[14] was first introduced by A. Guth[1] as an attempt to resolve the entropy, horizon, flatness and monopole problems of the standard hot big bang model. The key assumption is that the universe underwent a strongly first order phase transition during the first 10^{-35} sec; this, combined with the effects of general relativity, leads to a radically new cosmology.

A strongly first order phase transition is one in which the barrier separating the unbroken and broken symmetry phases is present not only at $T = T_c$ but also as the system supercools all the way to $T = 0$ (See Fig. 3). Thus, a system can be trapped in the unbroken phase even at $T = 0$ even though it has greater energy density than the broken phase due to the presence of the energy barrier. The energy density difference between the two phases will be called ϵ. In a condensed matter physics system, the transition would eventually be completed by nucleation and growth of bubbles, even if the activation energy is high at $T = 0$. Due to the effects of general relativity, though, the universe may never be able to complete such a transition.

To determine the effects of general relativity, it is first necessary to determine the zero of energy density. In most physical situations, the zero of energy density is an irrelevant concept since dynamics depends upon differences in energy density rather than the value itself. However, in general relativity, the energy-momentum tensor couples to the total energy, not just energy differences, and the zero of energy density must be determined. First, we can absorb the cosmological constant, Λ, in Eq. (9) into a redefinition of the energy density, $\rho \to \rho + \frac{3M_P^2\Lambda}{8\pi}$. Today, astrophysical observation tells us that the effective vacuum energy density (including the contribution of the cosmological constant) is zero in the present state: the symmetry breaking state near $T = 0$, as shown in Fig. 3. Once fixed, the rest of the scenario is determined.

As the universe supercools, the energy density in the *metastable* unbroken symmetry phase is given by $\rho_{meta} \approx aT^4 + \epsilon$ corresponding to the thermal energy density plus the constant energy density difference between the metastable phase and the stable broken symmetry phase. The value of ϵ is $O(T_c^4)$ in typical models. Thus, as T drops below T_c the value of ρ_{meta} approaches a constant, ϵ. (Note that for $T \gg T_c$ the energy density is dominated by the thermal contribution and the universe behaves as a radiation dominated FRW universe.) The solution to the Einstein equations, Eq. (9) with $\Lambda = 0$, is simple and dramatic:

$$R(t) = \exp(H_i t) \qquad (29)$$

where $H_i^2 \equiv 8\pi\epsilon/3M_p^2 = $ constant $(O(10^{10}\text{ GeV}))$. *The universe enters an epoch of exponential expansion with time constant H_i^{-1}*. The exponential expansion phase is nearly equivalent to the exponentially expanding de Sitter solution with $\rho = 0, \Lambda \neq 0$. The epoch of exponential expansion is referred to as the "inflationary" or "de Sitter" epoch. The exponential expansion represents a dramatic increase in the expansion rate $(\propto t^{1/2})$ up to that point.

The exponential expansion can be understood by the following intuitive argument. First, according to the conservation of energy equation, a phase with constant (positive) energy density has a negative pressure with magnitude equal to the energy density. That such a state has negative pressure is quite natural. If I placed a bubble of this phase in a background of zero energy density phase, I would expect the bubble to collapse since it leads to a lower energy state. The force responsible for the collapse must come from inside the bubble (since outside the bubble there is no energy density or pressure); thus, the phase inside the bubble has a negative pressure. In fact, as the conservation of energy equation indicates, the pressure is the negative of the energy density. Next, we note that this corresponds to a very strange state of matter since the positivity condition, $\rho + 3p > 0$ is violated in such a state. (Note that this means that the Hawking-Penrose singularity theorems do not apply to such cosmologies, either.) If we consider the second Einstein equation, Eq. (10), we see that when the positivity condition is violated the universe accelerates instead of decelerates. In this case, since ρ and p are constant, the acceleration is exponential. In short, the exponential expansion is the response of gravity to a very unusual state of matter with constant energy density and negative pressure.

If the de Sitter epoch continues for $t \gg H_i^{-1}$ and then the universe reheats to $T \approx M_G$ (an issue we will return to later), a number of cosmological problems can be resolved:

The entropy problem can be resolved because of the tremendous latent heat released after the transition to the stable phase is completed. The entropy density, $\sigma \propto T^3$, is roughly the same before and after the phase transition, but the volume, R^3 is increased exponentially! Beginning with total entropy $S_{before} \approx 1$ before the transition (a natural value), the entropy after the transition is:

$$S_{after} = S_{before} \frac{R^3(t_{after})}{R^3(t_{before})} \quad (30)$$

which can exceed the lower bound of 10^{86} computed above provided $R(t_{after}) > \exp(67) R(T_{before})$. Such expansion is easy to achieve in typical models; in fact, except for cases of special fine tuning, $t \gg 67 H_i^{-1}$ and the exponential prefactor is much greater. Note that this means that nearly all the entropy and matter we observe in our universe is produced by the inflationary transition.

The horizon problem is resolved because the observable universe, according to the model, was much smaller compared to a causal horizon volume before the phase transition. If S_{before} is O(1) or less before the transition (at $t < 10^{10} t_{Pl}$) then the observable universe has a radius $L(t)$ that is a causal horizon distance ℓ or less according to Eq. (22). That is, the region which evolves into our observable universe is much tinier before the transition than the equivalent in the big bang model. The exponential expansion epoch is necessary to expand such a tiny region into a region large enough to account for our present observable universe.

The flatness problem is resolved because an epoch of exponential expansion suppresses the spatial curvature of the universe. Recall that the Einstein equation (with $\Lambda = 0$) can be rewritten:

$$\Omega = 1 + \Delta(t); \text{ where } \Delta(t) \equiv \frac{k}{R^2 H^2} \quad (31)$$

where $\Delta(t)$ measures the deviation from spatial flatness. During inflation, $H \approx H_i$ remains nearly constant, but $R(t)$ grows exponentially. Even if $\Delta(t)$ is O(1) before the transition, after the transition it is exponentially suppressed. Thus, Ω is nearly unity. After the transition, the $\Delta(t)$ increases linearly with time (because the universe is in a FRW phase), but the coefficient of the linear term is so small ($\ll \exp^{-134}$) that even today $\Delta(t)$ is negligible ($\ll 1$). This result leads to the important prediction of the inflationary model that Ω *must be extraordinarily close to unity today*. This result can be tested by astrophysical measurement; the measurement of Ω has been a hot experimental project over many years now. The present best estimate of Ω is O(.1 - .2), but, with appropriate interpretation, the measurements do not exclude the possibility that $\Omega = 1$. In particular, all the measurements of Ω are based on measurements of the contribution of visible matter or matter gravitationally clumped about visible matter (clusters, superclusters. etc.). We already know that, if $\Omega = 1$, the contribution of ordinary particles (baryons, leptons, etc.) must be less than 10 – 20% and so there must be some form of "dark matter" — new particles — which account for the remaining energy density. There already exists some astrophysical evidence for dark matter and most recent particle physics models require some new particles that could serve as the source of the dark matter, so it does not seem so radical to suppose that that such matter could account for enough energy density that $\Omega = 1$. The only catch is that this dark matter cannot be distributed in the same way as ordinary matter, otherwise the measurements of Ω based on gravitational attraction to nearby superclusters would totally include their contribution. Since the dark matter has quite different physical properties from ordinary matter, it does not seem unreasonable that it may have followed a rather different history than ordinary matter and, thus, be distributed differently.[26]

From this author's perspective, it is difficult to accept the possibility, inflation or not, that $\Omega = .1$, as direct astrophysical measurements presently indicate. This would mean that the deviation of Ω from unity is $O(1)$ only, even though $\Omega = 1$ is a position of unstable equilibrium. Unless there is something special about the present cosmological epoch, we expect Ω to deviate greatly from unity or to be very close to unity (near a position of unstable equilibrium). Since we know of no reason why the present age $O(10^{10}$ yrs) should be special, we must suppose that, once the experimental situation becomes settled, we will find that $\Omega = 1$.

Guth's initial inspiration for developing the inflationary model came from an attempt to resolve the monopole problem, but this problem cannot really be resolved by his scenario. The inflationary epoch ends, according to Guth's proposal, with a transition to the broken symmetry phase by the nucleation and growth of bubbles (we shall discuss the feasibility of this process shortly). The bubbles, formed by a causal process, must have a radius of order a Hubble distance (H^{-1}) or smaller. Just as in the big bang model, when the bubbles coalesce, a high density of monopoles form with mean separation H^{-1} which dominate the universe (whose radius after inflation is $\gg H^{-1}$) with their mass independent of how much inflation has occurred. However, before considering this problem further, it is necessary to consider a more serious problem that undermines the whole scenario.

THE GRACEFUL EXIT PROBLEM: The problem with the Guth proposal, now referred to as the "old inflationary universe model," is that there does not exist a mechanism to successfully complete the transition to the broken symmetry phase and return the universe to a FRW expansion rate. The bubble nucleation mechanism is not feasible.

During inflation, $R(t) \propto \exp(H_i t)$. If a bubble is nucleated at time t_0 it grows at a velocity approaching the speed of light. The bubble wall travels a distance at time t given by:

$$\begin{aligned}\ell_b &= R(t) \int_{t_0}^t \frac{cdt'}{R(t)} \\ &= \frac{R(t)}{H_i}\left(\exp^{(-H_i t_0)} - \exp^{(-H_i t)}\right).\end{aligned} \quad (32)$$

If we divide out a factor of $R(t)$, we see that the coordinate distance is bounded above by H_i^{-1}; that is, there is a maximum coordinate distance beyond which the bubble wall does not travel even though it travels at the speed of light! Over time, more and more bubbles are nucleated, but each occupies a finite coordinate volume whose size decreases exponentially with time. It is simple to show that the bubbles do not percolate the universe even after an infinite time. That is, outside of a few clumps, the bubbles never complete the phase transition.

The resulting picture of our universe might be called a "cold swiss cheese universe." The universe contains isolated bubbles of broken symmetry phase with exponentially expanding unbroken symmetry phase. The bubble interiors are near zero temperature since the energy gained in transforming symmetric to symmetry breaking phase is stored in the bubble walls. Our universe, which by particle physics measurements corresponds to a symmetry breaking phase, could not correspond to the interior of one of the bubbles because there is insufficient energy and entropy. It could not correspond to one of the rare collisions between bubbles because such a region is too anisotropic.[15] In short, the cold swiss cheese model is an unacceptable model of our universe. As a result, the inflationary model nearly died.

3.2 NEW INFLATIONARY UNIVERSE MODEL

Several years after Guth's proposal, A. Linde[2] and, independently, A. Albrecht and P. Steinhardt[3] introduced a model with a new kind of phase transition that evades the problems of the old inflationary model and produces new benefits. The new model is commonly referred to as the "new inflationary universe model."

The new element is to consider a different kind of phase transition commonly referred to as a "slow rollover" transition.[3] In most cases, the slow rollover transition is a special kind of spinodal decomposition, as opposed to the Guth proposal which involved a homogeneous nucleation transition. The example discussed by Linde and Albrecht and Steinhardt involved a GUT theory in which the gauge symmetry is broken by radiative corrections to the effective potential corresponding to the "Coleman-Weinberg limit" discussed in Section 2.5. At zero temperature, the effective potential for the $SU(5)$ Coleman-Weinberg model is:[9]

$$V = B\phi^4[\ln(\phi^2/\sigma^2) - \frac{1}{2}] \quad (33)$$

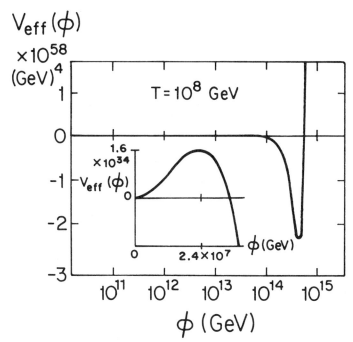

Figure 4: Free energy density for the Coleman-Weinberg model at $T \approx 10^8$ GeV. Barrier separating the symmetric ($\phi = 0$) and symmetry breaking ($\phi \neq 0$ phases can only be detected on small scales (inset).

where $B = 5625g^4/1024\pi^2$, g is the gauge coupling constant, and the scalar Higgs field Φ has been expressed as $\Phi = \phi(1, 1, 1, -3/2, -3/2)$. There is no term quadratic in the scalar field; the effective mass has been fine-tuned to zero. Also, there is no barrier between the symmetric ($\phi = 0$) phase and the spontaneous symmetry breaking (SSB) phase ($\phi \neq 0$). The finite temperature corrections, as discussed in Section 2.5, provide the only source of an energy barrier that can keep the universe trapped in a metastable symmetric phase as it cools from high temperatures.

There are several features that make the symmetry breaking transition in this model special. First, the finite temperature corrections keep the universe trapped in the symmetric phase even as the universe cools to very small T. See Fig. 4 for a cross-section of the potential near $T = 10^8$ GeV. As the temperature decreases below this value, the barrier becomes effectively negligible (the bubble nucleation rate approaches unity).[17] At this point, the formerly metastable phase becomes unstable and fluctuations can begin to drive the universe away from $\Phi = 0$ by a small amount. This condition is referred to by condensed matter physicists as "reaching the metastability limit" and it is characteristic of a spinodal decomposition transition. The subsequent evolution from Φ near zero to $\Phi = \sigma$ is governed by the semi-classical equation of motion:[3]

$$\ddot{\phi} + 3H\dot{\phi} + V'(\phi) = 0 \qquad (34)$$

where $H^2 \approx 8\pi\rho/3M_p^2$ is the Hubble constant, where ρ is the total energy density. After supercooling, the universe is dominated by the vacuum energy, $\epsilon \approx B\sigma^4$, of the symmetric phase. The thermal and matter energy density of the particles in the universe have been redshifted to a negligibly small value.

This semiclassical equation is analogous to the equation one would write for a ball with position ϕ rolling down a hill of shape V under the influence of a frictional drag force (the $\dot{\phi}$ term). If V is very flat, the $\ddot{\phi}$ term becomes negligible compared to the drag force. The ball (or ϕ) roll very slowly down the hill, restrained by the friction, until the potential gets steep. This friction dominated portion of the solution is referred to as the "slow roll."

The effective potential is so flat near $\phi = 0$ that the evolution towards $\phi = \sigma$ requires typically a time $t_{roll} = (100 - 1000)H^{-1}$ depending upon the choice of parameters.[2,3] During this "slow roll", the energy density of the universe is very nearly the same as that of the metastable phase ($\epsilon = V(\phi = 0)$) — essentially constant. Thus, for the same reason as in the old inflationary universe model, the universe begins to expand exponentially. In this case, however, the expansion only continues for a finite period of time as the universe slowly but inexorably evolves towards the stable phase. After the exponential expansion the scale factor is multiplied by a factor of $R(t) = \exp(Ht_{roll}) = \exp(100 - 1000)$.

The discussion has been oversimplified a bit because it appears in Fig. 4 that there is only a single SSB minimum, when in fact there are many of them associated with the spontaneously broken symmetry. Thus, when the fluctuations drive the universe away from $\phi = 0$, different parts of the universe are driven towards different spontaneous symmetry breaking minima. The universe breaks up into fluctuation regions or domains (see Fig. 5) inside of which ϕ evolves towards different SSB minima. A typical region is of a size of order the Hubble volume at that time (10^{-35} sec), around 10^{-25} cm. Where fluctuation regions come together there will generally be formed topological defects: monopoles where regions come together that differ by continuous symmetries, domain walls where regions differ by discrete symmetries (such as $\phi \to -\phi$). Then the slow evolution towards $\phi = \sigma$ begins and the inflation occurs. Each one of these regions grows by a factor of at least $\exp(100)$ so that a single fluctuation region is at least 10^{20} cm at the completion of the phase transition. Our observed universe at that time was about $1 - 10$ cm across. Thus, our observed universe fits easily inside a single fluctuation region.

The entropy, horizon, and flatness problems are solved, as in Guth's original proposal, because of the exponential expansion epoch. The problem of exiting

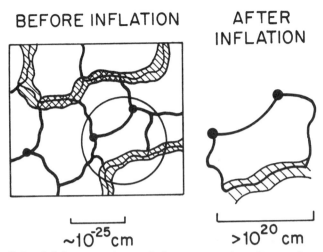

Figure 5: Spinodal decomposition of the universe into fluctuation regions and the subsequent inflation. Dark circles represent interesections of regions where monopoles form; hatched lines represent domain walls.

from the symmetric phase is resolved because the (important) exponential expansion occurs in the evolving stable phase, not in the trapped metastable phase. Monopoles and domain walls between separate fluctuation regions are formed; but in the inflationary picture our observed universe lies deep within a single fluctuation region and we should not have been able to observe even one such topological defect. (Under very artificial fine-tuned conditions, one can arrange a model in which a few monopoles are produced thermally within our observable universe just after the transition, but too few to have any undesirable effect on the evolution of the universe.)[16] The entropy, horizon, flatness, monopole and domain wall problems are resolved in a single stroke!

One must consider, however, what happens to the universe after the tremendous inflationary epoch. After a long period of exponential expansion, the density of all previously existing matter and energy is diluted by a huge exponential factor. The universe approaches essentially a vacuum state with the only significant contribution to the energy density coming from the vacuum energy density. A mechanism must be found for the universe to reheat in order to obtain a model corresponding to the universe in which we live: After the slow evolution over the flat part of the potential shown in Fig. 4, the scalar field expectation value begins to evolve very quickly and to oscillate about the SSB minimum. The scalar field can be thought of as a classical source coupled to quantum fields, and when the expectation value oscillates rapidly, it radiates particles. By this process, the vacuum energy of the symmetry phase is converted to matter and radiation.

The energy conversion process can be modeled as follows.[17] Suppose that the decay rate for the scalar field to ordinary matter and radiation is $\Gamma(\phi)$. The total energy density during reheating is given by $\rho_T = \frac{1}{2}\dot\phi^2 + V(\phi) + \rho_r$, where ρ_r is the radiation energy density (to which the scalar kinetic and potential energy is converted). The full equations of motion are then given by:

$$\ddot\phi + 3H\dot\phi + \Gamma\dot\phi + V'(\phi) = 0$$
$$H^2 = 8\pi\rho_T/3M_p^2 \qquad (35)$$
$$\dot\rho_r + 4H\rho_r = \Gamma\dot\phi^2.$$

For a very wide range of GUT models, the couplings of the Higgs field are such that the conversion of vacuum energy to radiation is very efficient; the temperature of the universe reheats close to the critical temperature, $T_c \approx M_G$.[17] This result is important for two reasons: Firstly, the large entropy that is observed in the universe today must be explained. Presumably, the universe has cooled adiabatically since the phase transition, so the large entropy (computed in Section 2.4) that is observed today must be generated through the efficient conversion of vacuum energy to radiation. Secondly, the observed excess of baryons over anti-baryons in the universe must be explained. After the exponential expansion, any excess baryon density before the transition is diluted by exponentially large factors. The subsequent production of radiation via the reheating makes the baryon excess to photon ratio negligibly small. However, since reheating to near T_c is achieved, color triplet mesons with masses near or below the GUT scale can be produced by the decay of the ϕ and/or subsequent particle interactions, and the mesons can then generate baryon asymmetry in much the same manner as in GUT models without inflation. Thus, the new cosmology leads to a universe in which the cosmological homogeneity, flatness, entropy, monopole, domain wall and baryon problems are resolved!

3.3 SOME REFINEMENTS FOR NEW INFLATION

Shortly after the introduction of the new inflationary universe model, several refinements of our understanding of the inflationary phase transition were discovered based upon the special properties of the exponential expansion phase. During the exponential expansion, the universe approaches a "de Sitter phase." One property of the de Sitter phase is that the curvature is not negligible and, in particular, there is generally a coupling of the curvature, R, to the scalar field of the form $\xi R\phi^2$. Such a term can seriously alter the flatness of the potential near $\phi = 0$ unless ξ is small. A simple way around this problem is to define a new "de Sitter mass" given by $m_{deS}^2 = m^2 + \xi R$, where m^2 is the bare mass of the scalar field. There is still only one fine-tuning of a mass term that is required, but it is tuning of a de Sitter space parameter rather than a flat space parameter.

A second feature of a metastable de Sitter phase is that the evolution away from the symmetric phase can be dramatically different from the case of flat space. Hawking and Moss[18] have analyzed the bubble nucleation process in de Sitter space and have found a very peculiar solution to the bubble nucleation field equations. They initially interpreted their solution as meaning that, if the barrier height is $O(H^4)$ or smaller, the total universe can evolve coherently out of the symmetric phase; in other words, all the universe evolves at once towards a single symmetry breaking phase. This does not in any substantive way alter the cosmological picture on the scale of our observable universe, but it does change our cosmological picture on very large scales which we cannot observe. For example, if the universe escapes the metastable phase via the Hawking-Moss solution, there is only one fluctuation region and no topological defects, even at very large distances from our observable universe. Probably, though, this interpretation of the Hawking-Moss solution is incorrect.[19] Causal physics, in this case a nucleation event, cannot act coherently on scales large than the Hubble radius H^{-1}. The Hawking-Moss solution actually involves scales $O(H^{-1})$ or smaller. Although the solution is totally uniform on these scales, one expects that, if the solution is properly extended to larger scales, it would approach the symmetric phase and appear as an ordinary bubble. At any rate, the Hawking-Moss solution is relevant for only a narrow range of parameters (small barrier heights) which cover only a very small region of the parameter space relevant to the inflationary model. In short, this detail can probably be safely ignored.

A third feature of a de Sitter Universe that can play an important role is the "effective event horizon". As we stated, causal physics cannot act coherently on scales that extend beyond the Hubble radius, which is a constant during inflation. Because the radius is constant, it represents in de Sitter space the distance from an observer to an event horizon. The de Sitter event horizon is observer dependent, unlike the physical event horizon about a black hole, but the de Sitter event horizon shares many properties of the black hole event horizon. One important property (discovered by Hawking) is that event horizons about black holes are the sources of fluctuations with a characteristic temperature, known as the "Hawking temperature," $T_H \equiv H/2\pi$. Similarly, (although different authors have different ways of treating this issue) These fluctuations not only further renormalize the de Sitter mass but can also drive the universe from the metastable to the stable phase at a faster rate than predicted by the semiclassical equation of motion, Eq. (34).[20] Such a process can prevent sufficient inflation from occuring; in fact, it has been shown that the coefficient B in Eq. (33) must be 10^{-3} times the value normally predicted by the simple GUTs model in order to suppress the fluctuations enough to obtain inflation. This is the first indication that inflation in the simplest GUT presents a difficulty.

More recently, there has been further criticism of the treatment of the inflationary phase transition. Essentially, the criticism is that in some phase transitions the mean field theoretic treatment of the phase transition presented in this discussion is not accurate — fluctuation effects become important. This criticism is valid when aimed at the inflationary models originally posed with couplings O(1).[21]

However, as we shall see, by far the worst problem for the GUT inflationary model derives from studying the fluctuations produced in the universe after the phase transition, as discussed in the next section. In order to obtain fluctuations that have a small enough amplitude to be consistent with astrophysical observation, we are forced to a model with a coupling many orders of magnitude weaker than a GUT coupling. This constraint is much more stringent than those imposed by de Sitter fluctuations driving the transition more rapidly and the couplings are so weak that one is well within the limits of mean field approximations.

Yet an additional criticism of the original new inflationary models is that it is incorrect to assume, in general, that the universe proceeds from the symmetric $SU(5)$ to the symmetry breaking phase corresponding to our present universe, $SU(3) \times SU(2) \times U(1)$, because there are many competing symmetry breaking minima in which the universe might first be trapped. In fact, without special choices of couplings, this is generally the case: the barrier to a competing minimum with different symmetry (e.g. $SU(4)$) disappears before the barrier to the desired $SU(3) \times SU(2) \times U(1)$ phase and the universe is trapped in the wrong phase.[22] Once again, though, the constraint from the density fluctuations essentially obviates this problem. The constraint requires that the coupling of the scalar field be much weaker than any typical gauge coupling. Therefore, it is most convenient to regard the scalar field as a gauge singlet, in which case the competing minimum problem can be easily avoided.

4 PRIMORDIAL FLUCTUATIONS AND GALAXY FORMATION IN THE INFLATIONARY UNIVERSE

The inflationary universe model was designed to solve the cosmological entropy, horizon, flatness, monopole, domain wall, and baryon asymmetry problems within a simple and efficient model. A few months after the model was introduced, though, yet another potential success of the model was discovered: the inflationary universe model predicts primordial fluctuations in the energy density of the universe with just the qualitative spectrum that many cosmologists believe is necessary to explain the origin of galaxies.[23,24] The model also leads to a prediction for the quantitative aspects of the spectrum, but these depend sensitively on the details of the particle physics model responsible for the inflationary phase transition. As we shall see, the quantitative predictions of the simple GUT models are in conflict with astrophysical constraints but it seems quite feasible that a somewhat different model can be designed that can meet the astrophysical constraints.

4.1 THE SCALE INVARIANT SPECTRUM

For many years cosmologists have debated the issue of what spectrum of fluctuations is necessary to account for the structure we observe in our universe: galaxies, galaxy clusters, superclusters, etc. Until the inflationary cosmology, the fluctuation spectrum could not be derived from first principles because the spectrum always depended upon the initial fluctuations that existed as the universe emerged from the big bang. To make matters worse, in order to form galaxies, coherent fluctuations have to exist on distance scales large compared to the Hubble radius (causal horizon distance) in the hot big bang model. Nevertheless, some cosmologists took the approach of working backwards — from observational data on the organization of galaxies in the universe, they hypothesized what kind of fluctuation spectrum is necessary to explain it without determining the source of those fluctuations. Of course, there is much debate about the interpretation of the observational data and whether or not a given fluctuation spectrum can really lead to a universe in agreement with the observational data. Strong opinions are formed and changed rapidly in this issue, and the problem is still far from being resolved, in my opinion. Nevertheless, it is fair to say that a large fraction of cosmologists have come to believe that a "scale invariant" spectrum of adiabatic fluctuations, as first proposed by Harrison and Zel'dovich, is what is required.[25]

The density fluctuation spectrum is the amplitude of the fractional energy density perturbation, $\delta\rho/\rho$, as a function of the wavelength, ℓ. An adiabatic fluctuation is one in which the fluctuation in the temperature and in the number of baryons is the same, as is expected in any GUT or microphysical theory. For wavelengths that are large compared to the Hubble radius, which sets the scale of the causal horizon, the fluctuation amplitude, $\delta\rho/\rho$, is not a coordinate (relativists use the word "gauge") invariant quantity — its numerical value depends upon how one fixes the hypersurfaces of constant time. For example, consider a universe in which the energy density is perfectly uniform in space but changing with time. By distorting the hypersurface of constant time, it will appear that the energy density is non-uniform. Obviously, this an uninteresting and unphysical non-uniformity. For wavelengths less than the Hubble radius, the fluctuation amplitude represents a physical (gauge invariant) quantity; for example, in a radiation dominated universe, $\delta\rho/\rho$ represents the amplitude of an acoustic wave. For this reason, $\delta\rho/\rho$ is only discussed on scales less than or equal to the Hubble horizon. Cosmologists, in fact, commonly define the spectrum of fluctuations by plotting $\delta\rho/\rho$ as a given wavelength just falls within the Hubble horizon (the horizon grows with time enclosing longer and longer wavelengths); this quantity we call $\delta\rho/\rho|_H$. In a FRW universe, the Hubble radius grows with time, so a plot of $\delta\rho/\rho|_H$ versus ℓ measures the fluctuations on different scales at different times (as each falls within the expanding Hubble radius). In this way, the amplitude $\delta\rho/\rho|_H$ represents the "virgin perturbation amplitude" before causal physics has had a chance to act coherently on the fluctuation and alter it.

Every physical scale, ℓ, at any given time in the universe can be characterized by a unique "comoving" scale in which the expansion of the universe has been divided out

$$\ell = R(t)\ell_{comoving}. \qquad (36)$$

So, for example, if we want to trace the history of the scale associated with galaxies, we trace an ℓ that varies as the universe expands or a unique $\ell_{comoving}$. When cosmologists refer to physics occuring on some scale, they are referring to the comoving scale. On the comoving scale associated with galaxies, we require that $\delta\rho/\rho|_H$ be greater than $10^{-(4-5)}$ in order to have big enough fluctuations to eventually condense and form galaxies. However, the fluctuation on the largest observable comoving scales cannot be greater than 10^{-4} or quadrupole anisotropies would have already been observed in the cosmic microwave background. The most recent experimental results indicate that the background is uniform in temperature (once the dipole anisotropy associated with the motion of the galaxy is subtracted) to at least one part in 10^4. Also, if $\delta\rho/\rho|_H$ is greater than unity on any small scale, many black holes would have been formed. The simplest spectrum that is consistent with all these constraints is one in which $\delta\rho/\rho|_H$ is nearly constant ($\approx 10^{-(4-5)}$) over all scales, which is the "scale invariant" spectrum proposed by Zel'dovich. In short, the spectrum means that the perturbation on any scale as it enters the horizon is the same independent of the time or scale.

Zel'dovich's motivation for proposing this simple spectrum was not based solely on this simplistic argument. He hoped to explain the organization and origin of galaxies from such a spectrum. Basically, he argued that, beginning from such a spectrum, fluctuations on small scales would be damped out by various dissipative mechanisms. The first surviving fluctuations would be on the galaxy cluster or supercluster scale and it is on this scale that condensations would first be formed. Zel'dovich argued that such fluctuations would condense in a hydrodynamically unstable mode, breaking up into subsections that collapse into flat, "pancake" surfaces which subsequently break up further into galaxies and galaxy clusters. One implication of such a "pancake theory" is that there should be large observable voids separating the pancakes, and already preliminary evidence for such voids has been reported.

The pancake theory is an example of what has been called a "top-down" theory, in which structure is first formed on large scales and then, when the large scale structures evolve. structure is formed on smaller (galactic) scales. For many years, a competing theory was the "bottom-up" theory, in which, as you may guess, structure is first formed on galactic scales and then galaxies clump gravitationally to form larger clusters and superclusters. The evidence from astrophysical measurement and computer simulations has not been totally decisive as to which theory works best. However, the inflationary prediction of a scale invariant spectrum seemed to strongly support Zel'dovich's notion of a top-down evolution.

The story takes a strange turn at this point, though. In order for the Zel'dovich spectrum to work with $\Omega = 1$, there has to exist dark matter in the form of a hot light particle species. The leading candidate at the time was a massive neutrino, which seemed to be allowed naturally in GUTs, whose finite mass was reported in some experiments, and which seemed to be so beautifully consistent with the pancake and inflationary theory. However, the bottom-up proponents soon realized that, if a different kind of particle accounted for the dark matter — a cold, weakly coupled particle such as an invisible axion — the pancake model would not follow. Instead, perturbations on small scales would not be dissipated and structure would first form on small scales. In short, *the scale invariant spectrum could lead to either the top-down or bottom-up picture depending upon the choice of dark matter particle!*[26] At the present time, GUT models (without some embellishment such as supersymmetry, superstrings, etc.) are somewhat in disfavor, there is still no concensus on the experimental measurement of the neutrino mass, and further computer simulations indicate that the neutrino dominated pancake theory is inconsistent with other astrophysical observations of the correlations in positions of galaxies. The best model, based on the computer simulations, is one in which the dark matter is cold (such as an invisible axion or a host of other candidates) and a bottom-up theory results. However, even here the model only works for $\Omega = 1$ if the dark matter is not distributed in the universe in the same way as ordinary matter. My own best guess is that probably the scale invariant spectrum is correct (or at least it can be made to be consistent with astrophysical observations and computer simulations), but that we are still far from understanding the details of how the spectrum evolves into the universe we observe.

4.2 INFLATION AND FLUCTUATIONS

Six months after the introduction the new inflationary universe model, four groups came together in an extemely exciting Nuffield Conference, organized by Stephen Hawking.[23] Each group was near completion of a computation of the fluctuation spectrum predicted by the inflationary model based on widely different computation methods. By the end of the conference, the results converged: the spectrum is scale invariant with a well-defined, though model-dependent, prediction of the fluctuation amplitudes.[24]

The result is remarkable because it is a success that was not anticipated as part of the inflationary universe program. Furthermore, it should be re-emphasized that this is the only cosmological model that has any hope of explaining our present universe and for which the spectrum of fluctuations can be derived from first principles.

The derivations of the four independent approaches are quite complicated and it is very difficult to see how one derivation is related to another. This is due to the fact that the different derivations are based on different choices of the hypersurfaces of constant time. The "history" of a fluctuation on a given scale appears to be quite different for different hypersurfaces, as we have argued. However, by the time the fluctuation re-enters the Hubble horizon in the FRW phase, the amplitude is the same independent of the choice of constant time hypersurfaces. Although I will not give the details of any one computation, I will attempt to provide a heuristic argument as to why the qualitative spectrum is found to be scale-invariant and a dimensional argument as to why, in the simplest $SU(5)$ Coleman-Weinberg model, the quantitative value of the amplitude is too large.

4.2.1 Why is the spectrum scale invariant?

The Hubble radius, H^{-1}, represents an important scale in the analysis of the creation and evolution of energy density perturbations. During the de Sitter or inflationary phase, H (and other parameters) are roughly constant. The scale factor, $R(t)$, grows exponentially with time undergoing one e-folding in a time interval $O(H^{-1})$. Microphysics can only operate coherently on physical scales less than $O(H^{-1})$.

Inflation has two important effects on the spectrum of fluctuations. First, it erases all previously existing fluctuations by expanding their wavelengths to such large scales that they are irrelevant for our observable universe. In this sense, inflation smooths the universe. If this were all that occurred, the model would be unacceptable because it would make the universe too smooth for the inhomogeneities in our universe to evolve. However, inflation also provides a natural source of new fluctuations coming from quantum fluctuations — a truly beautiful notion. Normally we consider quantum fluctuations to be important only on very small length scales; however, inflation can expand fluctuations on small scales to fluctuations on large scales of astrophysical interest.

The basis of the inflationary universe model is that all scales that we observe in our universe were much less than the the Hubble radius (the causal horizon radius) when the inflationary epoch began. Microphysics (quantum fluctuations, etc.) can create and alter the amplitude of perturbations on scales for which $H\ell \ll 1$. (In fact, in the de Sitter phase, the large curvature, or equivalently, the event horizon, automatically leads to large and calculable fluctuations on scales smaller than H^{-1}.) As inflation proceeds, the physical scale of a perturbation grows until $H\ell \approx 1$, and then onwards until $H\ell \gg 1$. Once the physical scale of the perturbation has grown such that $H\ell \approx 1$, microphysics cannot alter the amplitude of the physical perturbation: the perturbation "freezes out" at whatever amplitude it had at the time $H\ell \approx 1$. Since the inflationary epoch is essentially time translationally invariant, the same microphysics determines the amplitude on different scales as they expand beyond the Hubble horizon ($H\ell \approx 1$) at different times. Once the universe reheats and re-enters the FRW phase, the Hubble radius increases (at the rate characteristic of an FRW universe: $\propto t$) and the scale of perturbations expands with the universe, $\propto t^{1/2}$ or $t^{2/3}$. Thus, the Hubble radius gradually catches up with the scales that expanded beyond the horizon in the inflationary epoch; the perturbations are said to "re-enter the horizon." Since no microphysics could change the perturbation when $H\ell \gg 1$, and since the perturbations freeze out at (approximately) the same amplitude, each scale re-enters the horizon with the same amplitude of perturbation, $\delta\rho/\rho|_H$. Thus, the spectrum is scale invariant.

4.2.2 *What is the amplitude of the fluctuation spectrum?*

The computations of the fluctuation spectrum not only show that the spectrum is scale invariant, but also determine the amplitude of the spectrum. In fact, the methods used by Bardeen, et al. explicitly show that the amplitude is very insensitive to the equation of state during reheating and through the FRW epoch. I do not believe that there is a simple argument to explain the result, but the expression for the fluctuation amplitude is found to be:[24]

$$\delta\rho/\rho|_H = (4 \text{ or } \frac{2}{5})\frac{H\Delta\phi}{\dot{\phi}}|_{t_{freeze}} \quad (37)$$

where $\dot{\phi}$ is the time variation of the scalar field whose evolution signals the SSB phase and is responsible for the inflationary phase; $\Delta\phi$ is the fluctuation in ϕ during the inflationary epoch and the prefactor is 4 (or 2/5) if the universe is radiation dominated (or matter dominated) when a given scale re-enters the Hubble horizon in the FRW phase. The limits on $\delta\rho/\rho|_H$ obtained from the measured isotropy of the microwave background involve distance scales for which the (2/5) factor is appropriate. Note that the answer only depends upon the values of parameters at the time t_{freeze} when the given scale of interest expands beyond the Hubble radius during the inflationary phase, as seems sensible from our heuristic argument.

It is also clear from this result that the spectrum is not precisely scale invariant. During inflation $H\Delta\phi$ is essentially constant ($\approx H^2$), as determined by the quantum fluctuations of the scalar field (e.g. created the de Sitter equivalent of the Hawking temperature). The denominator, $\dot{\phi}$, is the classical time variation of the scalar field as determined by Eq. (34). During inflation, ϕ and $\dot{\phi}$ increase as the field rolls towards the SSB phase; thus, they have somewhat different values when perturbations on different scales freeze out as they expand beyond the de Sitter Hubble radius. The result is that the spectrum has scale dependence that is logarithmic. (The freeze-out scale depends only logarithmically on scale because the scale parameter grows exponentially with time.) The evaluation of the equation (using 2/5) for the Coleman-Weinberg GUT model is found to be:

$$\delta\rho/\rho|_H = 10\lambda^{1/2}[1 + \ln(10^{28} \text{ cm}/\ell)]^{3/2} \qquad (38)$$

where λ is the effective dimensionless quartic coupling. The value of λ is $O(10)$ during the inflationary epoch in the simplest $SU(5)$ GUT model and so $\delta\rho/\rho|_H$ is $O(10)$ on the largest scales that we observe today $O(10^{28}$ cm$)$. The result is roughly five to six orders of magnitude too big to be consistent with the observed isotropy of the microwave background. It might be pointed out that we have found that $\delta\rho/\rho|_H$ is greater than unity, so it is possible, in principle, that non-linear effects can save us. However, this is probably wishful thinking: if the non-linear effects do reduce the amplitude, eventually they reduce it to $O(1/10)$ or so when perturbation theory again becomes valid. The amplitude, according to the perturbative computation, does not become any smaller and we are still at least three orders of magnitude too large.

Although it is difficult to derive Eq. (37) explicitly, it is not difficult to explain why, on dimensional grounds, the answer we have obtained is order unity. The key is that, during inflation, ϕ lies very close to the top of the potential where the shape is nearly flat. The GUT scale, $\sigma \approx M_G$ appears only logarithmically in the potential and the only dimensional parameter that determines the evolution of ϕ and the perturbations (Hawking temperature $\approx H/2\pi$) is the Hubble parameter, H. Once ϕ grows much beyond H, its evolution is very rapid and the universe begins to reheat. The fractional perturbation, $\delta\rho/\rho|_H$, as we have argued, only depends upon the properties of the inflationary epoch. Because it is dimensionless and there is only relevant dimensional parameter, $\delta\rho/\rho|_H$ can only be a function of dimensionless quantities, such as λ. The dimensionless coupling, λ, is $O(1-10)$ for typical GUT models, so it is difficult to imagine obtaining the desired value of $\delta\rho/\rho|_H$, $O(10^{-4})$, that we need for the fluctuation spectrum to be consistent with the microwave background.

To obtain the desired value of $\delta\rho/\rho|_H$, the dimensionless coupling must be reduced by over *ten orders of magnitude* compared to its value in the simplest GUT model. By simply reducing λ and not other parameters, though, $H \propto \lambda$ is also reduced and then inflation is reduced. To keep H fixed while reducing λ, the GUT scale, σ, has to be increased. The severe constraint of the fluctuation spectrum forces parameters to lie in a very small range where $\sigma = 10^{(18-19)}$ GeV.

The result is a "squat and long" potential in which the flat "rollover" region of the potential is greatly increased. As a result, $\dot\phi$ is significantly greater than H^2 when observable scales in our universe expand beyond the Hubbles radius (during the last 60 e-foldings or so of inflation); this leads to a smaller value of $\delta\rho/\rho|_H$. Since there is no GUT model for which we expect the parameters to lie in the necessary range, the new inflationary model based on Coleman-Weinberg potentials is dead.

5 PRESCRIPTION FOR NEWER NEW INFLATION

There are three major problems with the inflationary model as formulated for Coleman-Weinberg GUT models. First and foremost, the density perturbations are too big in the simplest GUT models, which is a cosmological disaster. (Fixing the parameters to obtain acceptable fluctuation amplitudes would be more than sufficient to satisfy other criticisms of the model discussed in Section 3.3). Second, the model appears to require unnatural fine-tuning of parameters. In the Coleman-Weinberg model, this corresponded to setting $m_{deS}^2 = 0$ (to at least one part in 10^{10}). This is exacerbated by the new constraint from the density fluctuations that the effective dimensionless quartic coupling be $O(10^{-(12-16)})$, a very small value. Finally, if we do find a model with such a small effective scalar coupling, the scalar field could not be in thermal equilibrium before the transition and we must explain how the inflationary transition began.

In spite of these problems, the new inflationary universe model is very sound, even thriving. The key notion that was garnered from the studies of the Coleman-Weinberg model — the ideal of a "slow rollover" transition in which inflation occurs in evolving stable phase rather than in trapped metastable phase — can be carried over to a broad class of models. The notion of the slow rollover transition is crucial for two reasons. First, it insures that the transition can be completed. Second, unlike Guth's original picture where it was hoped that the transition would be completed by bubble nucleation and collision, the slow transition is a continuous process in which all physical quantities can be tracked in a smooth way. This last feature is crucial for analyzing the creation and growth of fluctuations, for example.

Thus, the hope is to incorporate the inflationary model within some other particle physics models with the hope that basic problems that occurred in the GUT models can be obviated. Already, as we shall mention in the next section, efforts have been made to incorporate inflation in supersymmetric and supergravity models, as well as modified GUT models. The problem really is that there is no good candidate for a particle field theory that satisfies the problems of particle physics, let alone inflation. Thus, for those attempting to formulate new models of particle physics, it is useful to codify the constraints that the

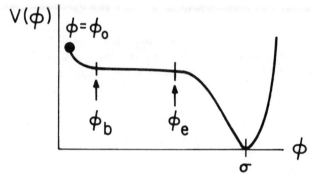

Figure 6: The de Sitter potential as a function of order parameter field.

inflationary model places on particle physics at very high energies.

The constraints from inflationary cosmology can be grouped into three categories. First, there are the *slow rollover constraints* — constraints on the detailed shape of the effective potential during the inflationary phase that guarantee sufficient inflation to solve the cosmological problems and density fluctuations with amplitudes in a range consistent with the observed isotropy of the microwave background. The prescription for these constraints has been discussed by M. Turner and the author.[27] A second category is the *cosmological constant* constraint — the constraint on the effective potential that the phase corresponding to the present universe have cosmological constant (or, equivalently, vacuum energy density) equal to zero. A third category is the *thermal constraint* — the condition that the energetically favored phase at high temperature correspond to the symmetric phase (or a phase near it) so that the universe supercools in that phase to begin the inflationary transition.[28]

5.1 SLOW ROLLOVER CONSTRAINTS

We shall consider a general effective (de Sitter) scalar potential of the form shown in Fig. 6 as a function of some scalar order parameter field (the field that drives the transition), ϕ. We shall assume that some dynamical process, such as high temperature effects, drives the universe to some value $\phi = \phi_0$ far from the stable phase, $\phi = \sigma$, as shown in the figure. We define the "flat region" of the potential as the interval $[\phi_b, \phi_e]$ where:

$$\begin{aligned} \phi_b &= \max\{\phi_0, \text{ minimum value of } \phi \text{ for which } V''(\phi) < 9H^2\} \\ \phi_e &= \text{maximum value of } \phi < \sigma \text{ where } V''(\phi) < 9H^2 \end{aligned} \quad (39)$$

For values of ϕ in this range, the $\ddot{\phi}$ term in Eq. (34) can be ignored compared to the drag term, $3H\dot{\phi}$, and ϕ rolls so slowly that the universe expands exponentially. We then have the following constraints:

(a): $\phi_e \leq \sigma \leq f M_p$, in order to guarantee that quantum gravity effects can be ignored. The "fudge" factor, f, should have a value such that the energy density stored in the scalar field is less than M_p^4.

(b): $\Delta\phi \equiv \phi_e - \phi_b \leq N^{1/2} H/2\pi$, where N is the total number of inflation e-foldings. This constraint guarantees that the flat region of the potential is long enough that quantum de Sitter fluctuations do not drive the transition too rapidly (which, as you might recall from Section 3.3, is a typical problem in GUT models). The fluctuations correspond roughly to a random walk of step-size $\delta\phi = H$ with one step every e-folding.

(c): $\int_{\phi_b}^{\phi_e} H dt \approx -\int_{\phi_b}^{\phi_e} 3H^2 d\phi/V'(\phi) \equiv N \geq 60$, the constraint that the number of e-foldings be sufficient to resolve the entropy, homogeneity, flatness, and monopole and domain wall problems. In practice, the other constraints force $N \gg 60$. That N is forced to be much greater than 60 is physically significant because some astrophysicists have proposed the possibility of "minimal inflation" — just enough to solve the cosmological problems but little enough that Ω can deviate by $O(1)$ from unity (so the model would be in agreement with present experimental measurements of Ω). However, the limit on the density fluctuation amplitudes requires a flatter potential and extra inflation. Minimal inflation is not possible.

(d): $\dot{\phi}_{60} = 10^4 H^2$, where $\dot{\phi}_{60}$ is the value of $\dot{\phi}$ 60 e-foldings before the end of inflation (corresponding to t_{freeze} for scale of our observable universe) in order to insure that $\delta\rho/\rho|_H$ is $O(10^{-4})$. See Eq. (37).

If these first four constraints are applied to a potential of the form:

$$V = V_0 - \beta\phi^3 + \lambda\phi^4 \tag{40}$$

where $V_0 = 27\beta^4/256\lambda^3$ is fixed to make the energy density zero at the minimum of the potential, then the constraints imply that: $\lambda < 4 \times 10^{-16}$ and $\beta \approx \lambda^{3/2} M_p$. The model leads to at least 10^4 e-foldings of inflation. That is, *it is not difficult to satisfy the inflationary constraints*.

(d): Reheating: $T_{RH} \approx \min\{(M_p H_e)^{1/2}, (\Gamma M_p)^{1/2}\}$ is the reheating temperature after the transition, obtained by solving the full equations of motion Eq. (35). If the decay rate, Γ, is fast compared to the Hubble expansion rate after inflation, the reheating temperature corresponds to the first value. With very weak couplings, though, the decay rate of the scalar field is likely to be much less than the Hubble expansion rate, and then

the second value applies. The reheating temperature must be at least 1 MeV in order to obtain the successful big bang predictions of primordial nucleosynthesis. The temperature probably must be much greater, $O(10^{12}$ GeV), in order for the baryon asymmetry to be generated dynamically after the phase transition.

(e): Sensible physics: These inflationary constraints must be incorporated into a viable particle physics model. As we shall see, there are cases when the inflationary constraints conflict or aid constraints due to particle physics. The most exciting possibility from my point of view is that inflationary cosmology, if correct, may help point the way towards new particle physics models that are useful for high energy physics.

5.2 COSMOLOGICAL CONSTANT CONSTRAINT

The cosmological constant or vacuum energy density constraints requires that $V(\phi = \sigma)$ be zero. The constraint is necessary because the cosmological constant is measured to be nearly zero in the present phase of our universe, which is supposed to correspond to the ground state, $\phi = \sigma$. In GUT models, this is a trivial constraint that can be satisfied simply by adding an overall constant to V and adjusting it so that V is equal to zero at the broken symmetry state corresponding to our present universe. However, in supersymmetry models, the constraint is more subtle because one cannot just add a constant term to V without explicitly breaking the supersymmetry, which destroys the purpose of introducing supersymmetry in the first place. A means of setting the cosmological constant to zero is to add an overall constant to the superpotential which generates V and then adjust it to make the cosmological constant zero. However, this not only changes the constant term in V, but also the coefficients of all the interactions due to the supersymmetry.[28] Thus, it is possible for the cosmological constant constraint to conflict with other inflationary or particle physics constraints. Supersymmetry connect two apparently unrelated sets of constraints: a constraint on the potential near $\phi = 0$ (the high temperature phase) and a constraint on the potential at large $\phi = \sigma$ (our present low temperature phase).

5.3 THERMAL CONSTRAINT

From the constraints that have already been discussed, it has been insured that the effective potential has the right shape between ϕ_0 and $\phi = \sigma$ for a successful inflationary universe scenario. However, one must explain how the universe approached the state $\phi = \phi_0$ in the first place. The conventional explanation has been that at high temperatures the ϕ_0 phase (or some phase near it) might be favored and then the universe supercools into that phase as inflation commences. If this is the case, one must check that the favored state at high T is ϕ_0 so that a transition will occur in which ϕ rolls over the flat portion of the potential that has been designed according to the slow rollover constraints. If ϕ near σ were favored at high temperatures (which is possible, especially in cases where the scalar fields couple only to themselves), then as the temperature decreases ϕ will never roll over the flat region of the potential. There will be no inflation even though the slow rollover constraints have been satisfied.

If ϕ couples to gauge fields, they tend to drive the universe towards the symmetric phase at high temperatures. In the symmetric phase, the maximum number of gauge fields are massless. Since the entropy is proportional to the number of massless degrees of freedom, the highest entropy state at high temperatures is the symmetric phase. The highest entropy state has the lowest free energy at high temperatures. However, from the constraints due to density fluctuations, we expect that the scalar field responsible for inflation is not strongly coupled to gauge fields since their couplings would lead to an effective quartic interaction coupling that is too large. If the scalar field is coupled only to itself (or other scalar fields) the finite temperature contribution is roughly:

$$\Delta V = \frac{\pi^2}{45} k^4 T^4 + \frac{1}{24} k^2 T^2 \mathrm{Tr} \left[\frac{\partial^2 V}{\partial \phi \partial \phi^*} \right] + \ldots \qquad (41)$$

where we have taken ϕ to be a complex singlet field and k is Boltzmann's constant. Insuring that the ϕ_0 is favored at high temperatures generally places a constraint on the signs of some of the couplings. For the case of supersymmetric models, this constraint eliminates a number of interesting candidates.[28,29]

The condition on the signs of the couplings is necessary but not sufficient to guarantee that the thermal constraint is satisfied.[30] It may be that, even though the $\phi = \phi_0$ phase is favored at high temperatures, the barrier separating it from the $\phi = \sigma$ phase is small compared to thermal fluctuations even when the temperature is quite high. This would mean that thermal fluctuations would carry the universe over the barrier at high temperatures before inflation could take place. This problem must be checked for each candidate model but is difficult to express in a prescriptive way.

Having introduced the thermal constraint and having indicated that it plays a role in eliminating some candidate models, I should now point out that it may not be necessary to take it into account. There is a potentially serious problem related to the thermal constraint that threatens the entire inflationary universe program.[31] The problem is that the usual assumption that the universe is driven to the false vacuum phase by high temperature effects is not feasible! Given the very weak couplings required for the inflation fields in order to obtain small density fluctuation amplitudes, it is straightforward to show that the inflation fields cannot have been in thermal equilibrium before the inflationary transition.[31] The conflict occurs in any inflationary model, supersymmetric or not.

What is desperately needed is a new approach for entering the inflationary phase transition. A first guess might be to suppose that, when our observable universe is contained within a single Planck radius at $t < 10^{-45}$ sec, strong gravitational interactions thermalize the distribution of particles. As the universe expands beyond the Planck size, the gravitational interactions become weak and the particles fall out of thermal equilibrium. However, just as with the microwave background, the particles maintain an effective thermal distribution and might remain trapped in the high temperature symmetric state up to the time of the inflationary transition.

Unfortunately, it does not make any sense to consider the notion of particles being in thermal equilibrium when the universe is within a Planck radius. The reason is that there is less than one particle per causal horizon during that epoch, so the notion of thermal equilibrium is hopeless. (The particle number density increases as T^3 but the causal horizon volume decreases as T^{-6}; the number of particles within the horizon becomes less than unity at a temperature slightly below the Planck scale, depending upon the number of particle species.) To assume that gravitational interactions can thermalize a distribution of particles on scales greater than the causal horizon appears to defeat one of the purposes of the inflationary model.

Several approaches to the dilemma seem possible: (a) Fluctuating Field Approach: Although almost no particles lie within the horizon when the observable universe is smaller than a Planck radius, the scalar field is continuously defined everywhere. It might be reasonable to suppose that, due to interaction with strong gravitational fluctuations, say, the field is left uncorrelated over distances greater than a causal horizon distance as the universe expands beyond the Planck radius. Depending upon the signs of the couplings of the fields to themselves and others, it may be energetically favorable for the fields to relax to a false vacuum state within the growing causal horizon because this may be the state of least gradient field energy density, although it is a state of higher potential energy density. In fact, just such a phenomenon occurs in those field theories which obey the thermal constraint. (b) Chaotic Inflation:[32] If we assume that the universe began in a globally chaotic state (not a big bang), random fluctuations might drive some rare regions into a phase near ϕ_0 and then such regions may inflate. Only those regions undergo inflation and we happen to live in such a region; it is not clear what fraction of the total universe such lucky regions occupy. No thermal constraint is required. In the chaotic inflationary model proposed by Linde, there need not even be a phase transition or metastable phase. Even for a pure $\lambda\phi^4$ theory, fluctuations can drive ϕ to very large values ($> M_p$), and then it can take so long for ϕ to relax back to the minimum (as computed using the slow roll equation, Eq. (34)) that sufficient inflation can take place. The problem with this approach is that the slow roll equation is only relevant and inflation will only occur if the gradient in ϕ is small over the region in question. (Note also that the region in question must be at least a Hubble horizon in size or else the FRW equations are not a good approximation for that region.) If we assume that ϕ undergoes a fluctuation, $\Delta\phi > M_p$, over a region on the scale of a causal horizon size, H^{-1}, then $\nabla\phi \approx HM_p$. The gradient energy density, $(\nabla\phi)^2$, must be less than the potential energy density. Recalling that $H^2 \approx 8\pi V/3M_p^2$, we see that this condition is met for $V = \lambda\phi^4$, say, provided $\phi < 3M_p$, which alas, is too small a value for there to be sufficient inflation according to the slow rollover equation. This inconsistency can be avoided if we assume that the fluctuation in ϕ is uniform on scales *larger* than H^{-1}. However, this assumption seems somewhat unnatural, especially since one of the goals of the inflationary model was to eliminate assumptions about uniformity on scales large compared to the horizon volume. In my opinion, the attractiveness of the chaotic inflation approach is at least somewhat questionable. (c) Universe Springs from Nothing:[33] Our universe tumbles from a state with no space or time directly into a false vacuum phase; it then inflates, reheats, etc. No thermal constraint is required. There does not really exist a mathematical formalism for describing such a transition, but there have been a number of clever attempts to develop a candidate formalism analogous to the usual bubble nucleation formalism in phase transitions. Although there is the suggestion that such an approach is feasible, it makes me uncomfortable to think that the viability of the inflationary model might rest on such an uncertain possibility.

My own point of view is that scheme (a) might be closest to the right approach. Based on this bias and since scheme (a) requires the same "thermal constraints" on the parameters, many continue to consider these constraints in building inflationary particle physics models.

6 PROSPECTS FOR "NEWER" NEW INFLATION

The search is on to find a viable particle physics model that satisfies the inflationary constraints discussed in the previous section. A simple approach is to take any acceptable GUT theory and then add a gauge singlet field that is very weakly coupled to itself and other fields such that the effective potential for the fields satisfies the constraints.[34] Such an approach demonstrates that inflation is at least compatible with particle physics, but the model is somewhat artificial.

Perhaps it would be better for the students in these lectures if I simply end the discussion at this point so that each can now search for his or her own favorite model based upon the constraints that I have discussed. However, given the growing literature on the subject, and my own prejudices, I will make a few remarks about attempts to incorporate inflation into supergravity models.

Supergravity inflation is certainly worthy of consideration by those who believe that supergravity may play an important role in high energy physics, either directly or effectively through some superstring theory. Even though supergravity and inflation appear to be quite independent of one another, they both depend critically on the properties of elementary particles at very high temperatures (energies). A careful examination should be made to determine whether or not supergravity is ultimately compatible with inflationary cosmology.

By far the most significant reason for considering supergravity inflation is that locally supersymmetric theories appear, at least naively, to produce the "slow rollover" phase transition necessary for inflation with little or no fine-tuning of parameters. The basic reason is that all couplings in a supersymmetric model are scaled by some power of (μ/M), where μ is the supersymmetry scale and $M \equiv M_p/\sqrt{8\pi} = 2.4 \times 10^{18}$ GeV. The power of (μ/M) varies from term to term in a prescribed way. Inflation requires that all couplings of the scalar field responsible for inflation be weak and sets constraints on the ratios of various couplings. In a supersymmetric model, (nearly) all these constraints can be met by setting just one parameter — the supersymmetry scale, $\mu \approx 10^{14}$ GeV. All dimensionless parameters can have values of $O(1)$.

The appearance that supersymmetry and inflation can be combined easily and naturally has caused many groups, including myself, to investigate supergravity inflation.[31] However, a more careful study of the supergravity models shows that almost all supergravity inflation models that have been proposed fail to satisfy one or more constraints for inflationary cosmology. In fact, except for some extreme possibilities, it can be shown that only a special class of supergravity models appears capable of satisfying all the constraints.[28] This conclusion seems depressing, since some may conclude that inflation and supergravity are perhaps rather incompatible. However, the particular class of supergravity models that satisfies the inflationary constraints has a number of attractive properties:[35] The models spontaneously break supersymmetry; set a supersymmetry breaking scale that is consistent with the mass hierarchy; produce baryon asymmetry reasonably consistent with astrophysical observations; and avoid the overproduction of entropy caused by the domination of the energy density by gravitinos and/or coherent field oscillations (a problem which occurs in nearly all supersymmetry models). The models, which have only one or two chiral fields, can act as supersymmetry breaking "hidden sectors" in conventional particle physics except now the same sector produces inflation at the same time that it spontaneously breaks the supersymmetry. All the specific models known at present have an unnatural fine-tuning of parameters, but the ubiquitous nature of these models encourages many of us to believe that supergravity inflation may yet be the best approach to achieving inflationary cosmology, as well as resolving some problems in particle phenomenology.

Recently, a group of us at Penn[36] studying supergravity inflation have noted yet one other feature of supergravity inflation, that may pertain as well to other inflationary models: quantum gravity corrections can play a major role in inflationary cosmology. Normally, we regard quantum gravity corrections as being negligible for macroscopic physics, yet here is an example where quantum gravity can play an important role on a cosmological scale. In particular, we find that the flat region of the effective potential required for inflation is so flat that the quantum graviton and gravitino corrections to the effective potential compete with the tree level terms. To make matters more interesting, we find in some cases that, beginning from a very simple tree level potential, the quantum gravity contributions provide just the sort of terms needed to enable the inflationary scenario. It seems somewhat ironic that quantum gravity corrections should have to be taken into account since, up to this point, inflationary cosmology involved energies safely below the Planck scale so that such contributions could be ignored. In the end, though, the constraints from the density perturbations forces a potential so flat that quantum gravity corrections cannot be ignored after all.

With the advent of superstring theories, the future of inflationary cosmology seems even harder to predict. Perhaps inflation is not needed at all! What does the horizon problem mean for a model in which the universe collapses from a higher dimensional manifold. Perhaps some feature of the collapse forces a spatially flat 3-manifold. Monopoles may exist in principle in the model, but perhaps they have a mass comparable to or larger than the reheat temperature after the collapse of the manifold and are there produced in very small numbers. In short, the status of the cosmological problems has to be re-evaluated in the context of the superstring cosmology. I suspect that remnants of some of these problems will remain and the primordial fluctuation problem will certainly remain: some mechanism is required to produce a scale invariant spectrum of perturbations stretching beyond the Hubble horizon after dimensional collapse. I believe, therefore, that some kind of inflationary model is still required. It may involve scalar fields in the effective four dimensional theory, in which case the study of supergravity inflation is directly relevant. Or it may occur as part of the dimensional collapse process itself, in which case the slow rollover transition does not involve ordinary scalar fields, but a more complicated order parameter that traces the evolution of the higher dimensional manifold. This second prospect, in my opinion, represents the most daring and challenging hope for inflationary cosmology.

ACKNOWLEDGEMENTS

I would like to thank T. Appelquist, F. Gursey, C. Sommerfield and the other organizers of TASI-1985 for their kind hospitality and support during my participation in the program. I wish to thank the students for their interest and challenging questions. I have been fortunate to have many outstanding collaborators during the course of my research on inflationary cosmology whose work forms the basis for this series of lectures. I would like to especially thank A. Albrecht, J. Bardeen, A. Guth, L. Jensen, P. Lindblom, B. Ovrut and M. Turner. This work has been supported in part by DOE Grant EY-76-C-02-3071 and by an Alfred P. Sloan Foundation Grant.

Bibliography

[1] Guth, A., *Phys. Rev.* **D23**, 347 (1981).

[2] Linde, A. D., *Phys. Lett.* **108B**, 389 (1982).

[3] Albrecht, A. and Steinhardt, P., *Phys. Rev. Lett.* **48**, 1220 (1982).

[4] The following include excellent discussions of standard hot big bang cosmology: Peebles, P. J. E., *Physical Cosmology* (Princeton: Princeton U. Press, 1971). Weinberg, S., *Gravitation and Cosmology* (New York: Wiley, 1972). Hawking, S. W. and Israel, W., eds., *General Relativity: An Einstein Centenary Survey*, (Cambridge: Cambridge U. Press, 1979).

[5] Yang, J., Turner, M. S., Steigman, G., Schramm, D. N., and Olive, K. A., *Astrophys. J.* submitted in 1983; *Astrophys. J.* **246**, 557 (1981).

[6] Rindler, W., *Mon. Not. R. Astron. Soc.* **116**, 663 (1956).

[7] Dicke, R. H. and Peebles, P. J. E., in *General Relativity: An Einstein Centenary Survey*, eds. Hawking, S. W. and Israel, W. (Cambridge: Cambridge U. Press, 1979).

[8] Preskill, J., *Phys. Rev. Lett.* **43**, 1365 (1979). Zel'dovich, Ya. and Khlopov, M., *Phys. Lett.* **79B**, 239 (1978).

[9] Coleman, S. and Weinberg, E., *Phys. Rev.* **D7**, 788 (1973).

[10] Kirshnitz, D. and Linde, A. D., *Phys. Lett.* **42B**, 471 (1972). Dolan, L. and Jackiw, R., *Phys. Rev.* **D9**, 3320 (1974). Weinberg, S., *Phys. Rev.* **D9**, 3357 (1974).

[11] Cahn, J., private communication.

[12] For an excellent review of GUTs and baryon asymmetry, plus extensive references, see Langacker, P., *Physics Reports* **72**, 187 (1981).

[13] Affleck, I., and Dine, M., *Nucl. Phys.* **B249**, 361 (1985).

[14] The following are some of the excellent reviews on aspects of inflationary cosmology: Linde, A. D., *Rep. Prog. Phys.* **47**, 925 (1984). Turner, M. S., in Proceedings of NATO ASI on Quarks and Leptons (Munich: 1983). Gibbons, G. W., Hawking, S. W. and Siklos, S. T. C, eds., *The Very Early Universe* - Proceedings of the 1982 Nuffield Workshop (Cambridge: Cambridge U. Press, 1983). Steinhardt, P., in *Birth of the Universe*, ed.

by Audouze, J. and Tran Than Van, J. (Editions Frontieres: 1982), p. 45; in *Particles and Fields - 1982*, ed. by Caswell, W. and Snow, G. (New York: AIP, 1983), p. 343.

[15] Guth, A. and Weinberg, E., *Nucl. Phys.* **B212**, 321 (1983).

[16] Lindblom, P. and Steinhardt, P., *Phys. Rev.* **D31**, 2151 (1985).

[17] Albrecht, A., Steinhardt, P., Turner, M. S., and Wilczek, F., *Phys. Rev. Lett.* **48**, 1437 (1982). Abbott, L., Farhi, E., and Wise, M., *Phys. Lett.* **117B**, 29 (1982). Dolgov, A. and Linde, A. D., *Phys. Lett.* **116B**, 329 (1982).

[18] Hawking, S. W. and Moss, I., *Phys. Lett.* **110B**, 35 (1982).

[19] Jensen, L. and Steinhardt, P., *Nucl. Phys.* **B237**, 176 (1984).

[20] Linde, A. D., *Phys. Lett.* **226B**, 335 (1982). Vilenkin, A. and Ford, L., *Phys. Rev.* **D26**, 1231 (1982).

[21] Mazenko, G. F., Wald, R. M. and Unruh, W. G., *Phys. Rev.* **D31**, 273 (1985). Evans, M. and McCarthy, J. G., *Phys. Rev.* **D31**, 1799 (1985).

[22] Breit, J., Gupta, S., and Zaks, A., *Phys. Rev. Lett.* **51**, 1007 (1983).

[23] Gibbons, G. W., Hawking, S. W. and Siklos, S. T. C, eds., *The Very Early Universe* - Proceedings of the 1982 Nuffield Workshop (Cambridge: Cambridge U. Press, 1983).

[24] Guth, A. and Pi, S.-Y., *Phys. Rev. Lett.* **49**, 1110 (1982). Hawking, S. W., *Phys. Lett.* **115B**, 295 (1982). Starobinskii, A. A., *Phys. Lett.* **117B**, 175 (1982). Bardeen, J., Steinhardt, P. and Turner, M. S., *Phys. Rev.* **D28**, 679 (1983).

[25] Zel'dovich, Ya., *Mon. Not. Roy. Astron. Soc.* **160**, 1P (1972). Harrison, E., *D1*, 2726 (1970).

[26] See the NATO ASI review article by M. S. Turner for excellent discussion and complete set of references.

[27] Steinhardt, P. and Turner, M. S., *Phys. Rev.* **D29**, 2162 (1984).

[28] Ovrut, B. and Steinhardt, P., *Phys. Lett.* **133B**, 161 (1983).

[29] Olive, K. A. and Srednicki, M., UCSB preprint 84-0519 (1984). Gelmini, G. B., Nanopoulos, D. V., and Olive, K. A., *Phys. Lett.* **131B**, 53 (1983).

[30] Lindblom, P., to be published.

[31] Steinhardt, P., in Proceedings of Inner Space/Outer Space Fermilab Conference, ed. by Kolb, E., Olive, K. A., Turner, M. S., and Seckel, D. (to appear 1985).

[32] Linde, A. D., *Phys. Lett* **129B**, 177 (1983).

[33] Vilenkin, A., *Phys. Lett.* **117B**, 25 (1982).

[34] Shafi, Q. and Vilenkin, A. *Phys. Rev. Lett.* **52**, 691 (1984).

[35] Ovrut B. and Steinhardt, P., *Phys. Lett.* **147B**, 263 (1984); *Phys. Rev. D30* **D30**, 2061 (1984); *Phys. Rev. Lett.* **53**, 732 (1984).

[36] Lindblom, P., Ovrut, B., and Steinhardt, P., to appear.

KALUZA-KLEIN THEORIES

Alan Chodos
Department of Physics
Yale University
New Haven, CT 06511

What follows is the written version of three lectures on Kaluza-Klein theories given at TASI II. This is not meant to be a review article, and consequently no attempt has been made to refer to the original literature, except where it was felt that the consultation of such references would be of genuine pedagogical benefit to the reader.

By now, however, there are a number of reviews of one kind or another which will both amplify the contents of this article and also provide a much more complete list of references. In particular, the reader may wish to have a look at:
I. "An Introduction To Kaluza-Klein Theories", H.C. Lee, ed; (World Scientific, Singapore (1984)). This is the proceedings of a workshop held at Chalk River, Ontario in the summer of 1983.
II. "Modern Kaluza-Klein Theories", T. Appelquist, A. Chodos, and P.G.O. Freund, eds. (Benjamin/Cummings, Menlo Park, (1985)). This is a reprint volume with a detailed introduction and extensive list of references.
III. "Kaluza-Klein Supergravity", by M.J. Duff, B.E.W. Nilsson, and C.N. Pope (Physics Reports, to appear). This is a treatise on the eleven-dimensional supergravity aspects of Kaluza-Klein theories, by members of a group that has been very active in this field.

PRELIMINARIES

We begin by trying to put forward a working definition of Kaluza-Klein theories. What are they? Perhaps the only properties that everyone agrees on are two: (i) they have something to do with the unification of gravity and the other forces of nature; and (ii) they involve the introduction of extra dimensions of space, i.e. they are formulated in $4+N$ dimensions, $N>0$.

Within this class of theories, I shall confine myself to field theories of pointlike particles. That is, even though string theories clearly meet criteria (i) and (ii), I shall leave the discussion of their properties to other lecturers in the school. This is not to say that the contents of this article are irrelevant for string theory; on the contrary, many of the topics we shall touch upon, such as cosmology, one-loop quantum effects, and the transmutation of spatial symmetries into gauge symmetries via dimensional reduction will have their counterparts in string theory. Much of the physics will be the same, and it is probably easier even for the rabid string enthusiast to absorb it first in the field-theoretic setting.

Finally, there is a further condition that one might wish to impose in defining Kaluza-Klein theories, namely that no gauge fields occur explicitly in the higher dimensional theory, but that they make their appearance only because of spontaneous compactification on an internal manifold with isometries (how this happens will be explained below). In light of recent developments, in particular the input from string theory and the difficulty in obtaining chiral fermions this way (see below), this condition is probably too restrictive. Nevertheless, the idea that gauge theories arise from geometrical symmetries in higher dimensions is an appealing one that has fueled much of the interest in Kaluza-Klein theories over the years. It is in some sense the "pure" or "original" Kaluza-Klein idea (or, in the language of soft-drink advertising, the Kaluza-Klein Klassic), and we shall devote some time to understanding how it works.

Before embarking on the physics, we shall spend a few moments going over some hopefully familiar mathematics, if for no other reason than to fix the notation. Since Kaluza-Klein theories by definition involve gravity, we review the basic facts about general relativity. Unlike some of the other lecturers, we shall do this the old-fashioned way, with tensors that have lots of indices and coordinates that enter explicitly in the formulas.

The fundamental dynamical variables of gravity in D dimensions are the elements of the metric tensor, $g_{AB}(x)$, on some pseudo-Riemannian manifold M. This tensor defines the infinitesimal line element

$$ds^2 = g_{AB}(x) \, dx^A dx^B.$$

From g_{AB}, one forms the Christoffel symbols

$$\Gamma^A_{BC} \equiv \frac{1}{2} g^{AD} [g_{DB,C} + g_{DC,B} - g_{BC,D}]$$

where $g^{AD} g_{DB} = \delta^A_B$ and the comma denotes ordinary differentiation. From the Γ's, one defines covariant differentiation:

$$\nabla_A V_C = V_{C,A} - \Gamma^B_{AC} V_B$$

$$\nabla_A V^C = V^C{}_{,A} + \Gamma^C_{AB} V^B$$

which has the property that

$$\nabla_A g_{BC} = 0 \, .$$

The Riemann curvature tensor is given by

$$[\nabla_A, \nabla_B] V_C = R^D{}_{CBA} V_D.$$

Explicitly,

$$R^D{}_{CBA} = \Gamma^D_{AC,B} - \Gamma^D_{BC,A} + \Gamma^E_{CA} \Gamma^D_{BE} - \Gamma^E_{CB} \Gamma^D_{AE} \, .$$

From the Riemann tensor, one constructs the Ricci tensor:

$$R_{CA} \equiv R^D{}_{CDA} = R_{AC}$$

and the scalar curvature

$$R = g^{CA} R_{CA} \, .$$

The Einstein-Hilbert action for general relativity in D dimensions is then

$$S = \frac{\eta}{16\pi G_D} \int d^D x \sqrt{|g|} \ (R + 2\Lambda) \ .$$

We have assumed that $g_{AB}(x)$ is dimensionless, so that (in units where $\hbar=1$) G_D has dimension $(\text{length})^{D-2}$ (because R contains two derivatives). The parameter η is ± 1; it is chosen to insure that, if one expands

$$g_{AB} = \eta_{AB} + h_{AB}(x)$$

where η_{AB} is the D-dimensional Minkowski metric, then the coefficient of $\dot{h}_{ij}\dot{h}^{ij}$ will be positive. Here i,j run over spatial indices only, and the dot denotes time differentiation. The reader is invited to determine η for our choice of convention.

Variation of this action yields the Einstein equations

$$R_{AB} - \frac{1}{2} g_{AB} R = \Lambda g_{AB} \ .$$

Of some importance to Kaluza-Klein theories is the notion of a symmetry of g_{AB}. We recall that under a general coordinate transformation

$$x'^A = x'^A(x^B)$$

the metric g_{AB} transforms as

$$g'_{AB}(x') = \frac{\partial x^C}{\partial x'^A} \frac{\partial x^D}{\partial x'^B} g_{CD}(x) \ .$$

This transformation is, of course, a symmetry of the theory, regardless of what g_{AB} is. But we are looking for something rather different, a way of defining symmetries of a particular metric. To proceed, we consider the infinitesimal case,

$$x'^A = x^A + \epsilon^A(x) \ .$$

To first order in ϵ, we have

$$g'_{AB}(x) + \epsilon^C g_{AB,C}(x)$$

$$= g_{AB}(x) - \epsilon^C{}_{,A} g_{CB} - \epsilon^D{}_{,B} g_{AD} \ .$$

Using $\nabla_C g_{AB} = 0$, we can write

$$g_{AB,C} = \Gamma^D_{AC} g_{DB} + \Gamma^D_{CB} g_{AD}$$

and hence

$$g'_{AB}(x) = g_{AB}(x) - (\nabla_A \varepsilon_B + \nabla_B \varepsilon_A) .$$

Our definition of symmetry will be that

$$g'_{AB}(x) = g_{AB}(x) .$$

i.e. that

$$\nabla_A \varepsilon_B + \nabla_B \varepsilon_A = 0 .$$

This is Killing's equation, and for every symmetry of the metric g_{AB} there is a Killing vector field $\varepsilon_A(x)$ obeying the equation. Note that it was essential that the symmetry condition involved comparing the old and new metrics at the same point x^A (and not, for example, $g'_{AB}(x') = g_{AB}(x)$). Only in this way does one obtain a covariant symmetry condition valid in any reference frame.

A useful construct is the Lie Bracket of two vector fields. Given two contravariant vectors ξ^A and η^A, one forms the operators $\xi = \xi^A \partial_A$ and $\eta = \eta^A \partial_A$ (∂_A is the ordinary partial derivative.) These are the vector fields usually defined in differential geometry.

The Lie Bracket is then just the commutator of the two operators:

$$\phi = [\xi,\eta] = (\xi^B \eta^A{}_{,B} - \eta^B \xi^A{}_{,B}) \partial_A .$$

If ξ and η are Killing vectors, then it is straightforward to show that ϕ is, too. In general, if one has a collection of Killing fields $\{\xi\}_\alpha$ on a manifold, then their Lie Brackets close to form an algebra

$$[\xi_\alpha , \xi_\beta] = -f^\gamma{}_{\alpha\beta} \xi_\gamma$$

with structure constants $f^\gamma{}_{\alpha\beta}$. The group associated with this algebra is then the group of isometries of the manifold. Of course, since there can be many groups with the same algebra, exactly what the group is will depend on the global structure of the manifold.

Finally, we study the question of what symmetries are possible on spaces that are (a) compact; (b) Riemannian, i.e. the metric has only positive definite eigenvalues; and (d) Einstein spaces, i.e. the metric obeys the equation

$$R_{AB} = \alpha\, g_{AB} \; .$$

(This is equivalent to Einstein's equations with a cosmological constant). We note that from Killing's equation

$$\nabla_A \xi_B + \nabla_B \xi_A = 0$$

we immediately have

$$g^{AB}\, \nabla_A \xi_B = 0 \; .$$

Now compute

$$0 = g^{AC} \nabla_C (\nabla_A \xi_B + \nabla_B \xi_A) = \nabla^2 \xi_B + g^{AC}[\nabla_C, \nabla_B]\xi_A$$

where the extra term in the commutator vanishes because ξ_A is divergenceless. Therefore

$$\nabla^2 \xi_B + g^{AC}\, R^D{}_{ABC}\, \xi_D = 0$$

But since

$$R_{EABC} = R_{AECB}$$

we have

$$g^{AC}\, R^D{}_{ABC} = R^D{}_B = \alpha \delta^D{}_B$$

Thus

$$\nabla^2 \xi_B = -\alpha \xi_B \; .$$

Now multiply this equation by $g^{AB}\xi_A$, integrate over the manifold M and integrate the left-hand side by parts. Since M is compact there are no surface terms. Hence

$$\int_M (dV)\, g^{AB} g^{CD} (\nabla_C \xi_A)(\nabla_D \xi_B) = \alpha \int_M g^{AB} \xi_A \xi_B (dV) \; .$$

Here (dV) stands for the integration measure $\sqrt{g}\, d^N x$.
Because M is Riemannian, both the integrated quantities are positive semidefinite. The left-hand side can vanish only if $\nabla_C \xi_A = 0$, and if $\alpha \neq 0$, the right-hand side can vanish only if $\xi_A = 0$. There are then three cases:

(i) $\alpha<0$. Both sides must separately vanish and $\xi_A=0$. There are no Killing fields.

(ii) $\alpha>0$. There are no restrictions from this analysis. Manifolds can exist with arbitrary isometry groups. (e.g. the N-sphere, whose symmetry group is $SO(N+1)$.)

(iii) $\alpha=0$. Then $\nabla_C \xi_A=0$. If there are two Killing fields ξ and η, then

$$[\xi,\eta]^B = \xi^A \partial_A \eta^B - \eta^A \partial_A \xi^B$$
$$= \xi^A \nabla_A \eta^B - \eta^A \nabla_A \xi^B \quad \text{(exercise: show this)}$$
$$= 0 .$$

Consequently, only Abelian symmetries are possible. An example of this type of manifold is the torus, which indeed admits only Abelian symmetries. More exotic examples are the Calabi-Yau manifolds in six dimensions, of recent interest in string phenomenology. These manifolds actually admit no continuous symmetries.

THE FIVE-DIMENSIONAL KALUZA-KLEIN THEORY

We consider pure Einstein gravity, without a cosmological constant, in five dimensions:

$$S = \frac{\eta}{16\pi G_5} \int d^5x \sqrt{-g}\ R .$$

For reasons that will become clear shortly, it is convenient to parametrize the metric in the form

$$g_{AB} = W(\phi) \left(\begin{array}{c|c} \gamma_{\mu\nu} + \phi A_\mu A_\nu & \phi A_\mu \\ \hline \phi A_\nu & \phi \end{array} \right)$$

where the indices μ and ν run over the four values of ordinary space-time. Note that if we do not restrict the dependence of $\gamma_{\mu\nu}$, A_μ and ϕ on x^μ and x^5, then there is no loss of generality in writing g_{AB} this way, because there are just as many functions among the γ's, A's and ϕ as there are among the g's. The advantage of this parametrization is exposed by considering the following special case of a general coordinate transformation in five dimensions:

$$x'^\mu = x^\mu$$
$$x'^5 = x^5 + f(x^\mu).$$

Using the general transformation law for the metric tensor, one readily discovers that

$$\gamma'_{\mu\nu}(x^\mu, x'^5) = \gamma_{\mu\nu}(x^\mu, x^5)$$
$$A'_\mu(x^\mu, x'^5) = A_\mu(x^\mu, x^5) - \partial_\mu f(x^\nu)$$

and

$$\phi'(x^\mu, x'^5) = \phi(x^\mu, x^5).$$

Thus A_μ undergoes what looks like an Abelian gauge transformation, whereas $\gamma_{\mu\nu}$ and ϕ remain the same. This leads us to suspect that the physical content of the theory involves a $U(1)$ gauge field $A_\mu(x)$, together with a gravitational field $\gamma_{\mu\nu}$ and possibly also a scalar field ϕ.

One obvious feature of our universe that needs to be incorporated in the model is that we apparently live in a four dimensional space-time, not a five-dimensional one. The way this is done is to assume that the topology of the five dimensional manifold has the product structure

$$M = M_4 \times S^1$$

where M_4 is topologically equivalent to four-dimensional Minkowski space and S^1 is a circle parametrized by a coordinate x^5:

$$0 \leq x^5 \leq 2\pi R.$$

Then any field in the problem, $\chi(x^\mu, x^5)$, (χ can be any of the γ's, A's or ϕ), may be expanded in a Fourier series:

$$\chi(x^\mu, x^5) = \sum_{n=-\infty}^{\infty} \chi^{(n)}(x^\mu) e^{in \frac{x^5}{R}}.$$

These expansions can be substituted in the action and, since the x^5 dependence is now explicit, the integration over x^5 can be performed. The result will be a complicated theory involving the mutual interaction of the infinite towers of four-dimensional fields

$\chi^{(n)}(x^\mu)$, that is fully equivalent to the original five-dimensional theory. "Dimensional reduction" is now achieved by keeping only the x^5-independent piece of the Fourier expansions, i.e. throwing away all but the n=0 terms. If one looks at the full action expanded about the background

$$ds^2 = \eta_{\mu\nu} dx^\mu dx^\nu + (dx^5)^2$$

one finds that the n<u>th</u> fourier components describe particles of mass m_n such that

$$m_n^2 = \frac{n^2}{R^2}$$

Therefore, if R is sufficiently small one can justify truncating the Fourier expansion on the grounds that the neglected pieces are relevant only at extremely high energies. It should be borne in mind, however, that the mass of a particle described by a quantum field in general will depend on the background about which the field is being expanded. In more complicated cases than the five-dimensional one, different backgrounds can lead to very different particle spectra, and one should be careful not to truncate the mode sum before the final choice of a background metric has been made.

The dimensionally reduced action can be shown to be

$$S_4 = \frac{\eta}{16\pi G_4} \int d^4x \sqrt{-\gamma} \left\{ R^{(4)}(\gamma) - \frac{1}{4} \phi F_{\mu\nu} F^{\mu\nu} \right.$$
$$\left. + \frac{1}{6} \frac{\partial_\mu \phi \partial^\mu \phi}{\phi^2} \right\}.$$

Here $G_4 = \frac{G_5}{2\pi R}$ is the effective Newton's constant, $R^{(4)}$ is the usual four-dimensional scalar curvature evaluated with the metric $\gamma_{\mu\nu}$, and $F_{\mu\nu}$ is, of course, the usual Abelian field strength

$$F_{\mu\nu} = \partial_\mu A_\nu - \partial_\nu A_\mu.$$

To obtain this form for S_4 we have made the particular choice $W(\phi) = \phi^{-1/3}$ in our parametrization of the five-dimensional metric. Other choices of W would lead to different-looking forms for S_4, but the physical content would be the same.

The form of the ϕ kinetic energy term shows that $\ln \phi/\phi_c$ plays the role of a traditional scalar field. Here ϕ_c is the constant background value of ϕ. The length of the circle in the fifth dimension is given by $L_5 = \int_0^{2\pi R} dx^5 \sqrt{g_{55}} = 2\pi R \, \phi_c^{1/3}$. Careful analysis shows that physical quantities always depend on L_5 and never on ϕ_c or R separately. It is convenient to set $\phi_c=1$ and then R really does have the interpretation of the radius of the circle in the fifth dimension.

We also see that in order to obtain the correct normalization for the electromagnetic piece, we must rescale A_μ. It is $\tilde{A}_\mu = (16\pi G_4)^{-1/2} A_\mu$ that is the conventionally normalized gauge potential.

It is of some interest to identify the gauge coupling constant in this model. Since the dimensionally reduced action contains only fields neutral under the gauge group, one has to be a little devious. One way to proceed would be to go back to the $n \neq 0$ modes and to study the effect of a gauge transformation on them. Equivalently, one can introduce an ad hoc scalar matter field Φ (having nothing to do with ϕ in S_4) which we shall minimally couple to the Kaluza-Klein metric:

$$\Box \, \Phi = 0 \; .$$

Here $\Box$ is the five-dimensional generalization of the d'Alembertian:

$$\Box \equiv g^{AB} \nabla_A \nabla_B \; .$$

To extract the charge, we can simplify our Ansatz for the metric, letting

$$g_{AB} = \begin{pmatrix} \eta_{\mu\nu} & (16\pi G_4)^{1/2} \tilde{A}_\mu \\ (16\pi G_4)^{1/2} \tilde{A}_\nu & 1 \end{pmatrix} \; .$$

This g_{AB} will solve the linearized Einstein equations in five dimensions, provided that $\tilde{A}_\mu$ is a solution of Maxwell's equations. We assume that Φ has the simplest non-trivial x^5 dependence:

$$\Phi = \phi_o(x^\mu) e^{iqx^5}$$

where $q = \dfrac{n}{R}$.

Then the equation of motion for Φ becomes

$$\Box \Phi + q^2 \Phi + 2iq(16\pi G_4)^{1/2} A_\mu \partial^\mu \Phi + iq(16\pi G_4)^{1/2} (\partial_\mu A^\mu) \Phi = 0 .$$

This is the standard equation for a scalar particle interacting with an electromagnetic field, provided that we identify the charge e_n of the particle as

$$e_n = \frac{\hbar}{c} (16\pi G_4)^{1/2} \frac{n}{R} .$$

Now there is really no compelling reason to identify e_n with a multiple of the fundamental unit of electric charge, especially since the mass of the particle in question is q^2, which will turn out to be extremely large. Nevertheless, to get an idea of the orders of magnitude involved, it is amusing to set

$$\frac{e_n^2}{4\pi \hbar c} = \frac{n^2}{137}$$

which then allows us to solve for L_5:

$$L_5 = 2\pi R = 2.38 \times 10^{-31} \text{ cm} .$$

Regardless of the precise numerical value, we remark that it is a general feature of Kaluza-Klein models that the gauge coupling constant g will always be given by a relation of the form

$$g = \text{(pure number)} \times \frac{L_p}{R}$$

where L_p is the Planck length:

$$L_p = \left(\frac{\hbar G_4}{c^3}\right)^{1/2} = 1.6 \times 10^{-33} \text{ cm.},$$

and R is a distance scale characteristic of the internal manifold. Given that typical values of g will be less than, but not too much less than, unity, we see that R will never be more than a few orders of magnitude bigger than L_p. It is non-trivial that Kaluza-Klein theories satisfy the consistency requirement that the internal manifold, which must be small to explain the fact that we don't see it, turns out to be extremely small indeed in order to accommodate the correct values for the gauge coupling constants.

BEYOND FIVE DIMENSIONS

Gravity and electromagnetism exhausted the list of known interactions in the days of Kaluza and Klein, but nowadays, of course, any unified theory must accommodate the full $SU(3) \times SU(2) \times U(1)$ gauge theory. To enlarge the group one allows for a more general internal manifold than the circle. We take M to be 4+N-dimensional, and of the form

$$M = M_4 \times B .$$

Here M_4 is the usual four-dimensional space-time, and B is N-dimensional, compact, Riemannian, and admits a set of Killing vectors $\{K\}_\alpha$, $\alpha = 1, 2 \ldots, m$ such that $[K_\alpha, K_\beta] = -f^\gamma{}_{\alpha\beta} K_\gamma$. By our earlier theorem, if the Ricci tensor of B vanishes then $f^\gamma{}_{\alpha\beta} = 0$. The obvious way to generalize the parametrization that we introduced in the five dimensional case is to write

$$g_{AB} = W(\phi) \begin{pmatrix} \gamma_{\mu\nu} + \phi_{k\ell} B^k{}_\mu B^\ell{}_\nu & \phi_{i\ell} B^\ell{}_\mu \\ \phi_{j\ell} B^\ell{}_\nu & \phi_{ij} \end{pmatrix}$$

whose inverse is easily determined to be

$$g^{AB} = W^{-1}(\phi) \begin{pmatrix} \gamma^{\mu\nu} & -B^i{}_\lambda \gamma^{\lambda\mu} \\ -B^j{}_\lambda \gamma^{\lambda\nu} & \phi^{ij} + B^i{}_\lambda B^j{}_\sigma \gamma^{\lambda\sigma} \end{pmatrix} .$$

One finds that $g \equiv \det g_{AB} = (\det \gamma)(\det \phi) W^D$ (exercise: show this) so that B^i_μ enters only polynomially in the Lagrangian $\sqrt{-g}\,R$, obviously a desirable circumstance if we are planning to extract gauge fields from B^i_μ.

Let x^μ denote coordinates on M_4, and let y^i denote coordinates on B. The Ansatz that corresponds to keeping only the n=0 modes in the five-dimensional case is

$$\gamma_{\mu\nu} = \gamma_{\mu\nu}(x)$$
$$\phi_{ij} = \phi_{ij}(y)$$

and

$$B^i_\mu(x,y) = K^i_\alpha(y)\, A^\alpha_\mu(x)\ .$$

Up to an overall normalization, $A^\alpha_\mu(x)$ will be the gauge fields valued in the Lie algebra generated by the Killing vectors K^i_α. There is one such field for each generator of the algebra. Since an N-dimensional manifold can have as many as $\frac{1}{2}N(N+1)$ Killing vectors, one can, in general, end up with many more gauge fields than extra dimensions.

To check the Ansatz, one can perform the following exercise: consider the infinitesimal coordinate transformation

$$x'^\mu = x^\mu$$
$$y'^i = y^i + \varepsilon^\alpha(x)\, K^i_\alpha(y)\ .$$

Show that

$$\gamma'_{\mu\nu}(x) = \gamma_{\mu\nu}(x)$$
$$\phi'_{ij}(y) = \phi_{ij}(y)$$

and

$$A^{i\alpha}_\mu(x) = A^\alpha_\mu(x) + f^\alpha{}_{\gamma\beta}\varepsilon^\gamma A^\beta_\mu - \varepsilon^\alpha{}_{,\mu}\ .$$

(Note: be sure to expand B^i_μ and B'^i_μ in terms of the <u>same</u> $K^i_\alpha(y)$).

One recognizes the change in $A^\alpha_\mu(x)$ as the infinitesimal version of the appropriate non-Abelian gauge transformation.

Superficially, it appears that the non-Abelian generalization of the Kaluza-Klein Ansatz is completely straightforward. But there are important differences. For D=5, (Minkowski space) x S^1 is a solution of the Einstein equations, because S^1 has no intrinsic curvature. For D>5, however, if we demand that a metric of the form

$$g_{AB} = W(\phi) \begin{pmatrix} \eta_{\mu\nu} & 0 \\ 0 & \phi_{ij} \end{pmatrix}$$

(which represents (Minkowski space) x B) solve the Einstein equations, we find, first of all, that the cosmological constant Λ must vanish, and second, that $R_{ij}(\phi) = 0$. From the latter property, we deduce $f^\gamma_{\alpha\beta} = 0$. So when D>5, we must abandon at least one of (a) Minkowski space; (b) non-Abelian symmetries; or (c) the Einstein field equations as an attribute of the ground-state metric of the theory.

SUPERGRAVITY IN ELEVEN DIMENSIONS

One motivation for the renewed interest in higher dimensions has been the fact that supergravity theories apparently prefer to be formulated in more than four dimensions. Indeed, a Lagrangian for N=8 supergravity in four dimensions was first discovered by the expedient of starting with a simpler (N=1) Lagrangian in eleven dimensions and then compactifying on a seven-dimensional torus.

It turns out that just as N=8 is the maximal extended supergravity in four dimensions, so eleven is the maximal number of dimensions in which supergravity can be consistently formulated. (This assertion is widely believed, although it has not been rigorously proved. To refute it, one would have to discover a consistent scheme for coupling fields with spins greater than two.) The eleven-dimensional supergravity Lagrangian is unique. Its degrees of freedom are: (a) the graviton, which in eleven dimensions has 44 physical degrees of freedom; (b) the gravitino, with 128 physical degrees of freedom; and (c) an anti-symmetric three-index Abelian gauge field $A_{\mu\nu\rho}$ with 84 physical degrees of freedom. No other fields or couplings are possible. Notice that, as always in a supersymmetric theory, the number of bosonic and fermionic degrees of

freedom are equal. This theory has an impressive list of symmetries:
(i) 11-dimensional general covariance; (ii) N=1 local supersymmetry;
(iii) SO(10,1) local Lorentz invariance; (iv) the Abelian gauge
invariance associated with $A_{\mu\nu\rho}$; and (v) assorted discrete
symmetries. Because of its uniqueness and its high degree of
symmetry, this theory has aroused considerable interest among those
whose ambition it is to unify all known interactions.

There is another, apparently completely independent motivation
for focussing on Kaluza-Klein theories in eleven dimensions. The
minimal gauge theory that will describe the observed low-energy
phenomena is, as we all know, SU(3)xSU(2)xU(1). If all the gauge
bosons of this group are to be generated by isometries of the internal
manifold, then one can show that at least seven internal dimensions
are required.

The counting is done as follows: for any group G, one can obtain
a manifold on which there exists a set of Killing vectors obeying the
algebra of G by constructing the coset space G/H (where H is a
subgroup of G). The dimension of G/H is dim G - dim H. The smallest
such manifold is therefore obtained by choosing dim H to be as large
as possible. For SU(3), the most favorable case is H=SU(2)xU(1),
yielding a four dimensional manifold. For SU(2), H=U(1); the
two-sphere is the resulting coset space. U(1) obviously requires one
dimension in the form of a circle. Hence for SU(3)xSU(2)xU(1) to be
realized as an isometry group of some manifold, the manifold must be
at least seven dimensional.

Thus eleven dimensions is both the most that one can have if
supergravity is to be incorporated in the theory, and the fewest one
can have if an SU(3)xSU(2)xU(1) gauge group is to emerge via the
Kaluza-Klein mechanism. This seeming coincidence has been taken by
some as a profound hint that Nature is really eleven dimensional, and
has certainly spawned a lot of literature on Kaluza-Klein theories in
eleven dimensions.

Let us look for candidate ground state configurations in eleven-
dimensional supergravity whose symmetries will dictate the observed

gauge group. Presumably the vacuum expectation value of the gravitino field $\psi_{\mu\alpha}$ is zero. (It has been conjectured that bilinears in $\psi_{\mu\alpha}$ may not have vanishing vacuum expectation values, but we shall ignore this possibility here.)

The field equations with $\psi_{\mu\alpha}=0$ are

$$R_{\mu\nu} - \frac{1}{2} g_{\mu\nu} R = \frac{1}{3} [F_{\mu\lambda\rho\sigma} F_\nu{}^{\lambda\rho\sigma} - \frac{1}{8} g_{\mu\nu} F_{\lambda\rho\sigma\tau} F^{\lambda\rho\sigma\tau}]$$

and

$$\nabla_\mu F^{\mu\nu\rho\sigma} = -\frac{1}{576} \varepsilon^{\mu_1\ldots\mu_8 \nu\rho\sigma} F_{\mu_1\ldots\mu_4} F_{\mu_5\ldots\mu_8}$$

where

$$F_{\mu\nu\rho\sigma} = \frac{1}{6} \nabla_{[\mu} A_{\nu\rho\sigma]}$$

is the field strength associated with $A_{\nu\rho\sigma}$.

For a vacuum solution, we seek a product manifold $M = M_d \times B_{11-d}$, where M_d is a d-dimensional non-compact maximally symmetric manifold representing space-time, and B is compact and admits non-Abelian symmetries.

If we set $F_{\mu\nu\rho\sigma}=0$, we are back to the case of ordinary gravity with vanishing cosmological constant. M_d must then be d-dimensional Minkowski space, and B must be Ricci-flat, which, as we have seen, precludes the possibility of a non-Abelian gauge group.

Freund and Rubin[1] made the observation that other interesting compactifications could be obtained by giving $F_{\mu\nu\rho\sigma}$ a vacuum expectation value as well. In order not to spoil the maximal symmetry of M_d, the only non-trivial possibility is to have

$$F_{\mu\nu\rho\sigma} = 3m\varepsilon_{\mu\nu\rho\sigma}$$

when μ,ν,ρ,σ take values on M_d. The crucial point here is that since $F_{\mu\nu\rho\sigma}$ has four indices, (which follows from the structure of 11-dimensional supergravity) then so must $\varepsilon_{\mu\nu\rho\sigma}$ have four indices which means in turn that the dimensionality of space-time must be four! That is, this type of compactification requires d=4, which is in startlingly good agreement with observation. (There is, actually, also the dual possibility of d=7 in which $*F \propto \varepsilon$, but this is clearly not as realistic.) If one follows Freund and Rubin and assumes in addition that $F_{\mu\nu\rho\sigma}=0$ if any of the indices refer to B_{11-d}, then one finds that the equation for $F_{\mu\nu\rho\sigma}$ is automatically satisfied, while the equation for $R_{\mu\nu}$ splits up into

$$R_{ab} = -12\,m^2\,g_{ab} \quad \text{(a,b indices on } M_4\text{)}$$

and

$$R_{ij} = 6\,m^2\,g_{ij} \quad \text{(i,j indices on } B_7\text{)}.$$

The good news about these equations is, first, as already mentioned, space-time is predicted to have dimension four, and second, B_7 is an Einstein space with the correct sign of the cosmological constant to insure that it is compact and can admit non-Abelian symmetries. The bad news is first, that M_4 is not Minkowski space but rather anti-de Sitter space with a huge cosmological constant, and second, that there are a variety of other phenomenological problems with this Ansatz, chief among them the fact that it is impossible to obtain chiral fermions in the massless spectrum. What this means and why it is true will be discussed shortly.

Notwithstanding these problems, a great deal of effort has gone into understanding the structure of seven-dimensional coset spaces that are also Einstein spaces. The most obvious such space is the seven-sphere $SO(8)/SO(7)$, which is maximally symmetric and has other intriguing mathematical properties. (For example S^7 is the manifold spanned by the unit octonions. Related to this is the fact that S^7 is the only manifold other than a Lie group that is absolutely parallelizable. That is, it admits an affine connection $\Gamma^\mu_{\nu\lambda} \neq \Gamma^\mu_{\lambda\nu}$ such that the Riemann tensor evaluated with this connection vanishes.) Unfortunately, the isometry group of S^7 is $SO(8)$, which does not contain $SU(3)\times SU(2)\times U(1)$ as a subgroup. This "round" S^7 preserves $N=8$ supersymmetry. There is another version of S^7, called the "squashed" seven sphere, which is also an Einstein space and whose isometry group is $SO(5)\times SU(2)$. In this case the supersymmetry is broken down to $N=1$. This possibility has been much studied by the

Imperial College group.[2] There is also an infinite class of
manifolds whose isometry group contains SU(3)xSU(2)xU(1). These would
seem to be more attractive phenomenologically, but in all cases the
fermionic spectrum is not realistic.

In the case of S^7, the Freund-Rubin Ansatz has also been
generalized to include non-vanishing $F_{ijk\ell}$, where these indices
refer to the internal space. The components of $F_{ijk\ell}$ are taken to
be proportional to the curl of $S_{ik\ell}$, the torsion tensor associated
with the absolutely-parallelizing connection mentioned above. In this
case the N=8 supersymmetry is broken completely.

THE CHIRAL FERMION PROBLEM

Hanging over all these attempts to derive low-energy
phenomenology from Kaluza-Klein theories is the extreme difficulty of
obtaining thereby a chiral spectrum of fermions.

The term "chiral fermions" refers to a particular property of
the observed families of leptons and quarks: the left-handed fermions
and the right-handed fermions transform under different
representations of the gauge group. The first question we must
address concerning this property is whether it is only a low-energy
manifestation, or whether it actually reflects the fermionic spectrum
of the underlying theory. If the former, then at some relatively low-
energy (compared to the energy scale of the fundamental theory, which
in our case is the Kaluza-Klein compactification scale) one would
expect the appearance of "mirror fermions", that is, families of
leptons and quarks in which the role of left-handed and right-handed
fermions is exactly interchanged. The combined spectrum of ordinary
and mirror fermions would then be non-chiral.

There are at least two compelling reasons for discounting the
possibility of mirror fermions. The first is that all the leptons and
quarks so far observed fall into ordinary families; no mirror
representations have yet been found. Second, the consistency of the
low-energy gauge theory demands that the anomalies that might spoil
the conservation of the gauge currents must cancel. This cancellation
would be automatic if mirror fermions existed, since each mirror

fermion would make the opposite contribution to any gauge anomaly from that of its ordinary partner. But, miraculously, the anomalies cancel already among the observed families of ordinary fermions. Nature seems to be sending the message that mirror fermions are not needed, and hence that they probably are not there.

If we then adopt the latter possibility, we must look for the appearance of chiral fermions at the level of the massless spectrum of the Dirac or Rarita-Schwinger operators on the internal manifold B. Let us look at the Dirac operator for simplicity. Because we are assuming a product manifold, the Dirac operator takes the form

$$\Gamma^A \partial_A = \gamma^\mu \partial_\mu + \gamma^i \partial_i$$

where μ refers to space-time and i to the internal space. Thus the Dirac operator on the internal space, $\gamma^i \partial_i$, plays the role of the mass operator in the dimensionally reduced theory. We want to study the solutions of

$$\gamma^i \partial_i \psi = 0$$

to see if the left-handed ones and the right-handed ones can transform differently under the isometry group of the internal manifold. The group-theoretical analysis of this problem has been done by Wetterich,[3] who shows that chiral fermions can exist only if the total dimensionality D satisfies D=2 mod 8. In particular, eleven (or any other odd number of dimensions) is ruled out if one insists upon getting chiral fermions from the Kaluza-Klein mechanism.

In fact, the situation is even worse. Recall that n, the number of left-handed minus the number of right-handed zero modes of a Dirac operator on a compact manifold is a topological invariant. Now the problem at hand is a little trickier, because we are not only interested in the absolute number n, but also in how these modes transform under the action of the isometry group of the manifold. Nevertheless, it is possible to define a more refined topological invariant that takes this subtlety into account. Furthermore, if it is possible to construct a particular manifold for which this topological invariant vanishes, it must vanish on all manifolds which can be obtained from the original manifold by continuous deformations that preserve the isometry group G. Using arguments along these lines, Witten[4] has shown that the Dirac operator never admits a chiral spectrum on any compact internal manifold in any dimension, whereas the Rarita-Schwinger operator does not have chiral fermions if the space is homogeneous. Thus only in the somewhat unlikely case of the Rarita-Schwinger operator on a non-homogeneous space could chiral fermions possibly exist.

This situation seems so unattractive and restrictive that various rather drastic proposals have been made to circumvent it. Perhaps the least drastic is the suggestion to add extra gauge fields to the original theory, thereby going beyond "pure" Kaluza-Klein. This is what one does in supersymmetric string theories, where the effective field theory in ten dimensions contains not only supergravity but also supersymmetric Yang-Mills. The addition of these extra gauge fields removes the correspondence between the gauge group and the isometry group of the internal manifold, and it has been shown that if these gauge fields assume topologically non-trivial background values then the spectrum of fermions can be chiral.

Another suggestion is that the internal manifold be taken to be non-compact, thereby relaxing one of the assumptions of the no-go theorem. Non-compact internal manifolds must be dealt with carefully: One may worry that the spectrum of excitations will be continuous, in which case the usual dimensional reduction will not work. This is related to the simple fact that if one is not careful, since the internal manifold is non-compact it will tend to be as large, in some sense, as the space-time manifold. It has been shown, however, that it is possible to construct non-compact internal manifolds of finite volume, whose excitation spectrum is at least partially discrete, which admit non-trivial isometry groups and chiral fermions.

Finally, Weinberg[5] has advocated a concept he calls "quasi-Riemannian" geometry. Normally, given a metric g_{AB} one defines the

vielbein e_A^a such that $e_A^a e_B^b \eta_{ab} = g_{AB}$. The index a refers to
the tangent-space group G_T, which is taken to be $SO(D-1,1)$ for a
D-dimensional manifold. In quasi-Riemannian geometry, one explores
the possibility that G_T is something other than $SO(D-1,1)$. This is
relevant to the problem of chiral fermions because spinors in a curved
space are taken to be world scalars; they are spinors with respect to
G_T. Modification of G_T might be expected to lead to the
modification of the spectrum of spinors, and Weinberg constructs
examples in which the spectrum is chiral.

COSMOLOGY

Because the Kaluza-Klein length scale is so small, the effects
of the extra dimensions on laboratory physics are at best indirect.
One might imagine, however, that the early history of the universe
could have been rather profoundly influenced by the higher
dimensionality of space. Also, as we shall see, the influence can
extend the other way as well: cosmology can, possibly, help to answer
one of the puzzles of Kaluza-Klein physics, namely why are the extra
dimensions so small?

Many papers have been written on Kaluza-Klein cosmology, but our
understanding of the subject is still rather murky. It is not
possible at this stage to give a logical and systematic account of
where things stand. Instead, I shall present two examples of the
kinds of questions that have been addressed in the literature. Much
work remains to be done before a coherent picture begins to emerge.

Cosmology in general relativity means letting the solutions
depend on time; the time development then represents the evolution of
the universe. The simplest time-dependent solution to the vacuum
Einstein equations in D dimensions was discovered in the same year
that Kaluza's paper was published. It is called the Kasner solution,
and in a suitable coordinate system it has the form:

$$ds^2 = -dt^2 + \sum_{i=1}^{D-1} \left(\frac{t}{t_o}\right)^{2p_i} \left(dx^i\right)^2$$

where, in order to solve $R_{\mu\nu}=0$, we must require

$$\sum_{i=1}^{D-1} p_i = \sum_{i=1}^{D-1} p_i^2 = 1 \quad . \tag{1}$$

The case where one of the p_i's is unity and all the rest vanish can be shown to be flat space in a funny coordinate system. In all other cases, the solution of eq. (1) requires at least one of the p_i's to be negative. That is, the Kasner solution describes a hybrid between an expanding universe ($p_i>0$) and a contracting one ($p_i<0$); if it were applied to the observable universe, then in some directions one would see a red shift and in others a blue shift.

This circumstance explains why the Kasner solution, despite its simplicity, has not found much application in standard cosmology. In Kaluza-Klein cosmology, however, it is just what we want. For example, when D=5, let us choose $p_1=p_2=p_3=1/2$, and $p_5=-1/2$. Then all three observed spatial dimensions expand at the same rate, while the fifth dimension contracts. The reason the fifth dimension is so small is simply that the universe is relatively old. In the early universe (i.e. at the time $t=t_0$) all four spatial dimensions were the same size, and correspondingly, the gravitational and gauge interactions described by the dimensionally reduced theory were merged into a unified interaction characterized by a single coupling strength.

This picture has been generalized to the case of eleven-dimensional supergravity by Freund[6], who has found solutions in which three spatial dimensions expand preferentially relative to the other seven. Just as in our discussion of the time-independent backgrounds, however, the generalization to an internal manifold B with curvature gives rise to new difficulties. For example, unlike the Kasner solution which begins with a singularity but then develops smoothly forever, the generic solution in the case where B is curved has a singularity both at t=0 and at some finite final time t_f.

Another example of the interplay between extra dimensions and cosmology has to do with the mechanism for entropy generation in the early universe. (For a discussion of the entropy problem in cosmology, see the lecture notes by Steinhardt in these proceedings.)

It was pointed out by Alvarez and Belén Gavela[7] that dimensional reduction itself gives rise to entropy production. At the most simple-minded level, if one assumes that dimensional reduction is adiabatic, then the entropy in a higher-dimensional space becomes compressed into effectively fewer dimensions, thereby giving rise to a huge increase in the entropy density.

A more detailed picture of what is happening was provided by Barr and Brown.[8] Before dimensional reduction, the entropy is spread out over many modes, corresponding to the full spectrum of the higher-dimensional theory. When the extra dimensions become small, the higher modes, whose masses depend inversely on the size of the extra dimensions, become very heavy, and can no longer be excited. The entropy then gets dumped into the massless modes, which are the ones that survive after dimensional reduction takes place. If the process is adiabatic, the temperature will rise, but not fast enough to re-excite the higher modes, so that the net effect is a vast increase in the entropy of the massless modes. What is not clear, however, is whether this increase is sufficient to account for all the entropy that is necessary on cosmological grounds.

Cosmology would appear to be one of the few available windows on the extra dimensions, but as of this writing the view from the window is still far from clear. Many interesting ideas have been proposed; what is hard to do is to assess how believable any of them is.

QUANTUM EFFECTS

So far our discussion has been entirely classical. It is also true that the theories we have been considering are non-renormalizable, so that we cannot hope to provide a full quantum description of the dynamics of Kaluza-Klein theories. Nevertheless, we can take the point of view that the Lagrangians we have been working with are effective theories valid at distance scales larger than the Planck length L_p. As long as the compactification radius is large compared to L_p, we can perform a loop expansion and keep only the lowest terms. The coupling constant G_D is the loop expansion parameter, because G_D^{-1} multiplies the action. Since G_D carries dimension (length)$^{D-2}$, the true dimensionless expansion parameter must be G_D/R^{D-2}, where R is a length scale characteristic of extra dimensions. Thus for

$$\frac{G_D}{R^{D-2}} = \left(\frac{L_P}{R}\right)^{D-2}$$

small, we may hope that the one-loop approximation is a valid guide to the quantum corrections.

The physics of these one-loop corrections is the same as that of the Casimir effect, first discussed by Casimir[9] in 1948. He considered two parallel, infinite, perfectly conducting plates separated by a distance a, and computed the zero-point energy of the electromagnetic field as a function of a. He predicted an attractive force between the plates, which has subsequently been seen experimentally.

For simplicity, let us consider instead the example of a free, massless scalar field subject to periodic boundary conditions:

$$\phi(t,x,y,z) = \phi(t,x,y,z + 2a).$$

The allowed frequencies are then $\omega^2(k,n) = k_x^2 + k_y^2 + \left(\frac{n\pi}{a}\right)^2$ and we want to compute $V = \frac{1}{2} \sum_{k,n} \hbar\omega(k,n)$ which is obviously infinite. However, if this sum is suitably regulated, a finite unambiguous part depending on a can be extracted. To proceed, we note the formula (we shall be working in Euclidean space throughout)

$$\int \frac{d^m k}{(k^2+\alpha^2)^s} = \pi^{m/2} \frac{\Gamma(s-m/2)}{\Gamma(s)} \alpha^{m-2s} .$$

One consequence of this formula can be derived by choosing m=1, differentiating with respect to s, and analytically continuing the result to s=0. One gets the strange-looking equation

$$\alpha = \int \frac{dk_0}{2\pi} \ln(k_0^2 + \alpha^2) . \tag{2}$$

Also, we shall compute not the energy V but rather $\tilde{V}=V/A$ where

$A = \int dxdy$ is the (infinite) area of the two-dimensional transverse space orthogonal to the compactified direction. We note that

$$A^{-1} = \frac{d^2k}{(2\pi)^2} \, .$$

Putting this together, and using Eq. (2) with $\alpha=\omega$, we have

$$V = \frac{1}{2} \int \frac{d^3k}{(2\pi)^3} \sum_{n=-\infty}^{\infty} \ln\left(k^2 + \left(\frac{n\pi}{a}\right)^2\right)$$

$$= -\frac{1}{2} \frac{d}{ds} \left\{ \sum_{n=-\infty}^{\infty} \frac{1}{(2\pi)^3} \pi^{3/2} \frac{\Gamma(s-3/2)}{\Gamma(s)} \left(\frac{n\pi}{a}\right)^{3-2s} \right\}_{s=0}$$

$$= -\frac{d}{ds} \left\{ \frac{\pi^{3/2}}{(2\pi)^3} \left(\frac{a}{\pi}\right)^{2s-3} \frac{\Gamma(s-3/2)}{\Gamma(s)} \zeta_R(2s-3) \right\}_{s=0}$$

where ζ_R is the Riemann zeta function:

$$\zeta_R(s) = \sum_{n=1}^{\infty} \frac{1}{n^s} \, .$$

The quantity in curly brackets is well-defined for $s>2$. To obtain a finite answer, we must analytically continue to a neighborhood of $s=0$. In this case, we simply quote a well-known result for ζ_R:

$$\zeta_R(1-z) = \frac{1}{(2\pi)^z} 2\Gamma(z) \cos\left(\frac{\pi z}{2}\right) \zeta_R(z) \quad (3)$$

and apply it with $1-z = 2s-3$. We note that since $\frac{1}{\Gamma(s)} \to 0$ as $s \to 0$, the answer is obtained by deleting the factor $\frac{1}{\Gamma(s)}$ and evaluating the remainder at $s=0$. We obtain

$$\tilde{V}(a) = \frac{-\pi^2}{720 \, a^3}$$

i.e. an attractive potential tending to make the separation a decrease. (Incidentally, formula (3) can be used to establish other interesting and useful facts, such as

$$\zeta_R(-1) = \sum_{n=1}^{\infty} n = -\frac{1}{12} \, .$$

(You have to know that $\sum_{n=1}^{\infty} \frac{1}{n^2} = \frac{\pi^2}{6}$).)

It is convenient to relate the Casimir energy to the one-loop effective potential familiar from field theory. We shall do this formally for the case of a free scalar field; the demonstration generalizes readily to other cases.

On the one hand,

$$\int D\phi e^{\frac{1}{2} \int d^4x \, \phi(\Box - m^2)\phi}$$

$$= \prod_k e^{-\frac{1}{2} \ln(k^2 + m^2)}$$

$$= e^{-\frac{1}{2} (\int d^4x) \int \frac{d^4k}{(2\pi)^4} \ln(k^2 + m^2)}$$

$$\equiv e^{-(\int d^4x) V_{eff}}$$

so that

$$V_{eff} = \frac{1}{2} \int \frac{d^4k}{(2\pi)^2} \ln(k^2 + m^2) .$$

On the other hand, from Eq. (2)

$$\frac{1}{2} \sum_k \hbar\omega = \frac{1}{2} \sum_k \int \left(\frac{dk_0}{2\pi}\right) \ln(k^2 + m^2)$$

$$= (\int d^3x) \frac{1}{2} \int \frac{d^4k}{(2\pi)^4} \ln(k^2 + m^2) = (\int d^3x) V_{eff} .$$

i.e. V_{eff} is the zero-point energy per unit volume.

To summarize: We have learned that the Casimir energy is related to the one-loop effective potential, and that each of these can be regulated using the zeta-function method, which essentially amounts to the observation that

$$\sum_i \ln \lambda_i = -\frac{d}{ds} \sum_i \left(\frac{1}{\lambda_i^s}\right) \Big|_{s=0}$$

where the sum on the right-hand side is to be defined near s=0 by analytic continuation.

What has all this to do with Kaluza-Klein? Whenever a quantum field is subjected to boundary conditions, dependent on a set of parameters $\{\alpha_i\}$, there will be an associated Casimir energy $V(\alpha_i)$ which will give rise to forces that will tend to make the system move in the parameter space. In the original Casimir effect, there was one parameter, namely the separation between the plates, and the Casimir energy produced an attractive force tending to bring the plates together.

In the simplest Kaluza-Klein setting, one has the classical background (four dimensional Minkowski space) x (a circle of radius r). Classically, r is undetermined. But the quantum fluctuations about this background will give rise to a potential $V(r)$ that will dynamically determine how r will behave.

The action is

$$S = \frac{1}{16\pi G_5} \int d^5 x \sqrt{|g|}\, R$$

and one writes

$$g_{\mu\nu} = g^{(c)}_{\mu\nu} + h_{\mu\nu}$$

where

$$g^{(c)}_{\mu\nu} = \left(\begin{array}{c|c} \eta_{\mu\nu} & 0 \\ \hline 0 & r \end{array}\right)$$

is the classical background and $h_{\mu\nu}$ is the quantum fluctuation. To compute the one-loop contribution to V, it suffices to keep only the quadratic terms in $h_{\mu\nu}$. One must choose a gauge to do the quantum computation. It turns out that the cylindrical gauge

$$h_{\mu 5, 5} = 0$$

picks out only the physical degrees of freedom, so that there is no contribution from the Faddeev-Popov ghost. The calculation is then very similar to the scalar case discussed above, once one takes into account the fact that one is in five dimensions and that the graviton therefore has five degrees of freedom. (In D dimensions, the graviton

has 1/2 D(D-3) physical degrees of freedom.)
The result is

$$V_{eff}(r) = - \frac{15}{4\pi^2} \zeta_R(5) \frac{1}{(2\pi r)^4} \quad (4)$$

(Here V_{eff} has been normalized to be energy per unit three-volume.) There is no stationary point to V_{eff}, so the quantum effects tend to make r shrink indefinitely. Of course, since the loop expansion only makes sense if $r > L_p$, all we can reliably say is that the circle will shrink until it becomes or order the Planck length, at which point the dynamics will be very much more complicated.

Another interesting example is due to Candelas and Weinberg.[10] They considered a background of (Minkowski space) x (the N-sphere S^N of radius r). This background is not a solution of the classical equations. They computed the Casimir energy due to the presence of f minimally coupled scalar fields. If f is large enough, they argued that the quantum contribution of the gravitational field itself could be ignored.

The action is, with D=4+N,

$$S = - \frac{1}{16\pi G_D} \int d^D x \sqrt{|g|} \, (R + 2\Lambda)$$

$$- \frac{1}{2} \int d^D x \sqrt{|g|} \, g^{AB} \sum_{i=1}^{f} \nabla_A \phi^i \nabla_B \phi^i \, ,$$

The computation yields

$$V_{eff} = V^{(0)} + V^{(1)}$$

where $V^{(0)}$ is the classical contribution that comes from two sources: (a) the curvature of S^N and (b) the cosmological constant Λ, and $V^{(1)}$ is the Casimir energy of the scalar fields. $V^{(0)}$ can be written down essentially by inspection:

$$V^{(0)} = - \frac{\Omega_N}{16\pi G_D} \left[N(N-1) r^{N-2} - 2\Lambda r^N \right]$$

(Ω_N is the volume of the unit N-sphere) whereas on dimensional grounds, $V^{(1)}$ must have the form

$$V^{(1)} = \frac{fc_N}{r^4}$$

(recall that the one-loop contribution is independent of G_D). The actual quantum computation reduces to the determination of C_N. Candelas and Weinberg performed this calculation for several values of N (N has to be odd for technical reasons), and also considered the cases of conformally coupled scalars and of spinors. In order to obtain a solution $r=r_0$ to the quantum corrected equations of motion one must impose two conditions:

1. $\left.\frac{dV}{dr}\right|_{r=r_0} = 0$

and

2. $V(r=r_0) = 0$.

The second condition insures that there is no net cosmological constant, so that Minkowski space is indeed a solution. Because the radius of S^N is the only adjustable parameter in the solution, in general one cannot satisfy the two conditions unless one also fine-tunes the input cosmological constant Λ. But if one does fine-tune Λ, one can indeed find examples of potentials with apparently stable minima satisfying conditions 1. and 2.

Of course the Candelas-Weinberg action is not a realistic model of a higher-dimensional theory. An additional problem is that C_N actually turns out to be unexpectedly small, of order 10^{-4} or 10^{-5}. (Part of the problem can be seen already in our expression, Eq. (4), for the five dimensional case. In addition to a coefficient that is about .4, there is also an extra $(2\pi)^{-4}$, making C_5 of order 2×10^{-4}.) Thus to get reasonable values for r one has to take $f \simeq 10^4$ or 10^5.

Given this very small Casimir energy of matter fields, the question naturally arises as to the hitherto neglected contribution of the graviton itself. After all, any Kaluza-Klein theory will contain the graviton. Unfortunately, the graviton contribution is much harder to deal with, for two reasons: (i) the spectrum of quantum fluctuations of the graviton is much more complicated; and (ii) the cosmological constant Λ now plays a dynamical role, in the sense that the previous form for $V^{(1)}$ is replaced by

$$\frac{F_N(2\Lambda r^2)}{r^4}$$

i.e. the constant C_N is replaced by a function of the radius which can be quite complicated. The reason for the appearance of Λ in $V^{(1)}$ is that the quantum fluctuations couple directly to Λ through the term $\Lambda \sqrt{|g|}$.

Despite these difficulties, the gravitational Casimir energy has been computed for the background $M_4 \times S^N$. The essential results are: (a) the graviton contribution is indeed 10^4 or 10^5 larger than that of equivalent numbers of matter fields, but (b) the graviton also gives rise to an imaginary part of V_{eff}, thereby rendering all otherwise interesting solutions unstable.

The program of including one-loop dynamical effects in the equations determining spontaneous compactification is thus at a crossroads. If it is to continue, the only hope for stable solutions appears to reside in broadening the class of backgrounds to something beyond $M_4 \times S^N$. This will require both inspired guesswork, and technical improvements to handle more complicated manifolds. Despite these obstacles, the potential payoff is great. We recall that in a Kaluza-Klein theory, the gauge group and the gauge coupling constants are determined by the shape and size of the internal manifold, respectively. The prospect of computing them from the dynamics of the underlying theory is an exciting one that one hopes will be realized, either within the pure Kaluza-Klein framework or in a related setting, in the not too distant future.

References

1. Freund, P.G.O., and Rubin, M.A., Phys. Lett. $\underline{97B}$, 233 (1980).
2. Duff, M.J., Nilsson, B.E.W., and Pope, C.N., "Kaluza-Klein Supergravity", Physics Reports (to appear) and references therein.
3. Wetterich, C., Nucl. Phys. $\underline{B222}$, 20 (1983).
4. Witten, E., in Proceedings of the Shelter Island Conference (MIT Press, Cambridge, Mass. (1985)).
5. Weinberg, S., Phys. Lett. $\underline{138B}$, 47 (1984).
6. Freund, P.G.O., Nucl. Phys. $\underline{B209}$, 146 (1982).
7. Alvarez, E., and Belén Gavela, M., Phys. Rev. Lett. $\underline{51}$, 931 (1983).
8. Barr, S., and Brown, L.S., Phys. Rev. $\underline{D29}$, 2779 (1984).
9. Casimir, H.B.G., Proc. Kon. Ned. Akad. v. Wet. $\underline{51}$, 793 (1948).
10. Candelas, P., and Weinberg, S., Nucl. Phys. $\underline{B237}$, 397 (1984).

PHYSICS RESULTS FROM THE CERN SUPER PROTON SYNCHROTRON PROTON-ANTIPROTON COLLIDER

S. Geer

Harvard University, Cambridge, Mass., USA

ABSTRACT

In the last four years the physics program of the CERN Super Proton Synchrotron proton-antiproton Collider has produced an impressive list of major results. In these lectures the results from three areas of the physics program are described and discussed: the physics of hadronic jet production; the study of the charged and neutral Intermediate Vector Bosons; and the physics of dimuon production at the Collider.

1. INTRODUCTION

When the CERN Super Proton Synchrotron p$\bar{\text{p}}$ Collider first became operational[1] in July 1981, hadronic interactions were recorded at a centre-of-mass energy ($\sqrt{s}$ = 546 GeV), that was almost an order of magnitude more than had previously been possible. In the last four years an impressive list of major results have come from the analysis of collider data: the measurement of the basic properties of typical events, such as particle multiplicity[2], rapidity and transverse-momentum distributions[3]; the observation of dramatic hadronic jet events[4], and the detailed measurement of jet production[5] and fragmentation[6] properties; the discovery of the charged[7] (W^+ and W^-) and neutral[8] (Z^0) Intermediate Vector Bosons (IVBs), the measurement of their masses, and their production and decay characteristics; and the first possible indication of the existence of the top quark[9] with a mass in the range 30 GeV/c^2 to 50 GeV/c^2. In addition to these classical results, several types of anomalous events have been observed (radiative Z^0 decays, electrons associated with jets and missing transverse energy, high-mass jet-jet events, monophotons, monojets, anomalous like-sign dimuon production + ... ?). These anomalies have provided the experimental and theoretical communities with much excitement and some inspiration over the last two years. Not all of the anomalies, however, will survive the test of time — indeed several of them have already been assigned to statistical fluctuations. If any of the remaining strange categories of events do survive, they will provide us with evidence for new and unexpected physics at the Collider, ensuring an interesting and exciting future for the collider program at CERN.

These lectures focus more on the established physics results coming from the p$\bar{\text{p}}$ Collider, than on the current zoo of anomalous events. Three topics are described and discussed in the following pages: i) hadronic jet physics; ii) W and Z^0 physics; and iii) dimuon physics. The majority of the results presented come from the UA1 detector, which is described in the following section.

2. EXPERIMENTAL DETAILS

The results described in these lectures have been made possible by the ability of the big collider experiments (UA1 and UA2) to detect and measure the properties of electrons, muons, neutrinos, and jets being produced in energetic p$\bar{\text{p}}$ collisions. A brief description of some of the relevant features of the UA1 detector is given in the following paragraphs. A more detailed description can be found elsewhere[10]. The corresponding details for the UA2 detector can also be found in the literature[11].

The UA1 detector has been designed to study, as systematically as possible, the phenomenology of p$\bar{\text{p}}$ collisions using a system of track chambers and of hermetic (4π solid angle) calorimeters that cover all polar angles greater than 0.2° from the beam direction. A schematic view of the UA1 detector is shown in Fig. 1. The main features of the detector and event reconstruction relevant to the physics results from the UA1 experiment are described in the following sections.

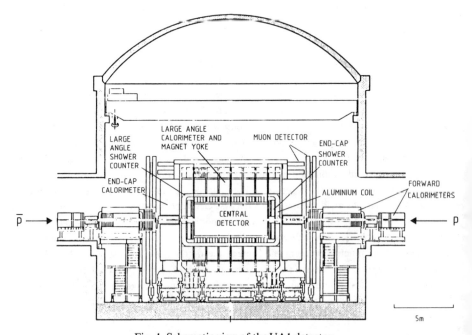

Fig. 1 Schematic view of the UA1 detector.

2.1 Charged-Particle Detection and Momentum Measurement

At the centre of the UA1 apparatus, there is a large-volume, cylindrical, central track detector (CD), 5.8 m in length and 2.3 m in diameter. The CD is a drift chamber containing 6110 sense wires, providing an average of $\sim$ 100 points per charged track coming from the interaction. The sign of the charges, and the momenta of charged particles emerging from an interaction, are determined by measuring the curvature of the associated CD tracks in the UA1 horizontal 0.7 T magnetic dipole field. The momentum accuracy for high-momentum tracks is limited by the localization error inherent in the system ($\leq$ 100 μm) and by the diffusion of electrons drifting in the chamber gas, which is proportional to $\sqrt{\ell}$ and amounts to about 350 μm after the maximum drift length ℓ = 19.2 cm. This results in a momentum accuracy of approximately $\pm$20% for a 1 m long track with 40 GeV/c momentum in the best direction with respect to the field.

2.2 The UA1 Calorimeters

A summary of the characteristics of the UA1 electromagnetic (e.m.) and hadronic calorimeters is given in Table 1. The e.m. calorimeters surround the UA1 CD. Large angles (between 25° and 155° with respect to the beam directions) are covered by the central e.m.

Table 1
A summary of the characteristics of the UA1 electromagnetic and hadronic calorimeters

Calorimeter	Angular coverage θ (deg.)	Thickness Radiation lengths	Thickness Absorption lengths	Cell size $\Delta\theta$ (deg.)	Cell size $\Delta\phi$ (deg.)	Sampling step (mm)		Segmentation in depth	Resolution
Barrel e.m.	25–155	26.4/sinθ	1.1/sinθ	5	180	Pb:	1.2	3.3/6.6/10.1/6.6X_0	0.15/$\sqrt{E}$
						Scint.:	1.5		
hadron	25–155	–	5.0/sinθ	15	18	Fe:	50	2.5/2.5λ	0.8/$\sqrt{E}$
						Scint.:	10		
End-caps e.m.	5–25	27/cosθ	1.1/cosθ	20	11	Pb:	4	4/7/9/7X_0	0.12/$\sqrt{E_T}$
						Scint.:	6		
hadron	155–175	–	7.1/cosθ	5	10	Fe:	50	3.5/3.5λ	0.8/$\sqrt{E}$
						Scint.:	10		
Forward e.m.	0.7–5	30	1.2	4	45	Pb:	3	4 × 7.5X_0	0.15/$\sqrt{E}$
						Scint.:	3		
hadron	175–179.3	–	10.2	–	–	Fe:	40	6 × 1.7λ	0.8/$\sqrt{E}$
						Scint.:	8		
Very forward e.m.	0.2–0.7	24.5	1.0	0.5	90	Pb:	3	5.7/5.3/5.8/7.7X_0	0.15/$\sqrt{E}$
						Scint.:	6		
hadron	179.3–179.8	–	5.7	0.5	90	Fe:	40	5 × 1.25λ	0.8/$\sqrt{E}$
						Scint.:	10		

calorimeters, consisting of two half-cylinders on either side of the beam, each made of 24 independent cells. Each cell subtends an angular interval $\Delta\theta = 5°$ and $\Delta\phi = 180°$, where θ is the polar angle and ϕ the azimuthal angle. A cell consists of a multilayer lead–scintillator sandwich, 26.6 radiation lengths (r.l.) thick, subdivided into four independent segments with thicknesses of 3.3, 6.6, 10.1, and 6.6 r.l. The scintillation light from each of the four segments is collected by wave-shifting plates on each side of the cell, and guided to photomultipliers located outside the detector. There are four photomultipliers looking at each segment, two at each end. Light division between the four photomultipliers is used to locate the position of the (e.m.) shower within a cell, with a resolution $\sigma_\phi \approx 0.3/\sqrt{E}$ radians and $\sigma_x \approx 6.3/\sqrt{E}$ cm, where E is the shower energy in GeV, ϕ the azimuthal angle in the plane perpendicular to the beam direction, and the x-axis is in the direction of the beams.

A second e.m. calorimeter, the end-cap calorimeter, covers the angular interval from 5° to 25° with respect to the beam direction. There are two end-cap calorimeters, one at each end of the CD. Each calorimeter is divided into 32 azimuthal sectors located 3 m from the interaction point. Each sector consists of a lead–scintillator sandwich, 27 r.l. thick, and segmented four times in depth. A position detector made of proportional tubes and located at a depth of 11 r.l., measures the position of e.m. showers with a precision of ~ 2 mm.

The energy of an electron can be measured in the UA1 e.m. calorimeters with a resolution $\sigma_E = 0.15\sqrt{E}$, where E is the electron energy in GeV. The absolute energy scale is obtained by illuminating the e.m. calorimeter cells at the beginnning and end of each data-taking period with an intense ^{60}Co source. A small number of e.m. cells have been calibrated directly with electron beams of known energy from the CERN SPS; the other e.m. cells are then adjusted so that their response to the ^{60}Co source is the same as that of the calibrated cells.

Surrounding the e.m. calorimeters is the iron yoke of the UA1 magnet. The yoke is laminated, and scintillator plates have been inserted between the iron plates to form the hadron calorimeter, which consists of 450 independent cells with a typical cell-size of 15° × 18° in the central region and 5° × 10° in the forward regions. The energy resolution of the hadron calorimeter $\sigma_{E_{had}} \approx 0.8\sqrt{E}$, where E_{had} is the hadronic shower energy in GeV.

2.3 Electron Identification

Isolated electrons with transverse energies in excess of 15 GeV can be clearly identified in the UA1 detector. Their experimental signature is illustrated in Fig. 2. The electron

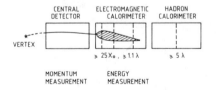

Fig. 2 Electron identification in UA1.

produces a charged track in the CD, and is totally absorbed in the e.m. calorimeters. In searching for an electron we therefore look for a track in the CD with a direction and momentum which matches the position and energy of an e.m. shower in the e.m. calorimeters. Furthermore, the e.m. shower must have the characteristics of an electron shower, namely its spatial extension (Δx, $\Delta \phi$) must be small, and its longitudinal profile in the four longitudinal segments of the calorimeter must be consistent with the expectations for an electron, with little or no energy deposited in the hadron calorimeter cells behind the e.m. shower.

2.4 Muon Identification and Measurement

Surrounding the hadron calorimeter, on the outside of the detector, are the UA1 muon chambers consisting of two drift chamber planes separated by a distance of 60 cm and covering the polar angular interval $5° < \theta < 175°$. Each of these planes is made from 50 drift chambers, $\approx 4 \times 6$ m^2 in size. Each chamber has four layers of drift tubes which define two orthogonal coordinates.

Muons with momentum in excess of ~ 3 GeV/c penetrate the whole UA1 apparatus. The experimental signature for a muon in the UA1 detector is illustrated in Fig. 3. A muon manifests itself in the CD as a charged track which matches, in position and angle, a corresponding track measured in the muon chambers on the outside of the detector. Between the two track chambers the muon leaves a characteristic (minimum ionizing) energy deposition in the calorimeter cells through which it passes. The muon momentum is measured in the CD. Using the known position of the interaction vertex together with the muon chamber track parameters, it is possible to obtain an independent measurement of the muon momentum through the determination of the muon deflection in the magnet yoke.

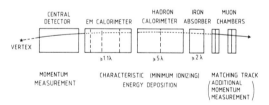

Fig. 3 Muon identification in UA1. A charged track in the CD matches, in position and angle, a charged track in the muon chambers on the outside of the detector.

2.5 Neutrino Emission

The presence of neutrino emission is inferred from an apparent lack of momentum conservation in the two components transverse to the beam direction. The detection and measurement of neutrino emission by this method is made possible in the UA1 detector by the nearly complete solid-angle coverage of the calorimeters down to 0.2° from the direction of the beams. In practice, we define an energy flow vector $\vec{\Delta E}$, adding vectorially the observed energy depositions in all the UA1 calorimeter cells. Neglecting particle masses, with an ideal calorimeter, momentum conservation requires $\vec{\Delta E} = 0$. If an energetic neutrino is emitted in the event, $\vec{\Delta E} = \vec{p}_\nu$, the neutrino momentum. This technique has been tested on typical (minimum bias) and jet-enriched events, for which energetic neutrino emission is not expected to occur. The transverse components ΔE_y and ΔE_z exhibit small residuals centred on zero, Gaussian in shape, and with an r.m.s. deviation given by $\sigma(\Delta E_{y,z}) = 0.4 \, (\Sigma_i |E_T^i|)^{1/2}$, where the sum is over all the transverse-energy contributions recorded in the UA1 calorimeter cells (the scalar transverse energy of the event), and all energies are in GeV. The scalar transverse energy is a measure of the 'temperature' of an event. Thus the missing transverse-energy resolution depends upon the temperature of the event (Fig. 4).

The longitudinal component of the energy flow vector ΔE_x has poor resolution for neutrino emission since ordinary particles can escape detection by travelling close to the x-axis, in the direction of the beam pipe. In practice we do not use ΔE_x for neutrino detection and measurement.

Finally, in constructing ΔE_y and ΔE_z, care must be taken to include the momentum components of any energetic muons reconstructed in the event, since high-energy muons penetrate the calorimeters without depositing a substantial amount of their energy in the detector.

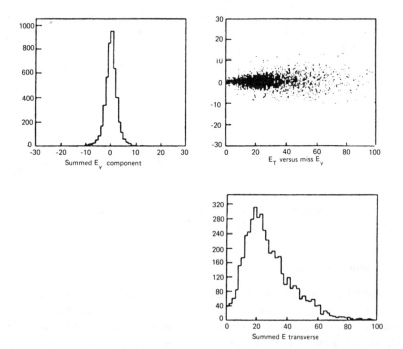

Fig. 4 Missing transverse energy resolution. The missing transverse energy in the y direction [ΔE_y (GeV)] is plotted versus the scalar sum of the transverse energies in all the UA1 calorimeter cells [E_T(GeV)], for a sample of typical (minimum bias) events.

2.6 The Trigger

In the period from 1981 to 1983 the Collider operated at a centre-of-mass energy $\sqrt{s} = 546$ GeV, and the UA1 experiment collected data from an integrated luminosity of 136 nb^{-1} (a process occurring with a cross-section of 1 nb would result in 136 events). In 1984 the centre-of-mass energy of the Collider was increased to $\sqrt{s} = 630$ GeV, and the UA1 experiment recorded a further 263 nb^{-1}. The total p$\bar{\text{p}}$ cross-section at the Collider has been measured by the UA4 experiment[12] to be $\sigma_{tot} = 61.9 \pm 1.5$ mb at $\sqrt{s} = 546$ GeV. Thus the UA1 data sample of 399 nb^{-1} corresponds to a total of $\sim 10^{10}$ interactions.

Clearly, it is only possible to record and analyse a small fraction of the interactions occurring at the Collider. The choice of which fraction of the events are to be recorded on magnetic tape and subsequently analysed is made with the help of a trigger. The trigger consists of a list of conditions which must be satisfied before an event will be recorded by the experiment. The basic triggers used by the UA1 experiment are the following:

a) *Electron trigger.* The presence of two adjacent e.m. calorimeter cells with a total transverse energy greater than a threshold value (typically 10 GeV).

b) *Muon trigger.* The presence of a track in the outer muon chambers that points back to the interaction vertex.

c) *Jet trigger.* The presence of a localized cluster of energy in the UA1 calorimeter cells with summed transverse energy greater than a threshold value (typically 15 GeV).

d) *Global E_T trigger.* The scalar sum of all the transverse energies measured in the UA1 calorimeter cells in excess of a threshold value (typically 40 GeV).

With the help of these triggers the data-taking rate is reduced to match the maximum tape writing speed, which corresponds to about three events per second. At the end of a typical three-months running period the UA1 experiment records several million events, which are subsequently analysed.

3. HADRONIC JET PRODUCTION

Hadronic jets are expected to be produced at the Collider as a result of the hard scattering of a parton from the proton with a parton from the antiproton. Each of the scattered partons (quarks, antiquarks, or gluons) fragments into a collimated jet of hadrons. A typical example of such an event observed in the UA1 detector is shown in Fig. 5. Two tightly collimated groups of charged tracks are seen in the central track detector (CD) associated with two large clusters of energy in the calorimeters. The two calorimeter clusters have roughly equal transverse energies (100 GeV and 80 GeV) and are back-to-back with each other in the plane perpendicular to the beam direction (transverse-momentum balance). The structure of the event can clearly be seen by looking at the transverse-energy flow plot (Fig. 6)

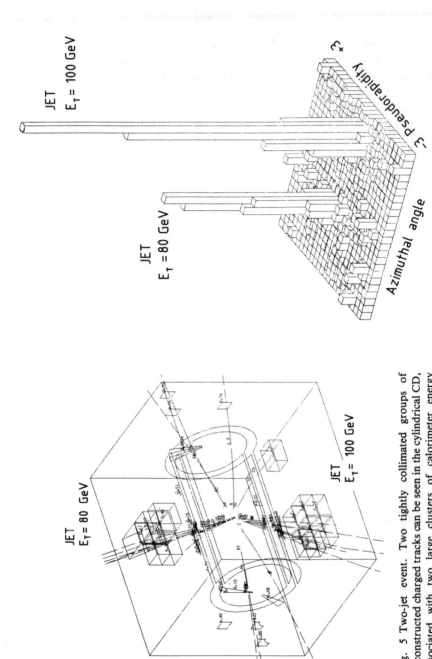

Fig. 5 Two-jet event. Two tightly collimated groups of reconstructed charged tracks can be seen in the cylindrical CD, associated with two large clusters of calorimeter energy depositions.

Fig. 6 Transverse-energy flow plot for the two-jet event shown in Fig. 5.

in which the transverse energy measured in the calorimeter cells is shown in the [pseudorapidity[*)], azimuthal angle]-plane. Two large towers of transverse energy are seen in the (η,ϕ)-plane at $\eta \sim 0$ (perpendicular to the beam direction), and separated from each other by an angle $\Delta\phi$ of roughly 180°. The two-jet system reconstructed from these clusters has a mass of 176 GeV/c². Thus, in this spectacular event almost one-third of the incoming proton and antiproton momentum has been thrown sideways.

3.1 Jet Reconstruction and Inclusive Jet Properties

In a typical two-jet event, hadronic jets manifest themselves in the UA1 detector as well-defined clusters of calorimeter energy. We can measure the size of a typical jet by studying the transverse-energy flow plots for many jets. This is illustrated in Figs. 7a to 7c, in

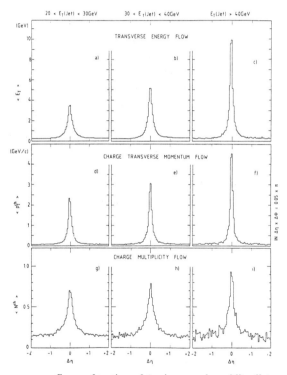

Fig. 7 a–c   Transverse-energy flow as function of $\Delta\eta$, i.e. pseudorapidity distance from the jet axis, for three slices of jet E_T.
d–e   Charged transverse momentum flow as function of $\Delta\eta$.
g–i   Charged multiplicity flow as function of $\Delta\eta$.

[*)] Pseudorapidity η is defined as $\eta \equiv -\ln(\tan\theta/2)$, where θ is the angle between the antiproton direction and the vector pointing from the interaction vertex to the centre of the calorimeter cell.

which the transverse-energy flow is shown along the pseudorapidity axis, summed over many jets by aligning all the jet axes at $\Delta\eta = 0$. The resulting average transverse-energy flow about the jet axis is shown for jets of transverse energy $E_T(jet)$ in the interval a) $20 < E_T(jet) < 30$ GeV, b) $30 < E_T(jet) < 40$ GeV, and c) $E_T(jet) > 40$ GeV. The full width at the base of the jet-associated transverse-energy flow peak is given by $\Delta\eta = \pm 1.0$, and is independent of E_T (jet). Defining the jet core to be the region $|\Delta\eta| < 0.2$, and the jet wings to be the region $0.2 < |\Delta\eta| < 1.0$, we find that the transverse energy contained in the jet core increases with increasing $E_T(jet)$, whilst the transverse energy in the jet wings is fairly independent of $E_T(jet)$. We see a similar behaviour if we look at the transverse-momentum flow of charged tracks around the jet axis (Figs. 7d to 7f) or the charged multiplicity flow around the jet axis (Figs. 7g to 7i).

In order to determine, on a jet-by-jet basis, the jet energy and direction, we need to use a jet reconstruction algorithm. The standard UA1 jet algorithm has been developed by studying the shape and characteristics of the jet transverse-energy clusters observed in the associated transverse-energy flow plots. Details of the UA1 jet algorithm can be found in Ref. 5. To reconstruct hadronic jets, a momentum vector is associated with each calorimeter cell. Its direction and magnitude are defined by the spatial position of the cell and the energy deposition within the cell. The cells with transverse energy in excess of a threshold value (typically 1.5 GeV) are grouped in clusters, and their associated momentum vectors are added if the distance between them in (pseudorapidity, azimuthal angle)-space ΔR is less than 1 $[\Delta R \equiv (\Delta\eta^2 + \Delta\phi^2)^{1/2}$, where ϕ is in radians]. Cells with lower transverse energy are then included in the cluster if they are within $\Delta R < 1$ with respect to the cluster axis. Jets are retained if they have a transverse momentum in excess of 5 GeV/c, and an axis within the pseudorapidity interval $|\eta| < 3.0$.

Defining jets in this way, we can measure the inclusive jet cross-section at the Collider, and compare our results with the QCD expectations for parton–parton scattering (Fig. 8).

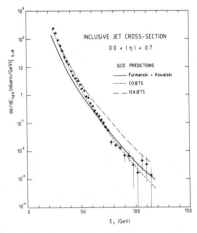

Fig. 8 Inclusive jet cross-section for jets produced in the pseudorapidity range $|\eta| < 0.7$. The curves are the predictions from various QCD Monte Carlo calculations.

The inclusive jet rate has been measured for jets with transverse energies from $\approx$ 25 GeV to $\approx$ 120 GeV. The measured jet cross-section falls by five orders of magnitude over this range, and is in excellent agreement with the QCD expectations.

3.2 The Two-Jet Cross-Section

The two-jet cross-section can be written in the form

$$\sigma(p\bar{p} \to cd) = \sum_{ab} P(p \to a) P(\bar{p} \to b) \, d\sigma(ab \to cd) , \qquad (1)$$

where we are considering a definite final state (cd), and summing over all allowed initial states (ab); $P(p \to a)$ is the probability of finding a parton of type a in the proton, and $d\sigma(ab \to cd)$ is the cross-section for the subprocess $ab \to cd$. In using expression (1) we assume factorization: that the probability of finding parton a in the proton is independent of the probability of finding parton b in the antiproton. We shall see that this is a reasonable assumption to make later on. To exhibit the kinematics more explicitly, we can rewrite expression (1) as

$$\sigma(p\bar{p} \to cd) = \sum_{ab} \{[F_a(x_1)/x_1]dx_1\} \{[F_b(x_2)/x_2]dx_2\} \, [d\sigma(ab \to cd)/d\cos\theta^*] \, d\cos\theta^* , \qquad (2)$$

where the flux factor $[F_a(x_1)dx_1]/x_1$ gives the probability of finding a parton of type a in the proton carrying a fraction of the proton's momentum between x_1 and $x_1 + dx_1$. The $F(x)$ are called structure functions, and describe the momentum distribution of the constituents inside the proton and antiproton. The subprocess cross-section $d\sigma(ab \to cd)/d\cos\theta^*$ has been written to display explicitly the angular dependence of the subprocess, θ^* being the angle between the incoming and outgoing partons in the subprocess centre-of-mass system. In practice the angular dependences of the various lowest-order QCD subprocesses are very similar. Furthermore, at present we do not experimentally distinguish between the various final states (cd). Therefore, summing over all two-jet final states, we can write the differential cross-section,

$$d\sigma/dx_1 dx_2 \, d\cos\theta^* \approx [F(x_1)/x_1] \, [F(x_2)/x_2] \, [d\sigma(gg \to gg)/d\cos\theta^*] . \qquad (3)$$

In practice the beam directions are known, and the four-vectors of the outgoing jets are measured. Thus, assuming massless partons, the complete kinematics of each two-jet event can be solved. The momenta x_1 and x_2 of the interacting partons are computed as follows:

$$\begin{aligned} x_1 &= [x_F + \sqrt{(x_F^2 + 4\tau)}]/2 \\ x_2 &= [-x_F + \sqrt{(x_F^2 + 4\tau)}]/2 , \end{aligned} \qquad (4)$$

where

$$x_F = (\vec{p}_{3L} + \vec{p}_{4L})/(\sqrt{s}/2) \tag{5}$$

$$\tau = (p_3 + p_4)^2/s .$$

In Eq. (5), $\vec{p}_3$ and $\vec{p}_4$ are the four-momenta of the final two jets and $\vec{p}_{3L}$ and $\vec{p}_{4L}$ are the longitudinal momentum components measured along the beam direction in the laboratory frame.

The c.m.s. scattering angle is computed in the rest frame of the final two jets [$(p_3 + p_4)$] relative to the axis defined by the interacting partons [$(\vec{p}_1 - \vec{p}_2)$] assumed to be massless and collinear with the beams:

$$\cos\theta^* = (\vec{p}_3 - \vec{p}_4) \cdot (\vec{p}_1 - \vec{p}_2)/(|\vec{p}_3 - \vec{p}_4||\vec{p}_1 - \vec{p}_2|) . \tag{6}$$

Thus we can extract the proton structure function F(x) and the angular dependence of the subprocesses $d\sigma/d\cos\theta^*$.

We begin with the structure function. Since we do not distinguish between the various incoming partons (quarks, antiquarks, and gluons) the structure function that is measured at the Collider is a mixture of the gluon structure function [G(x)] and the quark and antiquark structure functions [Q(x) and $\bar{Q}$(x)]. In QCD the relative strengths of the couplings at the ggg vertex and at the qqg vertex are given by the colour factors: $C_A = 3$ for the ggg vertex, and $C_F = 4/3$ for the qqg vertex. This being the case, the structure function that is actually measured at the Collider is weighted in favour of the gluons, and is given by (see Fig. 9)

$$F(x) = G(x) + (4/9) [Q(x) + \bar{Q}(x)] . \tag{7}$$

The UA1 measurement[5] of F(x) is shown in Fig. 10. The fraction of the proton momentum carried by partons with momentum x decreases exponentially with increasing x. This suggests a thermodynamical explanation for the parton distribution inside the proton. *A priori* QCD does not give a prediction for F(x). It does, however, predict the way in which F(x) changes with Q^2. In Fig. 10 the measured F(x) is compared with expectations based on QCD fits to deep-inelastic scattering data. The solid curve represents the structure function G(x) + (4/9)[Q(x) + $\bar{Q}$(x)] at $Q^2 = 20$ GeV^2 based on a parametrization of the CERN-Dortmund-Heidelberg-Saclay (CDHS) measurements[13] of G(x) and [Q(x) + $\bar{Q}$(x)] at low Q^2 (Q^2 = 2-200 GeV^2) using a neutrino beam on an iron target. The broken curve shows the expected modification of the structure function at the value of Q^2 appropriate to the UA1 experiment ($Q^2 = 2000$ GeV^2), due to QCD scaling violations (assuming $\Lambda = 0.2$ GeV). The expected contribution due to the quarks and antiquarks is shown separately. Subtracting this contribution we can extract from F(x) a measurement of the gluon structure function G(x) (Fig. 11). Thus the two-jet data measure directly, for the first time, the very large flux of gluons in the proton at small x (x $\leq$ 0.3), and also demonstrate the existence of QCD scaling violations at large Q^2.

661

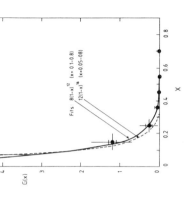

Fig. 11 The gluon structure function G(x), extracted from the effective structure function F(x) by subtracting the expected contribution from the quarks and antiquarks.

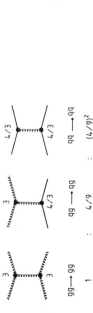

Fig. 9 Lowest-order QCD parton-parton scattering subprocesses. The colour factors, $C_A = 3$ and $C_F = 4/3$, determine the relative subprocess cross-sections.

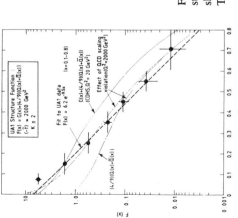

Fig. 10 The structure function F(x). The solid curve represents a QCD parametrization of the structure function at $Q^2 = 20$ GeV^2 based on the CDHS[13] measurements. The broken curves show the expected modification of F(x) at the values of Q^2 appropriate to the UA1 experiment. The expected contribution of quarks and antiquarks is shown separately.

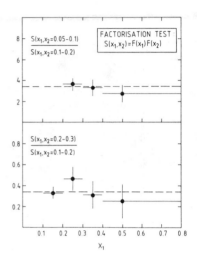

Fig. 12 The ratio $S(x_1,x_2')/S(x_1,x_2)$ shown as a function of x_1 for a) $x_2 = 0.1$–0.2, $x_2' = 0.05$–0.1, and b) $x_2 = 0.1$–0.2, $x_2' = 0.2$–0.3. In each case the data are consistent with a constant (broken line) independent of x_1.

We are now in a position to test the assumption of factorization that we made in writing expression (1). In Fig. 12 it is shown that the probability of finding a parton in the antiproton with momentum fraction x_2 is independent of the momentum fraction x_1 of the parton in the proton. The assumption we made in writing expression (1) is therefore justified.

We now consider the two-jet angular distribution. Many of the lowest-order QCD subprocesses have a direct analogy to an equivalent $e^+ e^-$ QED subprocess (an example is shown in Fig. 13). In both cases the interaction between two light spin-$\frac{1}{2}$ fermions is mediated by a massless spin-1 boson. It is therefore not surprising that the expected angular dependence of these lowest-order QCD subprocesses follows a QCD version of the Rutherford law,

$$d\sigma/d\cos\theta^* \approx (\alpha_s^2/\hat{s})(1 - \cos\theta^*)^{-2}, \qquad (8)$$

where $\sqrt{\hat{s}}$ is the subprocess centre-of-mass energy, taken to be the invariant mass of the two-jet system.

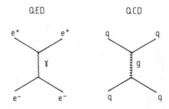

Fig. 13 A comparison between a QED and a QCD subprocess. Both interactions are mediated by a spin-1 massless boson. This results in a similar angular dependence for the subprocess cross-sections.

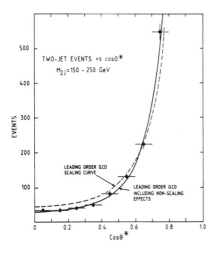

Fig. 14 Two-jet angular distribution. The data are close to the Rutherford law curve (broken line), but are better described by the QCD (solid) curve which takes into account scale-breaking effects.

Indeed the Rutherford law gives a good first description of the data[14] (Fig. 14). However, we would expect the angular distribution to be modified from the Rutherford law by higher-order QCD (scale-breaking) effects. To see this deviation from Rutherford scattering clearly, we define the angular variable[15]

$$\chi \equiv (1 + \cos\theta^*)/(1 - \cos\theta^*) \,. \tag{9}$$

This variable has the property that if Eq. (8) holds exactly, then the χ-distribution will be flat for $\chi \gtrsim 2$. The measured χ-distribution is shown in Fig. 15. There is a marked rise in this distribution with increasing χ, which is in agreement with the expected effect coming from QCD scale-breaking effects associated with the Q^2 dependence of α_s [through the α_s^2 factor in Eq. (8)], and the Q^2 dependence of the effective structure function. We conclude that the angular distribution observed for two-jet events at the Collider is in agreement with QCD expectations provided higher-order scale-breaking effects are taken into account.

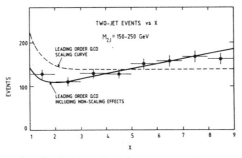

Fig. 15 The two-jet angular distribution plotted versus $\chi \equiv (1 + \cos\theta^*)/(1 - \cos\theta^*)$. The broken curve shows the leading-order QCD prediction suitably averaged over the contributing subprocesses. The solid curve includes scale-breaking corrections.

3.3 The Three-Jet Cross-Section

The highest scalar E_T events at the Collider are predominantly two-jet events. However, clearly defined three-jet events have also been observed. An example is shown in Fig. 16. The mass of the three-jet system is 236.7 GeV/c^2, which is 43% of the available p$\bar{\text{p}}$ energy.

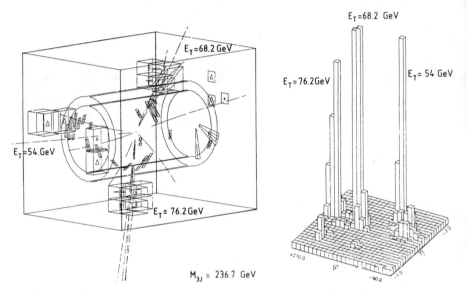

Fig. 16 A typical three-jet event in the UA1 detector and the associated transverse-energy flow plot.

For three final-state (massless) partons, the final-state parton configuration, at fixed subprocess c.m.s. energy, is specified by four independent variables. The leading-order QCD predictions for the subprocess cross-sections are given by

$$d^4\sigma/dx_3dx_4 \, d\cos\theta_3 \, d\psi = (1/32\pi^2)|M|^2 , \qquad (10)$$

where, neglecting constant and slowly varying factors, the spin- and colour-averaged matrix element squared $|M|^2$ may be written:

$$|M|^2 \approx (\alpha_s^3/\hat{s}) \, [x_{T_3}^2 x_{T_4}^2 x_{T_5}^2 \, (1 - x_3)(1 - x_4)(1 - x_5)]^{-1} . \qquad (11)$$

In Eqs. (10) and (11) the x_i ($i = 3,4,5$) are the energies (or momenta) of the outgoing partons, ordered so that $x_3 > x_4 > x_5$ and scaled to the total subprocess c.m.s. energy such that $x_3 + x_4 + x_5 = 2$; θ_i is the angle between the parton i and the beam direction ($x_{T_i} = x_i \sin\theta_i$), and ψ is the angle between the plane defined by partons 4 and 5 and the plane defined by the beam

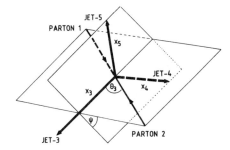

Fig. 17 The three-jet variables defined in the subprocess c.m.s. frame.

direction and parton 3 (see Fig. 17). The exact (leading-order) expressions for $|M|^2$ for the various incoming and outgoing parton combinations (e.g. gg → ggg, qg → qgg, etc.) have been given in a simple form by Berends et al.[16]. An important feature of these 2 → 3 subprocess cross-sections is that for three-parton final-state configurations which approach two-parton configurations, i.e. x_3, x_4 → 1 or x_{T_i} → 0, the cross-sections become large as a result of final- (or initial-) state bremsstrahlung processes. Naturally the comparison of theory with experiment must be restricted to a region of phase space where all the jets are well separated from each other and from the beams, and in which the theoretical cross-sections are finite and relatively slowly varying.

To ensure that all three jets are well separated from each other and from the beams, the following cuts are applied to define the three-jet sample:

i) $x_3 < 0.9$. This cut guarantees that jet-4 and jet-5 are well separated in angle in the subprocess c.m.s. frame. It is made to ensure that jet-4 and jet-5 will be resolved as separate jets, with full efficiency, by the jet algorithm.

ii) $|\cos \theta_3| < 0.6$, $30° < |\psi| < 150°$. These cuts guarantee that all three jets are well separated in angle from the incoming parton axis in the subprocess c.m.s. frame.

For three-jet events the total subprocess c.m.s. energy is taken to be the three-jet mass ($\sqrt{\hat{s}} = m_{3J}$), computed using the four-vectors of the three highest-p_T jets. To ensure good trigger and selection efficiency for the three-jet events, we require $m_{3J} > 150$ GeV/c^2. With these cuts the UA1 Collaboration has extracted a sample of 173 three-jet events. We can now see if the QCD single-bremsstrahlung formulae [Eqs. (10) and (11)] are able to describe the data.

The three-jet Dalitz plot (x_3 versus x_4) is shown in Fig. 18. The density of events in the Dalitz plot is significantly non-uniform over the range explored. In particular the density of events increases visibly with increasing x_4, for fixed x_3 ($x_3 \approx 0.85$), as the two-jet region ($x_3 \to 1$, $x_4 \to 1$) is approached. Less apparent, but also significant, is the increase in density of events with x_3, for fixed x_4 ($x_4 \approx 0.6$–0.8). This effect signals the contribution of quasi-collinear final-state bremsstrahlung processes which are expected to dominate for $x_3 \to 1$. The projections of the Dalitz plot onto the x_3 and x_4 axes are also shown. The solid curves, which have been normalized to the data, show the predicted distributions for

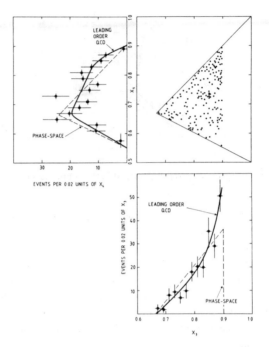

Fig. 18 The Dalitz plot (x_3 versus x_4) for the three-jet sample. The solid curves represent the predictions of the leading-order QCD bremsstrahlung formulae.

x_3 and x_4 based on the QCD single-bremsstrahlung formulae. The dominant subprocesses are predicted to have a very similar x_3 and x_4 dependence over this range, and the curves shown have been suitably averaged over the contributing subprocesses. The broken curves show the corresponding phase-space distributions computed assuming a constant matrix element. The data are clearly consistent with the predictions of the QCD single-bremsstrahlung formulae, and are inconsistent with the phase-space distributions.

The three-jet angular distribution (ψ versus $\cos \theta_3$) is shown in Fig. 19. The sign of $\cos \theta_3$ has been defined with reference to the direction of the 'fast' and 'slow' incoming partons. Since, in general, the magnitudes of the longitudinal momenta of the incoming partons are not equal in the laboratory frame, the direction of the longitudinal motion of the three-jet system in the laboratory frame uniquely identifies the 'fast' and 'slow' incoming partons. The sign of $\cos \theta_3$ has been defined to be positive when, in the c.m.s. frame, jet-3 points along the direction of the 'fast' incoming parton. Similarly, ψ is defined such that $\psi = 0°$ when jet-4 lies in the plane defined by jet-3 and the incoming parton axis, and points along the direction of the 'fast' incoming parton. The distribution in $\cos \theta_3$ shows a pronounced forward-backward peaking which is similar to the behaviour of the two-jet angular distribution (Fig. 14) and is in agreement with the predictions of the QCD bremsstrahlung formulae. Furthermore, the data also show a distinct ψ dependence. Configurations for which

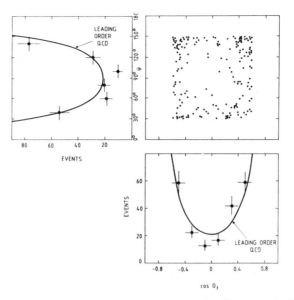

Fig. 19 The three-jet angular distribution: ψ versus $\cos\theta_3$. The theoretical curves are the predictions of the leading-order QCD bremsstrahlung formulae.

jet-4 and jet-5 lie close to the plane defined by jet-3 and the beams ($|\psi| \approx 30°, 150°$) are preferred to configurations for which $|\psi| \approx 90°$. This effect reflects the contribution of initial-state bremsstrahlung processes and is in agreement with the predictions of the QCD bremsstrahlung formulae. Finally, we note that the agreement between the data and the theoretical curves (Fig. 19) is only qualitative. In particular the measured angular distributions have a tendency to be steeper than the scaling curves. It is perhaps not unreasonable to suppose that, as in the two-jet case, the inclusion of scale-breaking corrections in the theoretical curves would improve the agreement between theory and experiment.

3.4 The Three-Jet/Two-Jet Ratio and the Determination of α_s

In QCD the relative yield of three-jet and two-jet events is directly related to the value of α_s. Integrating the differential cross-sections [Eqs. (8) and (10)] over the dimensionless variables in the region defined by the experimental cuts, gives

$$\sigma_{2J} = C_{2J}\alpha_s^2/\hat{s}, \qquad (12)$$

$$\sigma_{3J} = C_{3J}\alpha_s^3/\hat{s}, \qquad (13)$$

where C_{2J} and C_{3J} are dimensionless coefficients which depend on the subprocess, e.g. $gg \to gg$, $gg \to ggg$, etc. The numerical values of C_{2J} and C_{3J} appropriate to the experimental

Table 2

Summary of three-jet and two-jet cross-section coefficients [Eqs. (12) and (13)], calculated from the formulae of Berends et al.[16] and Combridge et al.[17]

	Cuts		
	3-jet: $x_3 < 0.9$, $|\cos\theta_3| < 0.6$ $30° < |\psi| < 150°$ 2-jet: $\cos\theta < 0.8$		
Parton combination[a]	C_{3J}	C_{2J}	C_{3J}/C_{2J}
gg	111.5	110.5	1.01
qg	38.1	48.2	0.79
$q\bar{q}$[b]	16.4	21.4	0.76

a) The cross-sections given refer to the elastic and single-gluon production processes only, e.g. gg → gg, gg → ggg, etc.

b) Calculated for identical flavours, e.g. $u\bar{u} \to u\bar{u}$, $d\bar{d} \to d\bar{d}$, etc.

cuts are listed in Table 2 for the various possible incoming parton combinations. The values given, which refer to the dominant elastic and single-gluon-production subprocesses only, have been calculated using the exact (leading-order) theoretical expressions for the subprocess cross-sections[16,17]. The two-jet subprocess cross-sections follow the well-known rule:

$$\sigma(gg) : \sigma(qg) : \sigma(q\bar{q}) \simeq 1 : (4/9) : (4/9)^2. \tag{14}$$

Remarkably, the corresponding three-jet cross-sections are seen to be in essentially the same proportions (to within ±25%). This has the important consequence that the three-jet to two-jet ratio is predicted to be almost independent of the incoming parton combination, and ensures that to a first approximation the same effective structure function is relevant for three-jet and two-jet production.

From Eqs. (12) and (13), averaging over the possible incoming parton combinations, the three-jet/two-jet ratio may be written:

$$\sigma_{3J}/\sigma_{2J} = \langle C_{3J}/C_{2J}\rangle \alpha_s. \tag{15}$$

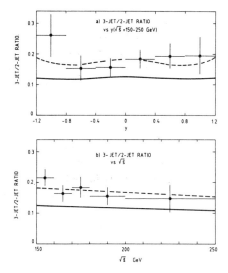

Fig. 20 The three-jet/two-jet ratio plotted a) versus the laboratory rapidity of the three-jet (two-jet) system, and b) versus subprocess c.m.s. energy (mass). The solid curves correspond to the choice of identical Q^2-scales for the three-jet and two-jet samples; the broken ones to the choice of a lower Q^2-scale for the three-jet sample.

The experimentally measured ratio of the three-jet to two-jet cross-sections is shown in Fig. 20a as a function of the laboratory rapidity y of the three-jet (or two-jet) system, and in Fig. 20b as a function of the subprocess centre-of-mass energy ($\sqrt{\hat{s}} = m_{3J}, m_{2J}$). The three-jet/two-jet ratio shows no significant dependence on rapidity or on subprocess centre-of-mass energy.

Further comparison between theory and experiment is made complicated by the theoretical uncertainty related to the freedom of choice of the Q^2-scale appropriate to three-jet and two-jet production. If the relevant Q^2-scale for two-jet production (Q^2_{2J}) is taken to be $\hat{t}$ (consistent with the data on the two-jet angular distribution), then for the two-jet sample we find $\langle Q_{2J} \rangle \simeq 0.45 m_{2J}$. Assuming that the Q^2-scales for the three-jet and two-jet samples are identical, i.e. $\langle Q_{3J} \rangle = 0.45 m_{3J}$, then Eq. (15) may be applied directly to predict the three-jet/two-jet ratio. In Fig. 20 the solid curve shows the expected value of the ratio calculated from Eq. (15) [assuming $\alpha_s(Q^2) = 12\pi/[21 \ln (Q^2/\Lambda^2)]$, i.e. six effective quark flavours, and taking $\Lambda = 0.2$ GeV]. If the Q^2-scales for three-jet and two-jet production are not identical, then Eq. (15) will acquire a correction due to the non-cancellation of the common factor α_s^2 in Eqs. (12) and (13) (and due to the non-cancellation of the effective structure function) as a result of the scaling deviations. By way of example, in Fig. 20 the broken curve represents the expected modification of the prediction using $\langle Q_{3J} \rangle = 0.33 m_{3J}$. On the basis of these comparisons, and assuming the validity of the expression for $\alpha_s^2(Q^2)$ given above, we conclude that the three-jet sample is probably characterized by a Q^2-scale which is lower than that for the corresponding two-jet sample at the same subprocess c.m.s. energy.

This result should not be considered surprising: indeed one might argue *a priori* that, for comparable angular acceptance, the Q^2-scale is naturally lower in the case where the available energy is shared among a larger number of final-state quanta (e.g. $\langle Q_{3J} \rangle \simeq 2/3 \langle Q_{2J} \rangle$).

Taking this into account, we can use the measured three-jet/two-jet ratio to obtain information on α_s, at a Q^2-value ($Q^2 \equiv Q^2_{2J} \equiv -\hat{t}$) of ≈ 4000 GeV^2. We find

$$\alpha_s (K_{3J}/K_{2J}) = 0.16 \pm 0.02 \pm 0.03,$$

where the first error is statistical, and the second one is the systematic error on the result. The unknown residual scale-breaking (and/or higher-order) correction appropriate to this three-jet/two-jet comparison is represented by the factor K_{3J}/K_{2J}. Assuming that the Q^2-scales for the three-jet and two-jet samples are matched (or almost matched), we anticipate that K_{3J}/K_{2J} will be close to unity. Clearly this factor cannot be eliminated until the next-to-leading-order corrections to the two-jet and three-jet cross-sections have been calculated theoretically.

3.5 Summary

Spectacular two-jet and three-jet events have been observed at the CERN p$\bar{\text{p}}$ Collider, demonstrating the importance of hard parton–parton scattering in hadronic collisions at high energies. The inclusive jet cross-section is in agreement with QCD predictions. Furthermore, the differential two-jet and three-jet cross-sections are in agreement with the QCD formulae, and provide confirmation of the need for higher-order (scale-breaking) QCD processes.

In the future we can go further, since examples of clean four-jet events have already been observed at the Collider (Fig. 21). These higher-order jet topologies clearly provide a challenge for the experimentalist to analyse and for the theorist to predict.

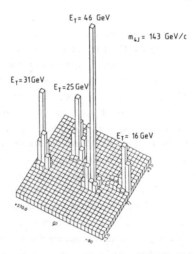

Fig. 21 Example of a four-jet event seen in the UA1 detector.

4. W± AND Z⁰ PHYSICS AT THE CERN pp̄ COLLIDER

The main motivation for transforming the SPS machine into a proton-antiproton collider[18] was the goal of producing and detecting the charged and neutral Intermediate Vector Bosons (IVBs) which mediate the weak interaction. In Section 4.1 we review the expected properties of the IVBs in the framework of the standard model[19]. The experimentally measured properties of the weak bosons, and the details of their production at the Collider, are then discussed in the following sections.

4.1 Expected Properties of the IVBs

The mass of the charged IVB, the W±, can be obtained from the known strength of the weak interaction between charged currents by using the relation

$$G_F/\sqrt{2} = g^2/8m_W^2, \tag{16}$$

where G_F is the Fermi coupling constant ($G_F \approx 1.02 \times 10^{-5}/m_p^2$, where m_p is the proton mass) and $g = e/\sin\theta_w$. Radiative corrections[20] modify this expectation for the W mass. If we take into account $O(\alpha)$ radiative corrections, we obtain

$$m_W = 38.65/\sin\theta_w, \tag{17}$$

where m_W is measured in GeV/c². Values of $\sin^2\theta_w$, obtained from the analysis of neutral-current neutrino interactions, fall typically between 0.21 and 0.24, corresponding to an expected range of m_W between 79 and 84 GeV/c².

The main decay modes of the W± can also be deduced from the standard model. The leptonic decay mode W⁻ → e⁻ν̄_e (or W⁺ → e⁺ν_e) has a partial width given by

$$\Gamma(e\nu_e) = G_F m_W^3/6\pi\sqrt{2}. \tag{18}$$

The partial widths of all other decay modes into a fermion-antifermion pair are related to $\Gamma(e\nu_e)$. Neglecting all fermion masses one finds

$$\Gamma(e\nu_e) = \Gamma(\mu\nu_\mu) = \Gamma(\tau\nu_\tau) \tag{19}$$

and

$$\Gamma(q\bar{q}')/\Gamma(e\nu_e) = 3|U_{qq'}|^2, \tag{20}$$

where the factor of 3 in Eq. (20) reflects the three possible colour states of the quarks, and the quantity $U_{qq'}$ takes into account the mixing between quarks which occurs in the presence of the weak interaction[21]. For example, in the decay W⁻ → dū, $|U_{qq'}|$ is just given by $\cos\theta_C$, where θ_C is the Cabibbo angle[22], whereas in the decay W⁻ → sū, $|U_{qq'}| = \sin\theta_C$. For a quark q belonging to a given fermion generation, the sum $\Sigma_{q'}|U_{qq'}|^2$ extended over all possible generations is equal to 1.

As a consequence of Eqs. (19) and (20), and neglecting all fermion masses, the branching ratio for the leptonic decay mode $W^\pm \to e^\pm \nu_e(\bar\nu_e)$ is given by

$$B_{e\nu} = \Gamma(e\nu_e)/\Gamma_W = \Gamma(\mu\nu_\mu)/\Gamma_W = \Gamma(\tau\nu_\tau)/\Gamma_W = 1/4N_G , \qquad (21)$$

where Γ_W is the total width of the $W^\pm$, and N_G is the total number of fermion generations. At present we know of three generations. Setting $N_G = 3$ in Eq. (21), we find $B_{e\nu} \approx 0.08$. Rewriting Eq. (21) as

$$\Gamma_W = 4N_G \Gamma(e\nu_e) \qquad (22)$$

gives $\Gamma_W \approx 3$ GeV/c^2, using the numerical value $\Gamma(e\nu_e) \approx 260$ MeV/c^2 obtained from Eq. (18).

We now consider the neutral IVB, the Z^0. In the standard model with only one physical Higgs particle, the Z^0 mass is related to the $W^\pm$ mass by the formula

$$m_Z^2 = m_W^2/\cos^2\theta_w , \qquad (23)$$

which predicts the value of m_Z to fall within the range 90-95 GeV/c^2.

The Z^0 decays mainly into a fermion-antifermion pair $f\bar f$ (f = ν, e^-, μ^-, τ^-, or any quark q). The coupling of the Z^0 to such a pair may be parametrized by an effective Lagrangian given by [23]

$$\mathcal{L}_{\text{eff}} = -(m_Z/\sqrt 2)(G_F/\sqrt 2)\, \bar f \gamma_\mu (V_f - A_f \gamma_5) f Z^\mu , \qquad (24)$$

where

$$V_f = A_f = 1 \text{ for neutrinos}; \qquad (25a)$$

$$A_f = \pm 1 \text{ and}$$

$$V_f = \pm(1 - 4|Q_f|\sin^2\theta_w) \text{ for fermions f with charge } Q_f \lessgtr 0 . \qquad (25b)$$

We note that for charged leptons ($|Q_f| = 1$) the vector coupling V_f is very small because $\sin^2\theta_w$ has a value close to 0.25.

The partial decay widths of the Z^0 into a lepton-antilepton pair $\ell\bar\ell$ are given by

$$\Gamma(\ell\bar\ell) = (G_F m_Z^3/24\pi\sqrt 2)(V_\ell^2 + A_\ell^2) , \qquad (26)$$

and the corresponding width for the Z^0 decay into a $q\bar q$ pair by

$$\Gamma(q\bar q) = 3(G_F m_Z^3/24\pi\sqrt 2)(V_q^2 + A_q^2) , \qquad (27)$$

where, as usual, the factor of 3 results from the three possible colour states of the quarks. Quark-antiquark pairs are expected to belong to the same generation because flavour changing is suppressed in the weak interaction between neutral currents[24].

Quantitatively one finds

$$\Gamma(\nu\bar{\nu}) : \Gamma(e^+e^-) : \Gamma(q\bar{q}) = 2 : [1 + (1 - 4\sin^2\theta_w)^2] : 3[1 + (1 - b\sin^2\theta_w)^2] , \quad (28)$$

where $b = 8/3$ (4/3) for $q = u$ (d).

Setting $\sin^2\theta_w = 0.215$ one obtains

$$\Gamma(\nu\bar{\nu}) : \Gamma(e^+e^-) : \Gamma(u\bar{u}) : \Gamma(d\bar{d}) = 2 : 1.02 : 3.55 : 4.53 , \quad (29)$$

from which we can calculate the branching ratio for the decay mode $Z^0 \to e^+e^-$. Assuming three fermion generations ($N_G = 3$):

$$B_{ee} = \Gamma(e^+e^-)/\Gamma_Z = \Gamma(\mu^+\mu^-)/\Gamma_Z = \Gamma(\tau^+\tau^-)/\Gamma_Z \approx 1.02/11.1\, N_G \approx 3\%, \quad (30)$$

where Γ_Z is the total width of the Z^0. Using the numerical value $\Gamma(e^+e^-) \approx 90$ MeV, which can be obtained from Eq. (26), the value of Γ_Z turns out to be of the order of 3 GeV/c². The measurement of Γ_Z can be considered as a method for estimating the total number of neutrino types with $m_\nu \ll m_Z/2$, because each decay channel $Z^0 \to \nu\bar{\nu}$ contributes ~ 180 MeV/c² to Γ_Z. Before the Collider experiments, the upper limit on the number of light neutrinos from particle physics experiments[25] was $N_\nu \lesssim 10^5$. As we shall see later, this limit has been dramatically improved by the observation of the Z^0 decaying into e^+e^- at the CERN $p\bar{p}$ Collider.

The production of $W^\pm$'s and Z^0's at the CERN $p\bar{p}$ Collider is expected to occur as the result of $q\bar{q}$ annihilation [the Drell-Yan mechanism[26]]. The basic subprocesses are

$$\begin{aligned} u + \bar{d} &\to W^+ , \\ d + \bar{u} &\to W^- , \\ u + \bar{u} &\to Z^0 , \\ d + \bar{d} &\to Z^0 . \end{aligned} \quad (31)$$

Consider the subprocess $u + \bar{d} \to W^+$. The cross-section is given by

$$\hat{\sigma}(x_1,x_2) = \sqrt{2}\pi G_F m_W \cos^2\theta_C \delta(x_1 x_2 s - m_W^2) , \quad (32)$$

where $x_1\sqrt{s}/2$ ($x_2\sqrt{s}/2$) is the longitudinal momentum of the incoming u quark ($\bar{d}$ antiquark) in the proton (antiproton), and we have neglected the W^+ width. To obtain the inclusive cross-section for W^+ production we must perform a convolution integral, which takes into account the x distributions (structure functions) of the u and $\bar{d}$ quarks in the incident hadrons, and sum over all relevant subprocesses. We obtain

$$\sigma(p\bar{p} \to W^+ + ...) = \int_0^1 dx \int_0^1 dx_2 \hat{\sigma}(x_1,x_2) W^+(x_1,x_2) \quad (33)$$

$$= \sqrt{2}\pi G_F \tau \cos^2\theta_C \int_\tau^1 (dx/x) W^+[x_1,(\tau/x)],$$

where $\tau = m_W^2/s$ and

$$W^+(x_1,x_2) = (1/3)[U(x_1)D(x_2) + D(x_1)U(x_2)], \quad (34)$$

$U(x)$ and $D(x)$ being the x distributions of the u and d quarks in the incident protons.

In complete anology, the cross-section for inclusive Z^0 production in $p\bar{p}$ collisions is given by

$$\sigma(p\bar{p} \to Z^0 + ...) = 2G_F\pi\sqrt{2}\tau \int_\tau^1 (dx/x)Z[x,(\tau/x)], \quad (35)$$

where

$$Z(x_1,x_2) = (2/3)\,U(x_1)U(x_2)\,[(1/4) - (2/3)\sin^2\theta_w + (8/9)\sin^4\theta_w]$$

$$+ (2/3)\,D(x_1)D(x_2)\,[(1/4) - (1/3)\sin^2\theta_w + (2/9)\sin^4\theta_w]. \quad (36)$$

The resulting numerical estimates for the IVB production cross-sections at the $p\bar{p}$ Collider ($\sqrt{s} = 540$ GeV) are

$$\sigma(p\bar{p} \to W^+ + ...) = \sigma(p\bar{p} \to W^- + ...) \approx 1.8 \times 10^{-33} \text{ cm}^2$$

$$\sigma(p\bar{p} \to Z^0 + ...) \approx 10^{-33} \text{ cm}^2.$$

Non-leading QCD terms have the effect of multiplying these cross-section values by a factor k. Such an effect has been observed in the production of lepton pairs by hadron collisions at lower energies, where the value $k \approx 2$ has been measured [27]. For inclusive $W^\pm$ and Z^0 production at the $p\bar{p}$ Collider the value $k \approx 1.5$ has been predicted [28]. At the present time the Collider experiments UA1 and UA2 have recorded a few hundred nb^{-1} of data (a 1 nb cross-section process would result in a few hundred events recorded on tape). Thus, in the framework of the standard model, we would expect a few hundred $W \to e\nu$ decays and a few tens of $Z^0 \to e^+e^-$ decays to have been observed at the Collider to date. In the following subsections we will see that this is indeed the case.

4.2 $W \to e\nu$ Decays

The first IVB decays to be observed at the Collider [7] were $W^\pm \to e^\pm \nu_e$ decays. Experimentally, in searching for this decay, we look for an isolated energetic electron in coincidence with an energetic neutrino (missing transverse energy). This signature provides us with an essentially background-free sample of W decays in the UA1 detector, provided we

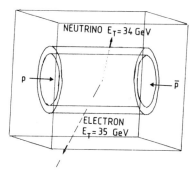

Fig. 22 Graphic display of a $W^+ \to e^+ \nu_e$ decay seen in the UA1 detector.

require that both the electron and the neutrino have transverse energies in excess of 15 GeV. Currently the UA1 Collaboration have a sample of 172 W decays satisfying these requirements. More details of the selection requirements used in extracting this sample of 172 W decays are given in subsection 4.4. An example of an event is shown in Fig. 22. An energetic isolated electron with a transverse energy of 35 GeV is seen recoiling against a neutrino (missing transverse energy) with roughly equal transverse energy ($E_T^\nu = 34$ GeV). This is what we would expect for the two-body decay of a massive object into an electron and a neutrino. In Fig. 23 the correlation between the electron and neutrino transverse energies is shown for the sample of 172 W events. We can see, once again, that the electron and neutrino tend to have equal (and in fact opposite) transverse momenta. The transverse-energy spectra for both leptons have end-points around 40 GeV. We might already guess, therefore, that we are seeing

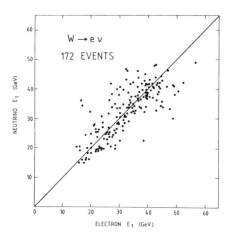

Fig. 23 Correlation between the electron and neutrino transverse energies for the sample of 172 $W^\pm \to e\nu_e$ decays seen in the UA1 experiment.

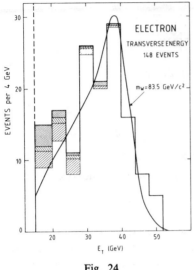
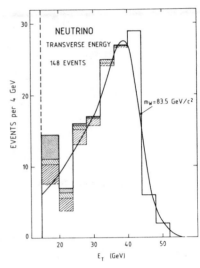

Fig. 24   Fig. 25

The electron (Fig. 24) and neutrino (Fig. 25) transverse-energy distributions for the UA1 sample of well-measured $W \to e\nu_e$ decays. The shaded parts show the expected background contributions from jet–jet fluctuations (cross-hatched) and $W \to \tau\nu_\tau$ decays with $\tau \to$ hadrons (top left to bottom right hatching) and $\tau \to e\nu_e\nu_\tau$ (top right to bottom left hatching). The curves show the predictions for the background-subtracted distributions (normalized to the data), corresponding to a W with a mass of 83.5 GeV/c².

the decay of an object with a mass around 80 GeV/c². To make the best determination of the W mass, only those events are used in which both the electron and the neutrino are well measured. The electron and neutrino transverse momentum spectra for the resulting sample of 148 $W \to e\nu$ decays are shown in Figs. 24 and 25, respectively. The expected contribution from background processes is also shown in these figures (shaded parts). The background contribution is small, and it is essentially zero if we require both the electron and the neutrino to have transverse energies in excess of 30 GeV. The distribution that we fit to determine the W mass is the transverse-mass distribution, defined as

$$m_T(e\nu) \equiv [2E_T^e E_T^\nu (1 - \cos \phi_{e\nu})]^{1/2}, \qquad (37)$$

where E_T^e and E_T^ν are the electron and neutrino transverse energies, and $\phi_{e\nu}$ is the angle between electron and neutrino in the plane transverse to the beam direction. This variable is chosen because we can fit it in a model-independent way (it is not necessary to make a strong assumption about the underlying W transverse momentum distribution, for example). In fact in order to be completely background-free we use the 'enhanced' transverse mass distribution

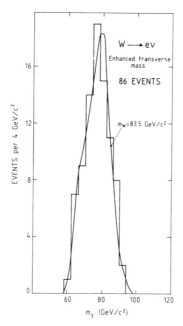

Fig. 26 The enhanced transverse-mass distribution (see text) for the ($e\nu_e$) system in the UA1 sample of 89 well-measured W → $e\nu_e$ decays where both the electron and the neutrino have transverse energies in excess of 30 GeV. The curve shows the result of fitting for the W mass.

(Fig. 26), which is defined as the electron–neutrino transverse mass for those events in which both leptons have a transverse energy in excess of 30 GeV. The resulting measurement for the W mass is

$$m_W = 83.5 ^{+1.1}_{-1.0} \pm 2.8 \text{ GeV}/c^2 ,$$

where the second error is the systematic error on the energy scale of the experiment. We can also put an upper limit on the W width from the fits. We find

$$\Gamma_W \leqslant 6.5 \text{ GeV}/c^2 \quad (90\% \text{ c.l.}) .$$

These results are in excellent agreement with the expectations of the standard model. Furthermore, the standard model with the above values for the W mass and width gives a good description of the lepton transverse-energy spectra (Figs. 24 and 25) and the enhanced transverse-mass distribution (Fig. 26). However, before concluding that the observed W boson is really the charged IVB of the standard model, we note that the standard model W should couple only to left-handed quarks and leptons. Since a W is produced at the Collider from the fusion of a left-handed quark from the proton with a right-handed antiquark from the

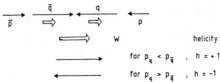
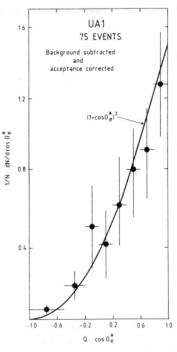

Fig. 27 a) $W^\pm$ production: polarization of the $W^\pm$ due to the (V − A) coupling of the W to the quarks. b) $W^\pm$ leptonic decay: preferred direction of leptons. The μ^+ from W^+ preferentially follows the $\bar{p}$ direction.

Fig. 28 The angular distribution of the electron emission angle θ^* in the rest frame of the W after correction for experimental acceptance.

antiproton (see Fig. 27), W's are produced fully polarized. Furthermore, the weak decay of the W acts as a polarization analyser, resulting in a strong forward-backward asymmetry in which electrons (positrons) from the W^- (W^+) decay prefer to be emitted in the direction of the proton (antiproton) beam. To study this effect, we look at the decay angular distribution of the charged lepton from W decays, specifically at the distribution of the emission angle θ^* of the electron (positron) with respect to the proton (antiproton) direction in the W rest frame. In (V − A) theory this distribution should be of the form $(1 + \cos\theta^*)^2$, which is in excellent agreement with the observed distribution (Fig. 28). Furthermore, it has been shown by Jacob[29] that for a particle of arbitrary spin J, one expects

$$\langle \cos\theta^* \rangle = \langle\lambda\rangle\langle\mu\rangle/J(J + 1), \qquad (38)$$

where $\langle\mu\rangle$ and $\langle\lambda\rangle$ are, respectively, the global helicity of the production system (u$\bar{\text{d}}$) and the decay system (e$\bar{\nu}$). For (V − A) one then has $\langle\lambda\rangle = \langle\mu\rangle = -1$, and $J = 1$, leading to the maximal value $\langle \cos\theta^* \rangle = 0.5$. For $J = 0$ one obviously expects $\langle \cos\theta^* \rangle = 0$, and for any other spin value $J \geqslant 2$, $\langle \cos \theta^* \rangle \leqslant 1/6$. Experimentally we find $\langle \cos \theta^* \rangle = 0.43 \pm 0.07$, which supports both the $J = 1$ assignment and maximal helicity states at production and decay. Note that the choice of sign $\langle \mu \rangle = \langle \lambda \rangle = \pm 1$ cannot be separated, i.e. right- and left-handed currents, at both production and decay, cannot be resolved without a polarization measurement.

Finally, we note that the UA1 experiment has also observed other leptonic decays of the W: $W^\pm \to \tau^\pm \nu_\tau$ and $W^\pm \to \mu^\pm \nu_\mu$ (Ref. 7).

4.3 $Z^0 \to e^+ e^-$ Decays and the Standard Model Parameters

The signature for the decay $Z^0 \to e^+ e^-$ is the presence of two isolated energetic electrons. Once again, it is easy to obtain an essentially background-free sample at the Collider. To select $Z^0 \to e^+ e^-$ decays we begin by selecting all those events in which there are two isolated electromagnetic (e.m.) clusters with transverse energy in excess of 8 GeV. We then require that at least one of the two clusters has transverse energy in excess of 15 GeV and passes all the isolated electron quality cuts used to obtain the $W \to e\nu$ event sample. With the current data sample the UA1 Collaboration have a sample of 39 events passing these cuts. The two-cluster mass distribution for these events is shown in Fig. 29. There are 21 events with $m(e^+e^-) < 50 \text{ GeV}/c^2$, no events in the interval $50 < m(e^+e^-) < 80 \text{ GeV}/c^2$ and a cluster of 18 events with $m(e^+e^-) > 80 \text{ GeV}/c^2$, centred on a mass of $\sim 95 \text{ GeV}/c^2$. The shape of the low-mass part of the distribution is consistent with our expectations for background from two-jet fluctuations, although it is not excluded that, in some of these events one or both electrons could be genuine. We identify the high-mass events with the decay $Z^0 \to e^+ e^-$. An example of one of these Z^0 events is shown in Fig. 30. The two electrons have approximately equal transverse energies and are more or less back-to-back in the transverse plane, suggesting the two-body decay of a massive object. We obtain the best value for the Z^0 mass by

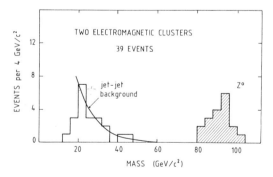

Fig. 29 The invariant-mass distribution for two isolated e.m. clusters seen in the UA1 experiment. The curve shows the expected shape of the contribution from events arising from jet–jet fluctuations. The high-mass Z^0 peak is clearly separated from the background.

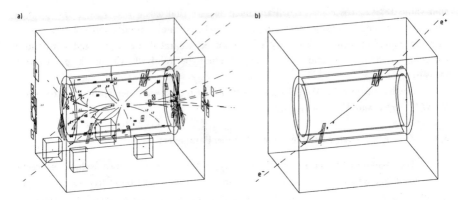

Fig. 30 Graphic display of a $Z^0 \to e^+e^-$ decay seen in the UA1 detector. a) All reconstructed vertex-associated tracks and all calorimeter hits are displayed. b) The same, but thresholds are raised to $p_T > 2$ GeV/c for charged tracks and $E_T > 2$ GeV for calorimeter hits.

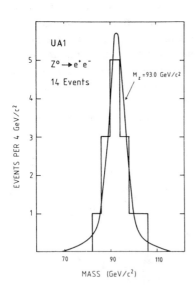

Fig. 31 Invariant-mass distribution for the 14 well-measured $Z^0 \to e^+e^-$ decays. The curve shows the prediction corresponding to a Z^0 mass of 93 GeV/c².

restricting ourselves to those $Z^0 \to e^+e^-$ decays for which both electrons have well-measured energies. The mass distribution for the resulting sample of 14 events is shown in Fig. 31. A fit to this distribution yields

$$m_Z = 93.0 \pm 1.4 \pm (3.2) \text{ GeV/c}^2,$$

$$\Gamma_Z \leqslant 8.0 \text{ GeV/c}^2 \quad (90\% \text{ c.l.}).$$

Once again we appear to be in excellent agreement with the expectations of the standard model. However, in order to make a more complete comparison with the standard model it is useful to define the standard model parameters:

$$\sin^2 \theta_w \equiv (38.65/m_W)^2 \,, \tag{39}$$

$$\varrho \equiv (m_W/m_Z \cos\theta_w)^2 \,. \tag{40}$$

The analysis of neutral-current neutrino interactions implies that $\sin^2 \theta_w$ is in the range 0.21 to 0.24, and from Eq. (23) we anticipate that $\varrho = 1$. Experimentally, the UA1 measurements of m_W and m_Z imply that

$$\sin^2 \theta_w = 0.214 {}^{+\,0.005}_{-\,0.006} \pm 0.15 \,,$$

$$\varrho = 1.025 \pm 0.041 \pm 0.021 \,,$$

in excellent agreement with the standard model. We conclude that the W and Z^0 bosons observed at the CERN SPS $p\bar{p}$ Collider are indeed the weak IVBs of the SU(2) × U(1) standard model.

4.4 The W and Z^0 Production Cross-Sections, an Example of the Details Involved in an Experimental Analysis, and the Number of Light Neutrinos in the Universe

We have seen that the masses and decay properties of the W and Z^0 are entirely consistent with the predictions of the standard model. Furthermore, the number of W and Z^0 events that have been observed are consistent with the numbers we had anticipated in subsection 4.1: a few hundred W → $e\nu$ decays and a few tens of Z^0 → e^+e^- decays. We will now look at the IVB production cross-sections more carefully. To illustrate the method employed in deducing the production cross-sections from the data, we must go into some of the experimental details [a more complete description of these details can be found elsewhere[30]]. Here we concentrate on the W production cross-section and begin by describing the selection of W events in the UA1 experiment. We then assess the number of background events we expect in our sample and the efficiency with which we have selected W → $e\nu$ decays. Knowing the integrated luminosity of the experiment, we can then calculate the production cross-section.

The trigger used throughout the data taking required the presence of an e.m. cluster (one or two adjacent e.m. calorimeter cells) with a transverse energy in excess of 10 GeV, and at an angle of more than 5° with respect to the beam direction. After complete off-line reconstruction, about 5×10^5 events have at least one e.m. cluster with transverse energy E_T > 15 GeV. To extract an essentially background-free sample of $W^\pm$ decays, we apply a number of cuts to the data. On the basis of Monte Carlo studies together with a visual inspection of the events on a high-resolution graphics display, these selection cuts have been designed to minimize the background in the final event sample whilst maintaining a resonable efficiency for the selection of W → $e\nu$ decays. The cuts that have been applied to the data are as follows:

i) *Cluster validation*. To ensure that we have a reliable reconstruction for the e.m. cluster, we require that, after removing the central detector (CD) track with the highest transverse momentum (the electron candidate), the sum of the transverse momenta of all other tracks entering the two calorimeter cells associated with the cluster be less than 3 GeV/c. We further require that the centroids of the energy depositions in the four longitudinal samplings of the e.m. calorimeter be consistent with a single e.m. cluster.

ii) *Associated track requirement*. If the e.m. cluster is at an angle of more than 15° with respect to the horizontal, where the CD has good efficiency, we require the presence in the CD of a charged track associated with the cluster, and having transverse momentum consistent with being in excess of 15 GeV/c.

iii) *Loose isolation*. To exclude two-jet events in which one jet fakes the electron signature, we demand that the electron be isolated in a cone in (pseudorapidity, azimuthal angle)-space with radius $\Delta R = 0.4$, where $\Delta R \equiv (\Delta\eta^2 + \Delta\phi^2)^{1/2}$ and ϕ is in radians. The isolation conditions within this cone are: a) that after removing the track with the highest transverse momentum, the sum of the transverse momenta of all other tracks in the cone is less than 10% of the e.m. cluster transverse energy; b) that the cluster transverse energy is at least 90% of the total transverse energy (e.m. calorimeter plus hadronic calorimeter) inside the cone.

iv) *Tight isolation*. To obtain further rejection against two-jet events, we require additional isolation inside a larger cone with radius $\Delta R = 0.7$. The isolation conditions within this cone are: a) that after removing the track with the highest transverse momentum, the sum of the transverse momenta of all other tracks in the cone is less than 3 GeV/c; b) that the sum of the transverse energy depositions in the calorimeter, after excluding the e.m. cluster, is less than 3.2 GeV.

v) *Electromagnetic shape*. To reject hadronic showers faking an electron signature, we require that the energy E_{had} deposited in the hadron calorimeter cell immediately behind the e.m. cluster be either less than 2% of the e.m. cluster energy or less than 0.6 GeV (1.5 GeV) for clusters with pseudorapidity $|\eta| < 1.5$ ($|\eta| > 1.5$). We further require that the quality parameter χ_R^2 be less than 60, where χ_R^2 measures the goodness of fit to the electron hypothesis; χ_R^2 has been developed from a study of electron test-beam data, making full use of the longitudinal profile of the shower measured in the four samplings of the electromagnetic calorimeter, and taking into account the electron energy and angle of incidence.

vi) *Neutrino emission.* Finally, we require that the missing transverse energy in the event be in excess of 15 GeV after validation during which obvious instrumental problems are removed. If the electron cluster is within 15° of the vertical direction there must be no reconstructed jet in the event (with transverse energy in excess of 7.5 GeV in the calorimeters, or transverse momentum in excess of 5 GeV in the central detector) which is also going in the vertical direction (within 15°).

After these criteria have been imposed on the data, we are left with a sample of 172 events for further analysis, of which 59 events come from collisions at $\sqrt{s}$ = 546 GeV and 113 events from collisions at $\sqrt{s}$ = 630 GeV.

We must now estimate the possible sources of background in our event sample:

i) *Hadronic interactions.* Hadronic jets can fake an isolated electron signature if a) they fragment in such a way that one energetic charged pion overlaps with one or more neutral pions, or b) they contain a genuine energetic electron arising from heavy-flavour decays which overlaps with one or more neutrals. In both these cases we would expect a two-jet topology with little or no missing transverse energy in the event. To estimate the background from these sources, we drop the last requirement imposed in the selection, namely the presence of a missing transverse energy $\Delta E_m > 15$ GeV, and examine the missing transverse energy for the enlarged sample. For those events with $\Delta E_m < 15$ GeV, the ΔE_m distribution is well described by the experimental resolution in the measurement of this quantity (Fig. 32), and furthermore almost all these events have a two-jet

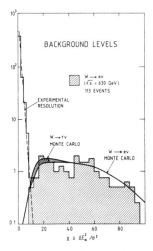

Fig. 32 Background from two-jet fluctuations. The missing transverse-energy squared ΔE_m^2 divided by the experimental resolution for events which come from the W selection procedure after removing the requirement that $\Delta E_m > 15$ GeV. The contribution from W events (hatched sub-histogram) agrees with the Monte Carlo prediction. The contribution from events with $\Delta E_m < 15$ GeV (non-hatched) is well described by the experimental resolution (dashed) curve.

topology. Extrapolating the resolution curve in the region $\Delta E_m > 15$ GeV, we estimate a background in the $W^\pm \to e^\pm \nu_e$ sample of 3.4 ± 1.8 events at $\sqrt{s} = 546$ GeV and 1.9 ± 0.6 events at $\sqrt{s} = 630$ GeV.

ii) $W \to \tau \nu_\tau$. This background is evaluated using the ISAJET [31] Monte Carlo together with a full simulation of the UA1 detector. We estimate a background of 11.8 ± 0.6 events from this source, of which 2.7 events are associated with the decay $\tau \to \pi^\pm(\pi^0)\nu_\tau$ and 9.1 events are associated with the decay $\tau \to e\nu_e\nu_\tau$.

Finally, before we can evaluate the production cross-section, we must calculate the efficiency of our selection procedure for obtaining the W sample. Two techniques have been used to estimate the efficiency of the selection criteria: i) the ISAJET Monte Carlo together with a full simulation of the UA1 detector, and ii) a randomized W-decay technique in which the 172 real W events are used, but the electron is replaced by a generated electron coming from a random (V–A) decay of the produced W. The two methods give consistent results, which are shown in Table 3. The efficiency of the trigger and selection requirements is 0.69 ± 0.03 for the $\sqrt{s} = 546$ GeV data sample and 0.62 ± 0.03 for the $\sqrt{s} = 630$ GeV sample.

We now know both the background in our sample of 172 W events and the efficiency with which we have selected W decays. We can therefore evaluate the production cross section. The integrated luminosity of the experiment is 136 nb^{-1} at $\sqrt{s} = 546$ GeV and 263 nb^{-1} at $\sqrt{s} = 630$ GeV, known to about ±15% uncertainty. The resulting cross-section is

$$(\sigma \cdot B)_W = 0.55 \pm 0.08\,(\pm 0.09)\text{ nb}$$

at $\sqrt{s} = 546$ GeV, where the last error takes into account the systematic errors and B is the branching ratio for the decay $W \to e\nu$. This result is not far from the prediction of Altarelli et al.[32], $0.38\,^{+0.12}_{-0.05}$ nb, where we take the branching ratio B = 0.089.

The corresponding experimental result for the 1984 data at $\sqrt{s} = 630$ GeV is

$$(\sigma \cdot B)_W = 0.63 \pm 0.05\,(\pm 0.09)\text{ nb}.$$

This is in agreement with the theoretical expectation of $0.47\,^{+0.14}_{-0.08}$ nb. We note that the 15% systematic uncertainty on these results disappears in the ratio of the two cross-sections. We obtain the result

$$\sigma_W(\sqrt{s} = 630\text{ GeV})/\sigma_W(\sqrt{s} = 546\text{ GeV}) = 1.15 \pm 0.17,$$

in agreement with the theoretical expectation of 1.24.

Thus, after following the analysis details, we find that the number of observed $W \to e\nu$ decays is completely consistent with the expected rate predicted by the SU(2) × U(1) standard model plus quantum chromodynamics (QCD). An analysis precedure similar to the one used to extract the W cross-section is also used for the Z^0 cross-section. The results are

$$(\sigma \cdot B)_Z = 40 \pm 20 \, (\pm 6) \, \text{pb} \quad (\sqrt{s} = 546 \, \text{GeV})$$

and

$$(\sigma \cdot B)_Z = 79 \pm 21 \, (\pm 12) \, \text{pb} \quad (\sqrt{s} = 630 \, \text{GeV}).$$

These are in good agreement with the theoretical expectations[32)] of 41^{+13}_{-7} pb at $\sqrt{s}$ = 546 GeV, and 51^{+16}_{-8} pb at $\sqrt{s}$ = 630 GeV. Once again we have agreement between experiment and theory.

Table 3

Selection efficiency. The efficiency for each cut in the selection leading to the sample of 172 $W^\pm \to e^\pm \nu_e$ events is shown, evaluated by the two methods described in the text. The efficiency given for each requirement is for those events passing all preceding requirements.

Cut	Efficiency			
	$\sqrt{s}$ = 546 GeV		$\sqrt{s}$ = 630 GeV	
	Events with random decay	ISAJET	Events with random decay	ISAJET
Electron E_T > 15 GeV	0.86	0.87	0.85	0.84
i) Cluster validation	0.97	0.96	0.96	0.95
ii) Associated track requirement	0.97	0.98	0.95	0.99
iii) Loose isolation (in cone ΔR = 0.4)	0.98	0.99	0.98	0.98
iv) Tight isolation (in cone ΔR = 0.7)	0.98	0.98	0.93	0.98
v) Electromagnetic shape: a) hadronic energy b) χ^2_R	0.95 0.99	0.95 0.98	0.95 0.99	0.95 0.99
vi) Neutrino emission	0.98	0.97	0.97	0.96
Trigger efficiency	0.96	0.96	0.96	0.96
Overall efficiency	0.69 ± 0.03	0.68 ± 0.03	0.62 ± 0.03	0.66 ± 0.03

To compare the W and Z^0 production cross-sections it is convenient to define the ratio

$$R \equiv (\sigma \cdot B)_W/(\sigma \cdot B)_Z . \tag{41}$$

At $\sqrt{s} = 546$ GeV we obtain

$$R = 13.9 \, ^{+\,10.9}_{-\,5.7} ,$$

and at $\sqrt{s} = 630$ GeV

$$R = 8.0 \, ^{+\,2.9}_{-\,2.1} .$$

Averaging over the two energies

$$R = 9.3 \, ^{+\,2.6}_{-\,1.9} ,$$

in excellent agreement with the standard model plus QCD expectation[33] of 9.2 ± 0.6.

The reason why R is an interesting quantity to measure is because it is sensitive to the number of light neutrinos in the Universe. In a modified standard model in which there are further (> 3) generations, the expected value for R would be different. If there were no associated new charged leptons or quarks lighter than the W, we would expect R to increase with the number of generations because of additional $Z^0 \rightarrow \nu\bar{\nu}$ decays. The relationship between R and the number of light neutrinos in the Universe N_ν has been calculated by Deshpande et al.[33]. Numerically, they arrive at the expression

$$N_\nu = (1.73 \pm 0.10)R - 12.55 , \tag{42}$$

where the error on the coefficient of R comes from a lack of precise knowledge of the proton structure functions. Using Eq. (42) together with the experimental result for R, we obtain the interesting result that $N_\nu \leqslant 9$ at 90% c.l.

4.5 W Production Properties

Having established the cleanliness of the W data sample, we can use the W events to study the way in which W's are produced. The production of W's at the $p\bar{p}$ Collider proceeds by the Drell–Yan mechanism[26] in which a quark from the proton annihilates with an antiquark from the antiproton. In a QCD-improved picture of the production mechanism the annihilating quark and antiquark are coloured, and there are higher-order corrections to the bare Drell–Yan process in which one or more gluons are radiated from the incoming partons. This initial-state gluon bremsstrahlung is expected to give rise to i) a long tail in the transverse-momentum distribution of the weak bosons, and ii) the occasional observation of one or more hadronic jets produced in association with the higher transverse-momentum

IVBs. In the region in which QCD perturbation theory is applicable (for sufficiently large transverse momenta), the W transverse momentum distribution, together with the rate of occurrence and properties of hadronic jets produced in association with the W's, provides us with a quantitative test of the theory.

The W transverse momentum $p_T^{(W)}$ is obtained by adding the electron and neutrino transverse-momentum vectors. The resulting distribution for $p_T^{(W)}$ is shown in Fig. 33. The distribution peaks at a value $p_T^{(W)} \sim 4$ GeV/c (primarily reflecting the experimental resolution of the measurement of the missing transverse energy in the event), and has a long tail extending to $p_T^{(W)} \sim 40$ GeV/c. The shape of the measured W transverse-momentum distribution is in reasonable agreement with the expectations of the QCD-improved Drell-Yan model. As an example of the QCD predictions, Fig. 33 shows the curve from Altarelli et al.[32], using the structure functions of Glück, Hoffmann and Reya[34] with $\Lambda = 0.4$ GeV. This gives an excellent description of the data.

The QCD-improved Drell-Yan mechanism predicts that the highest transverse-momentum W's are produced in association with one or more gluons radiated off the incoming annihilating quarks. If these gluons are sufficiently energetic and are emitted at sufficiently large angle to the beam direction, they will be observed as hadronic jets balancing the transverse momentum of the W and recoiling against the W in the plane transverse to the beam direction. The UA1 jet algorithm has been used to search for jet activity produced in association with the weak bosons. Hadronic jets are indeed found in those events containing the highest transverse momentum W's (Fig. 33). In 38% of the W sample, one or more

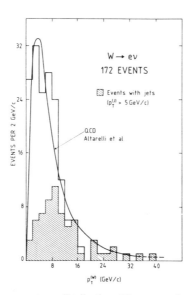

Fig. 33 The W transverse-momentum distribution. The curve shows the QCD prediction of Ref. 32. The hatched sub-histogram shows the contribution from events in which the UA1 jet algorithm reconstructs one or more hadronic jets with transverse momentum > 5 GeV/c.

Table 4

Rate of occurrence of jets in W events

$\sqrt{s}$ (GeV)	Number of events						Total number of jets
	Total	≥ 1 jet	1 jet	2 jet	3 jet	4 jet	
546	59	17	15	1	1	0	20
630	113	48	35	7	4	2	69
TOTAL	172	65	50	8	5	2	89

hadronic jets with transverse-momentum $p_T^{(J)}$ in excess of 5 GeV/c are observed. The number of observed events (Table 4) decreases with increasing jet multiplicity by a factor of roughly 0.3 per additional jet. Defining[35] $R^{(W)}$,

$$R^{(W)} \equiv \sigma[W + \text{jet(s)}]/\sigma(W), \tag{43}$$

where $\sigma(W)$ is the cross-section for producing a W with no observed jets, we find $R_W = 0.61 \pm 0.10$. A similar preliminary analysis for Z^0 events gives $R_Z = 0.7 \pm 0.3$, and the ratio of these ratios $R \equiv R_Z/R_W = 1.1 \pm 0.5$. Clearly the rate of jet activity observed in Z^0 events is consistent with the rate observed in W events.

In order to make a quantitative comparison with the predictions of perturbative QCD, we restrict ourselves to the region in which the reconstruction efficiency for jets within the acceptance of the detector is 100%, namely $p_T^{(J)} > 20$ GeV/c. At $\sqrt{s} = 546$ GeV we have one event with one jet passing this cut. After correcting the jet reconstruction efficiency for the geometrical acceptance of the detector (0.9) the resulting cross-section is $0.012 ^{+0.014}_{-0.007}$ (± 0.002) nb, in agreement with the QCD prediction of 0.010 ± 0.001 nb. This prediction is essentially that of Ref. 35 recalculated to take into account the cuts used in defining the present UA1 $W^\pm \to e^\pm \nu_e$ sample. At $\sqrt{s} = 630$ GeV we have six events with one jet with $p_T^{(J)} > 20$ GeV/c, corresponding to a cross-section of 0.041 ± 0.015 (± 0.006) nb—once again in agreement with the QCD prediction of 0.016 ± 0.002 nb. The rate of jet activity produced in association with W's is therefore in agreement with the expectations for initial-state gluon bremsstrahlung jets. Furthermore, the kinematics of the observed jets also suggests an initial-state bremsstrahlung origin.

The jet transverse-momentum distribution is shown in Fig. 34. This distribution essentially reflects the $p_T^{(W)}$ distribution and is well described by the expectation[35] from QCD perturbation theory. Furthermore, as expected, the angular distribution of these jets is

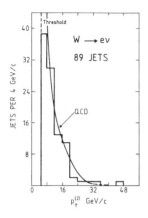

Fig. 34 Jet transverse-momentum distributions for all jets produced in association with the W. The curve shows the QCD prediction of Ref. 35 normalized to the tail of the distribution (the region in which we expect to have good reconstruction efficiency for the jets).

strongly peaked in the beam directions. This can be seen in Fig. 35 where the distribution of $\cos\theta^*$ is shown, θ^* being the angle between the jet and the average beam direction in the rest frame of the W and the jet(s). In the region in which the experimental acceptance is reasonably constant ($|\cos\theta^*| < 0.95$) the shape of the angular distribution is well described by the QCD expectation for bremsstrahlung jets, which is basically $(1-|\cos\theta^*|)^{-1}$.

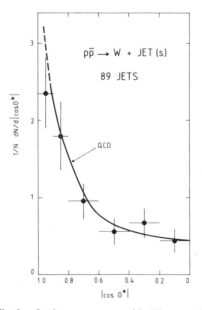

Fig. 35 The angular distribution for jets reconstructed in W events. The distribution of $\cos\theta^*$ is shown, θ^* being the angle between the jet and the average beam direction in the rest frame of the W and the jet(s). The curve shows the QCD prediction from Ref. 35.

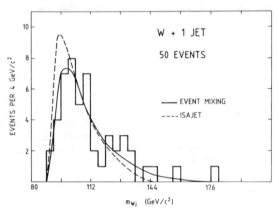

Fig. 36 The (W + jet)-mass distribution for events in which one jet is reconstructed by the UA1 jet algorithm. The curves show the predictions of ISAJET (dashed curve) and event mixing (solid curve).

Finally, we examine the invariant mass of the (W + 1 jet)-system for those events in which one jet has been reconstructed by the UA1 jet algorithm (Fig. 36). The shape of this mass distribution is a little broader than the expectation from ISAJET. It is, however, well described by a simple Monte Carlo in which the observed W four-vectors are randomly associated with the four-vectors from our sample of jets, suggesting that the shape of the (W + 1 jet) mass plot is controlled more by the proton structure functions than by the QCD matrix element. We conclude from this mass distribution that there is no evidence for the production of a massive-state X which subsequently decays into a W plus one jet. For $m_X > 180$ GeV/c^2, with X subsequently decaying into a W and a single hadronic jet, we place an upper limit on the production cross-section for X of

$$\sigma \cdot B_{(X \to W + jet)}/\sigma_W < 0.013 \qquad \text{(at 90\% c.l.)}.$$

We have seen that the transverse momentum of the produced W's, and the associated production of bremsstrahlung jets with the highest transverse-momentum W's, is well described by the QCD-improved Drell-Yan mechanism. The longitudinal momentum distribution of the W bosons is also of some interest, since it is expected to reflect the structure functions of the incoming annihilating partons, predominantly u-quarks and d-quarks. Unfortunately we do not measure the longitudinal momentum of the W directly since we do not measure the longitudinal component of the neutrino momentum. We can overcome this difficulty by imposing the mass of the W on the electron–neutrino system. This will yield two solutions for the longitudinal component of the neutrino momentum, one corresponding to the neutrino being emitted forwards in the W rest frame, the other corresponding to the neutrino being emitted backwards. Hence we have two solutions for x_W, the Feynman x for the W. In practice, in one-third of the events one of these two solutions for x_W is trivially

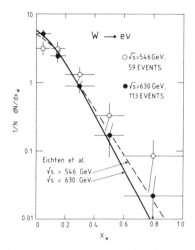
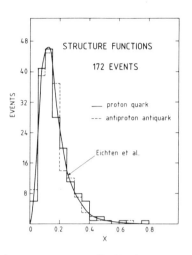

Fig. 37 Feynman x-distribution for W's produced in pp̄ collisions at √s = 546 GeV (open circles) and √s = 630 GeV (full circles). The curves show the predictions using the structure functions of Eichten et al.

Fig. 38 Feynman x-distributions for the proton and antiproton partons making the W. The curve shows the Eichten et al. prediction.

unphysical ($x_W > 1$), and in another one-third the ambiguity is resolved after consideration of energy and momentum conservation in the whole event. In the cases where the ambiguity in x_W is resolved, it tends to be nearly always the lowest of the two x_W solutions which is chosen. In our analysis the lowest of the two solutions is used for all the events. Taking 84 GeV/c² for the mass of the W, the resulting x_W distribution is shown in Fig. 37 for the √s = 546 GeV and √s = 630 GeV data samples separately. There is some indication of the expected softening of the W longitudinal momentum distribution with increasing √s. The data are in reasonable agreement with the expectation resulting from the structure functions of Eichten et al.[36] with $\Lambda = 0.2$ GeV. Using the well-known relations

$$x_p x_{\bar{p}} = m_W^2/s \quad \text{(energy conservation)} \tag{44}$$

and

$$x_W = |x_p - x_{\bar{p}}| \quad \text{(momentum conservation)}, \tag{45}$$

we can determine the parton distributions in the proton and antiproton sampled in W production. The results are shown in Fig. 38. The proton- and antiproton-parton distributions are consistent with each other, and are well described by the structure functions of Eichten et al. For those events in which the charge of the electron, and hence the W, is well determined, we can identify the proton (antiproton) parton with a u- (d̄)-quark for a W⁺, or a

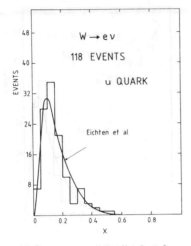 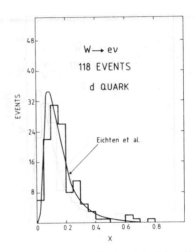

Fig. 39 Feynman x-distribution for u-quarks sampled in W production. The curve shows the Eichten et al. prediction.

Fig. 40 Feynman x-distribution for d-quarks sampled in W production. The curve shows the Eichten et al. prediction.

d-($\bar{u}$)-quark for a W^-. There are 118 events in which the charge of the electron track is determined to better than 2 standard deviations (from infinite momentum). The resulting u- and d-quark distributions for these events are shown in Figs. 39 and 40 respectively. Once again there is agreement with the predictions of Eichten et al.

4.6 Summary

The existence of the $W^\pm$ and Z^0 IVBs is well established, and their properties have been found to be in excellent agreement with the predictions of the SU(2) × U(1) standard model. We can now use the IVBs as a laboratory, not only for testing the SU(2) × U(1) part of the standard model but also for testing the perturbative QCD predictions for $W^\pm$ and Z^0 production at the Collider. Finally, we have seen that the rate at which we observe $W^\pm \to e^\pm \nu_e$ and $Z^0 \to e^+ e^-$ decays provides us with a powerful tool for counting the number of neutrinos in the Universe. In the coming few years we expect a factor of approximately 30 increase in the number of W and Z^0 decays recorded at the Collider. With this increase in statistics we can hope to improve our knowledge of the number of light neutrino species in the Universe, from the present limit of $N_\nu \leq 9$ (at 90% c.l.) to a measurement of ± 0.5 neutrino species.

5. DIMUON PHYSICS AT THE CERN p$\bar{\text{p}}$ COLLIDER

Muons are produced in hadronic events through electromagnetic (e.m.) and weak interaction processes. The most copious sources of single muons at the Collider are pion and kaon weak decays. The probability for these mesons to decay inside the volume of the central track detector (CD) is $0.02/p_T$ and $0.11/p_T$ for pions and kaons, respectively, where p_T is the transverse momentum of the meson measured in GeV/c. If we assume a charged-particle mixture of 50% $\pi^\pm$ and 25% $K^\pm$ measured by the UA2 and UA5 Collaborations[37], the resulting probability for a muon to come from meson decay in the CD volume is $0.04/p_T$ per produced hadron. We are interested in muons with relatively high transverse momenta, $p_T > 5$ GeV/c. Thus we expect about 8×10^{-3} muons per produced charged track with $p_T \geq 5$ GeV/c from meson decays, and $\sim 6 \times 10^{-5}$ dimuons per charged track with $p_T > 5$ GeV/c from this same source. This is already a small number. Since we are not interested in dimuons coming from two independent pion and/or kaon decays, we would like to recognize and completely exclude dimuon events coming from this source. This can be done by exploiting the information recorded in the CD.

A meson decaying in the volume of the track chamber produces a track which, at the decay point, will in general change momentum and direction. For most decays the resulting 'kink' in the CD track is easily recognizable, enabling us to identify and remove muons originating from meson decays. The small remaining dimuon background from unidentified decays can be estimated from events containing a single high-p_T muon by consideration of the decay probability of all other high-p_T tracks in these events. The probability that the 'kink' in the second high-p_T track will not be recognized as such has been studied by detailed Monte Carlo simulation, enabling a numerical estimate of the dimuon background to be made as a function of the reconstructed muon transverse momenta. In the total dimuon data sample, which corresponds to an integrated luminosity of 378 nb^{-1}, we expect 25 fake dimuon events to come from two randomly associated meson decays which produce two muons with $p_T > 3$ GeV/c, at least one of which has a $p_T > 5$ GeV/c. In the current UA1 data sample we have a total of 222 dimuon events. We conclude that there is a significant dimuon signal in the UA1 data, which does not come from—and can be cleanly separated from—two muons coming from pion and/or kaon decays.

5.1 Sources of Dimuon Events

Having eliminated dimuons coming from pion and kaon decays, we now consider more interesting sources of dimuon events at the Collider.

5.1.1 $Z^0 \to \mu^+\mu^-$. Neutral Intermediate Vector Boson (IVB) decays in the e^+e^- channel were described in subsection 4.3. We would expect [and observe[8]] a similar number of Z^0 decays to be seen in the $\mu^+\mu^-$ channel. The resulting high-mass dimuon system will consist of two oppositely charged, well-isolated energetic muons. Therefore Z^0 decays have a unique signature and are easily distinguishable from other sources of dimuons.

5.1.2 Drell–Yan production. The Drell–Yan mechanism, which is responsible for Z^0 production at the Collider, can also produce a 'heavy' photon which subsequently 'decays' into a $\mu^+\mu^-$ system:

$$q\bar{q} \to \gamma^* \to \mu^+\mu^- . \tag{46}$$

The resulting dimuon system will be very similar in appearance to dimuons coming from Z^0 decay (i.e. two oppositely charged isolated muons). However, the continuum mass spectrum of the $\mu^+\mu^-$ system from Drell-Yan production falls like $\sim m^{-3}$, producing muon pairs with a mass far below that of the Z^0. In addition to continuum $\mu^+\mu^-$ production, we also expect some resonance production of the vector $c\bar{c}$ and $b\bar{b}$ mesons (J/ψ and Υ, respectively) which can subsequently decay into isolated $\mu^+\mu^-$ pairs (with branching ratios of 7.4% and 2.9%, respectively). Observation and measurement of the properties of both continuum and resonant isolated $\mu^+\mu^-$ production will therefore help us to understand the quark-antiquark annihilation process in which a heavy photon γ^* is produced.

5.1.3 Heavy-flavour production. The Drell-Yan mechanism produces muons via the e.m. interaction. Muons can also be produced by the weak interaction decays of heavy-flavour quarks c, b, and, if it exists, t. Heavy quarks are produced at the Collider in quark-antiquark pairs. The strong interaction, for example, can produce $Q\bar{Q}$ pairs (where $Q = c, b,$ or t) by gluon-gluon fusion to give

$$gg \to Q\bar{Q}. \tag{47}$$

Alternatively, the e.m. interaction can produce $Q\bar{Q}$ pairs by the Drell-Yan mechanism,

$$q\bar{q} \to \gamma^* \to Q\bar{Q}, \tag{48}$$

or the weak interaction can produce heavy-flavour quarks by IVB decays

$$W^+ \to c\bar{b}, t\bar{b}, c\bar{s}, \tag{49}$$

$$Z^0 \to Q\bar{Q}. \tag{50}$$

In all these cases the heavy quarks can decay semileptonically into muons with a typical branching ratio of $\sim 10\%$. Hence if we are dealing with a $Q\bar{Q}$ pair, the probability for both quarks to decay semileptonically is $\sim 1\%$. We take as an example $b\bar{b}$ production:

$$p\bar{p} \to b\bar{b}(b \to \mu^- \bar{\nu}_\mu c)(\bar{b} \to \mu^+ \nu_\mu \bar{c}). \tag{51}$$

We end up with two oppositely charged muons which we expect to be close to the hadronic debris from the c and $\bar{c}$ decays. Typically we would see two oppositely charged muons which are recoiling back-to-back with each other in the plane transverse to the beam direction, and which are close to, or actually buried inside, two hadronic jets. Here we have considered only the first-generation decays of the heavy quarks. Clearly, in our $b\bar{b}$ example we could have a second-generation decay in which the c (or $\bar{c}$) quark decays semileptonically to produce a muon. However, since the semileptonic branching ratio is $\sim 10\%$, second-generation decays are suppressed by an order of magnitude with respect to first-generation decays. If a second

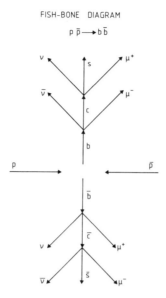

Fig. 41 'Fish-bone' diagram showing the process $p\bar{p} \to b\bar{b}$ with subsequent semileptonic decays of the heavy-flavour quarks.

generation decay occurs, we again end up with two non-isolated muons, but this time they can have either like-sign or opposite-sign electric charge, and can be either back-to-back with each other in the transverse plane, or be close to each other depending on whether they were both produced from the same Q ($\bar{Q}$) decay cascade or one from the Q and one from the $\bar{Q}$ cascade. The possibilities can be seen from the so-called 'fish-bone' diagram (Fig. 41) which provides a graphic representation of the heavy-flavour cascades.

5.2 The Dimuon Data Sample, and Drell-Yan Production

The transverse momentum threshold for a muon to penetrate the entire UA1 detector (~ 8 absorption lengths) and be seen in the muon chambers on the outside of the detector is ~ 3 GeV/c. We therefore require that both muon candidates have a transverse momentum in excess of 3 GeV/c. For two muons which are back-to-back in the transverse plane, this cut will correspond to a minimum mass for the dimuon system of ~ 6 GeV/c^2. After applying technical cuts to remove cosmic rays, leakage of hadronic showers, and kinks (pion and kaon decays), we are left with 222 dimuon events, from an integrated luminosity of 378 nb^{-1}. The dimuon mass plot is shown in Fig. 42. The mass spectrum decreases rapidly from threshold to dimuon masses of almost 40 GeV/c^2. There is a big gap in the spectrum between 40 GeV/c^2 and 80 GeV/c^2, and a cluster of 10 events, centred on the Z^0 mass, in the region m($\mu^+\mu^-$) > 80 GeV/c^2. The high-mass events are entirely consistent with the decay $Z^0 \to \mu^+\mu^-$. An

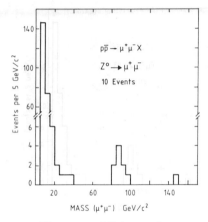

Fig. 42 Dimuon mass spectrum. The group of high-mass ($m_{\mu^+\mu^-} > 80$ GeV/c^2) dimuons are identified with the decay $Z^0 \to \mu^+\mu^-$ (the highest mass event is poorly measured and is also consistent with a Z^0 decay).

example is shown in Fig. 43. There are two energetic oppositely charged isolated muons, emerging roughly back-to-back with each other in the transverse plane. A preliminary fit to the mass of the Z^0 from the sample of 10 $Z^0 \to \mu^+\mu^-$ decays gives $m_{Z^0} = 88.8 ^{+5.5}_{-4.6}$ GeV/c^2, and the cross-section × branching ratio is

$$\sigma \cdot B(Z^0 \to \mu^+\mu^-) = 51.4 \pm 17.1 \pm 8.6 \text{ pb}.$$

These results are completely consistent with the corresponding results from the $Z^0 \to e^+e^-$ data sample described in Section 4.

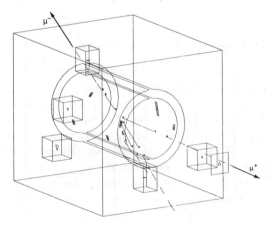

Fig. 43 Graphic display of a $Z^0 \to \mu^+\mu^-$ decay seen in the UA1 detector.

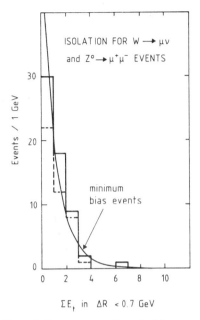

Fig. 44 Muon isolation. The calorimeter energy deposited in a cone of $\Delta R = 0.7$ (see text) around the muon direction is shown for muons from IVB decays. The distribution is consistent with the expected distribution for isolated muons based on the energy flow in minimum bias events (curve).

We remove the 10 Z^0 events from our data sample, and consider the properties and origin of the remaining 212 dimuon events.

The properties that we use to distinguish between heavy-flavour and Drell-Yan processes are the muon isolation, the charge assignments (like-sign muons or unlike-sign muons), and the dimuon (and muon-muon-jet-jet) mass. We begin by defining an isolation variable E_I for each muon, where E_I is the sum of the transverse energies of all the calorimeter cells inside a cone centred on the muon. The isolation cone is in (pseudorapidity, azimuthal angle)-space, and has a radius $\Delta R \equiv (\Delta \eta^2 + \Delta \phi^2)^{1/2} = 0.7$, where ϕ is in radians. A study of minimum-bias data and of the isolation in W and Z^0 decays (Fig. 44) suggests that, to define an isolated muon, a cut of $E_I < 4$ GeV is reasonable. The isolation of the 424 muons from the 212 low-mass dimuon events is shown in Fig. 45. The data are well described by a mixture of truly isolated muons and, based on the expectations of the EUROJET Monte Carlo[38], non-isolated muons coming from heavy-flavour decays. In order to study the muon isolation in our sample of 212 low-mass dimuon events, the events have been separated into two subsamples: unlike-sign dimuons (150 events) and like-sign dimuons (62 events). We now define a dimuon isolation variable,

$$E_I^{\mu\mu} \equiv (E_I^{\mu 1})^2 + (E_I^{\mu 2})^2 . \tag{52}$$

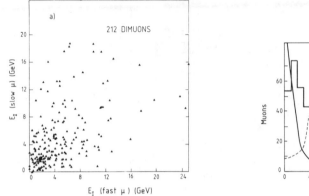
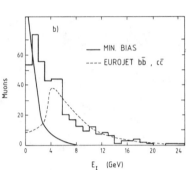

Fig. 45 Dimuon isolation. a) The calorimeter energy deposited (in a cone $\Delta R < 0.7$) around the muon is shown for the fastest muon versus the slowest muon. b) The isolation variable E_I for all muons in the dimuon sample. The curves show the expectations for truly isolated muons (min. bias) and muons from heavy-flavour decays.

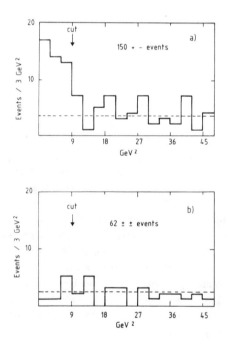

Fig. 46 Dimuon isolation variable $E_I^{\mu\mu}$ (see text) for a) unlike-sign dimuons, and b) like-sign dimuons.

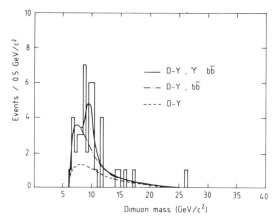

Fig. 47 Dimuon mass distribution for isolated $\mu^+\mu^-$ events, showing the expected shapes of the contributions from continuum Drell-Yan production, Υ production, and heavy-flavour decays.

Truly isolated dimuons should have $E_I^{\mu\mu} \leq 9$ GeV^2. Comparing the dimuon isolation for the 150 unlike-sign dimuons with the corresponding isolation for the 62 like-sign dimuons shows that (Fig. 46) there is a clear excess of isolated $\mu^+\mu^-$ events above the level expected for the heavy-flavour process. No corresponding excess is seen in the like-sign dimuon events.

We are now in a position to extract the Drell-Yan production cross-section from the data. To do this we make a fit to the dimuon mass plot for the 44 isolated $\mu^+\mu^-$ events satisfying the requirement $E_I^{\mu\mu} < 9$ GeV^2 (Fig. 47). Injected into the fit is the expected shape for the $\mu^+\mu^-$ mass spectrum from continuum Drell-Yan production (normalized to the tail of the mass distribution $m_{\mu^+\mu^-} > 11$ GeV/c^2), and the expected shapes for dimuons from the Drell-Yan production of Υ, Υ', and Υ'', and finally for dimuons from $b\bar{b}$ production and subsequent semileptonic decay. The result of this mass fit gives

i) for the continuum Drell-Yan production, a total of 19.6 ± 6.5 events with $m(\mu^+\mu^-) > 6$ GeV/c^2,

$$\sigma_{DY} (m > 11 \text{ GeV/c}^2) = 264 \pm 104 \text{ pb},$$

$$(d^2\sigma/dmdy)|_{y=0, m=0} = 17 \pm 6 \text{ pb};$$

ii) for the resonance Drell-Yan production of Υ, Υ', and Υ'', a total of 5.7 ± 2.8 events; taking B·σ for $\Upsilon : \Upsilon' : \Upsilon''$ to be 1 : 0.3 : 0.15, we obtain

$$B\cdot\sigma(\Upsilon + \Upsilon' + \Upsilon'') = 375 \pm 192 \text{ pb},$$

$$B[d\sigma(\Upsilon)/dy]_{y=0} = 63 \pm 34 \text{ pb}.$$

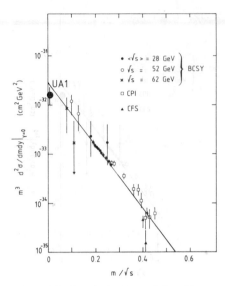
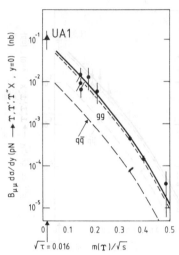

Fig. 48 Scaling behaviour of the dilepton system from Drell–Yan production. The UA1 measurement, at low m/√s, is seen to lie on the extrapolation of the world data[39].

Fig. 49 Scaling behaviour of Υ production. The UA1 measurement at low m/√s lies on the extrapolation of the world data.

iii) The contribution from heavy-flavour events is 19.5 ± 6.5 events, consistent with the guess we would have made on the basis of the dimuon isolation plot (Fig. 46).

We can compare the UA1 measurement of continuum and resonance Drell–Yan production cross-sections with the world data[39] (Figs. 48 and 49). Expressed as the variable $m^3 d^2\sigma/dmdy$, the continuum cross-section falls exponentially with increasing $\sqrt{\tau} \equiv m/\sqrt{s}$, which essentially just reflects the quark and antiquark structure functions. The UA1 point, at large √s and therefore small √τ, falls on the extrapolation of the world data. This is also true for the Υ cross-section (Fig. 49). We conclude that the observed Drell–Yan cross-sections are consistent with our expectations for the Collider.

5.3 Non-Isolated Dimuons and B^0–$\bar{B}^0$ Mixing

We now concentrate on the non-isolated dimuons. There are 161 dimuon events in which the muons fail the isolation cut $E_T^{j\mu} < 9$ GeV². Of these, 106 events have unlike-sign dimuons, and the remaining 55 events have like-sign dimuons, giving the ratio

$$R \equiv [N(\mu^+\mu^-) + N(\mu^-\mu^-)]/N(\mu^+\mu^-) = 0.46 \pm 0.10.$$

Fig. 50 The B^0-$\bar{B}^0$ mixing can take place via the emission of two virtual charged IVBs. The strengths of the couplings at the vertices are described by the KM matrix (see text).

We would expect these non-isolated dimuon events to be produced from heavy-flavour decays. However, remembering that the like-sign dimuons can only come from second- (or higher) generation decays which should be suppressed by an order of magnitude, the ratio R seems to be too big. In fact, when a detailed Monte Carlo study is performed using the EUROJET program[38], the expected value of R of 0.32 (preliminary) is not far from the observed value, but is a little lower. If, with more data, the excess of like-sign dimuons persists, we must ask ourselves what mechanism is generating the extra $\mu^+\mu^+$ and $\mu^-\mu^-$ events. One obvious and exciting possibility is that, analogous to K^0-$\bar{K}^0$ mixing, there is B^0-$\bar{B}^0$ mixing which proceeds via the diagram shown in Fig. 50. In this way a b-quark ($\bar{b}$-antiquark) from the initial $b\bar{b}$ pair turns itself into a $\bar{b}$-antiquark (b-quark) before decaying semileptonically, enabling like-sign dimuons to come from first-generation decays. In particular, an inspection of the K–M matrix

$$\begin{array}{c} \\ u \\ c \\ t \end{array} \begin{array}{ccc} d & s & b \\ \left[\begin{array}{ccc} 0.97 & 0.23 & < 0.01 \\ -0.22 & 0.96 & -0.06 \\ -0.01 & 0.06 & 0.98 \end{array} \right] \end{array} \quad (53)$$

which describes the weak couplings between the quarks, shows that the most likely states to mix are B_s^0 with $\bar{B}_s^0$ (see Figure 50). If the probability for a B^0 to turn itself into a $\bar{B}^0$ before decaying is Δ, then

$$R \equiv [N(\mu^+\mu^+) + N(\mu^-\mu^-)]/N(\mu^+\mu^-) \approx 2\Delta/(1 + \Delta^2) \ . \quad (54)$$

A measurement of R, and hence of Δ, can therefore determine the strength of the mixing. Clearly, if it turns out that mixing is taking place in the B^0-$\bar{B}^0$ system, we will have a new and exciting place to look for CP violation. Before leaving this topic we note that a preliminary study with the EUROJET Monte Carlo indicates that the expected value for R is 0.53 with full mixing of the B_s^0-$\bar{B}_s^0$ system. Finally, we also note that, if we exclude the contribution from the Drell–Yan process, the isolated dimuon events—which we expect to come predominantly from $b\bar{b}$ production and semileptonic decay—also yield a similar value for R of 0.45 ± 0.10.

Returning to the origin of the non-isolated dimuon events, in the above we have been assuming that these come mostly from $b\bar{b}$ and $c\bar{c}$ decays. The evidence for this is the following:

i) The four-body system (muon + muon + jet + jet) has a mass distribution which is consistent with the expected distribution from the EUROJET Monte Carlo for heavy-flavour production (Fig. 51).

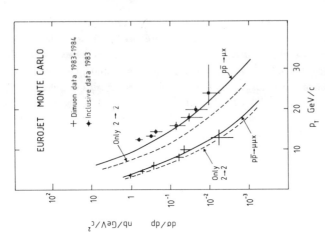

Fig. 52 The differential cross-sections $d\sigma/dp_T$ for inclusive muon production (upper data points) and dimuon production (lower data points) are well described by the EUROJET Monte Carlo for the $b\bar{b}$ and $c\bar{c}$ production (solid curves).

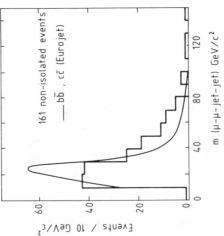

Fig. 51 The four-body (μ-μ-jet-jet) invariant mass distribution for non-isolated dimuon events, together with the $b\bar{b}$ + $c\bar{c}$ prediction from the EUROJET Monte Carlo (curve).

ii) The muon and dimuon transverse-momentum distributions are also well described by the EUROJET Monte Carlo.

iii) The differential cross-section $d\sigma/dp_T$ for dimuon production is well described by the EUROJET Monte Carlo, with the overall normalization fixed by the observed inclusive single-muon distribution (Fig. 52). This is true only if $2 \to 3$ processes are included in the Monte Carlo calculation, showing the importance of higher-order QCD corrections.

iv) An enhanced strange-particle content (K^0, $\bar{K}^0$, Λ^0, and $\bar{\Lambda}^0$) is seen in the non-isolated events. Taking V^0's with $p_T > 0.3$ GeV/c, $|\eta| < 2.0$, and decay length > 5 mm, the uncorrected rate of observed V^0's per event is

minimum-bias data	0.44 ± 0.09
isolated $\mu^+\mu^-$	0.64 ± 0.14
non-isolated $\mu^+\mu^-$	1.45 ± 0.21
non-isolated $\mu^\pm\mu^\pm$	0.91 ± 0.20

The excess of strange particles can be interpreted as arising from the hadronic end-products of the heavy-flavour cascades.

Thus both the rate and the features of the non-isolated dimuon events indicate a heavy-flavour origin which is well described by the EUROJET Monte Carlo. The resulting measurement of the $b\bar{b}$ cross-section at the Collider is

$$\sigma(p\bar{p} \to b\bar{b}) = 1.3 \pm 0.1\,(\pm 0.2)\,\mu b$$

for $\quad p_T(b), p_T(\bar{b}) > 5$ GeV/c, and $|\eta| < 2.0$.

5.4 Isolated Like-Sign Dimuons

We are left with one last category of events to understand: those having isolated like-sign dimuons. If we employ our tight dimuon isolation cut $E_t^{\mu\mu} < 9$ GeV^2, seven like-sign dimuons survive. The mass distribution for these seven events is shown in Fig. 53. The events are at low dimuon masses and can be understood as being in the tail of the non-isolated

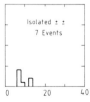

Fig. 53 Invariant dimuon mass distribution for isolated like-sign dimuons.

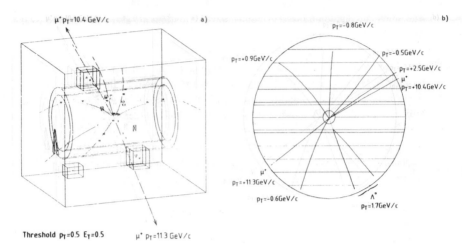

Fig. 54 An example of a semi-isolated high-mass $\mu^+\mu^-$ event: a) graphic display of the event in the UA1 detector, and b) transverse view of the charged tracks in the central track detector.

dimuon spectrum coming from heavy-flavour decays. Apparently, then, we have been able to explain all the observed dimuon events in terms of Drell–Yan and heavy-flavour processes. There is, however, an indication of a possible anomaly if we relax the muon isolation cuts and look at like-sign dimuon events where both muons pass the single-muon isolation cut $E_T^i <$ 4 GeV. We find[40] a handful of semi-isolated like-sign dimuon events with dimuon mass ~ 12 GeV/c^2, which are not completely understood in the framework of the EUROJET Monte Carlo. An example of one of these anomalous semi-isolated like-sign dimuon events is shown in Fig. 54. Note that there is a Λ^0 in the event; in fact the majority of the high-mass semi-isolated like-sign dimuons seem to have strange-particle activity in them. The origin of these events is left as an open question at this point in time. Hopefully, in the near future, more data will clarify this issue.

5.5 Summary

Dimuon events provide a fairly clean tool for studying Drell–Yan and heavy-flavour production at the Collider. Using the muon charges (like-sign or unlike-sign), the dimuon mass, and the dimuon isolation, we are able to unfold the contributions from the different production processes, and find that both the Drell–Yan events and the heavy-flavour events have rates and properties consistent with our expectations. There are two open questions coming from the data:

1) Does B^0–$\bar{B}^0$ mixing occur?
2) What is the origin of the high-mass semi-isolated like-sign dimuons?

In order to answer these questions, and to exclude the possibility that one or both effects are due to a statistical fluctuation, we need to accumulate more data, which we expect to acquire in the near future.

6. CONCLUSIONS

In these lectures only a few of the numerous physics results that have emerged from the CERN SPS $p\bar{p}$ Collider in the last few years have been described. The pioneering discovery work that resulted in hadronic jets being observed as spectacular naked-eye phenomena, and the Nobel Prize-winning discovery of the $W^{\pm}$ and Z^0 bosons that mediate the weak interaction, has been completed. We are now using these new tools (jets, $W^{\pm}$'s, and Z^0's) to probe the details of the QCD plus SU(2) × U(1) standard model, and are finding an impressive agreement between theory and experiment almost everywhere we look. The future of the Collider program looks bright. There are a number of very fundamental measurements that can be performed and questions that can (hopefully) be answered in the coming years; for example: i) How many light neutrinos are there in the Universe? ii) Is there mixing in the B^0-$\bar{B}^0$ system? iii) Are there any more charged leptons beyond the τ?; and finally iv) Do any of the anomalies mentioned in the Introduction require new physics beyond the standard model?

Obviously the most exciting note on which to close these lectures is the possibility that perhaps there is still something unexpected and fundamental to be discovered at the CERN SPS $p\bar{p}$ Collider.

REFERENCES

1) The staff of the CERN proton-antiproton project, Phys. Lett. **107B**, 306 (1981).

2) Arnison, G. et al. (UA1), Phys. Lett. **107B**, 320 (1981) and **123B**, 108 (1983).
Alpgård, K. et al. (UA5), Phys. Lett. **107B**, 315 (1981).
Alner, G.J. et al. (UA5), Phys. Lett. **138B**, 304 (1984).

3) Arnison, G. et al. (UA1), Phys. Lett. **118B**, 167 (1982).
Banner, M. et al. (UA2), Phys. Lett. **122B**, 322 (1983).

4) Arnison, G. et al. (UA1), Phys. Lett. **118B**, 173 (1982) and **123B**, 115 (1983).
Banner, M. et al. (UA2), Phys. Lett. **118B**, 203 (1982).
Bagnaia, P. et al. (UA2), Z. Phys. **C20**, 117 (1983).

5) Arnison, G. et al. (UA1), Phys. Lett. **132B**, 214 (1983) and **136B**, 294 (1984).
Bagnaia, P. et al. (UA2), Phys. Lett. **144B**, 283 (1984).

6) Arnison, G. et al. (UA1), **132B**, 223 (1983).
Bagnaia, P. et al. (UA2), Phys. Lett. **144B**, 291 (1984).

7) Arnison, G. et al. (UA1), Phys. Lett. **122B**, 103 (1983), **129B**, 273 (1983), **134B**, 469 (1984), and **135B**, 250 (1984);
Banner, M. et al. (UA2), Phys. Lett. **122B**, 476 (1983).

8) Arnison, G. et al. (UA1), Phys. Lett. **126B**, 398 (1983) and **147B**, 241 (1984).
Bagnaia, P. et al. (UA2), Phys. Lett. **129B**, 130 (1983) and Z. Phys. **C24**, 1 (1984).

9) Arnison, G. et al. (UA1), Phys. Lett. **147B**, 493 (1984).

10) UA1 Collaboration, Proposal CERN/SPSC/78-06/P92 (1978).

11) Mansoulié, B., Proc. 3rd Moriond Workshop on $p\bar{p}$ Physics (1983) (Éditions Frontières, Gif-sur-Yvette, 1984), p. 609.

12) Bozzo, M. et al. (UA4), Phys. Lett. **147B**, 392 (1984).

13) Abramowicz, H. et al. (CDHS), Z. Phys. **C12**, 289 (1982).

14) Arnison, G. et al. (UA1), preprint CERN-EP/85-98 (1985), submitted to Physics Letters B.

15) Combridge, B.L., and Maxwell, C.J., Nucl. Phys. **B239**, 428 (1984).

16) Berends, F. et al., Phys. Lett. **103B**, 124 (1981).
Gottschalk, T., and Sivers, D., Phys. Rev. **D21**, 102 (1980).

17) Combridge, B. et al., Phys. Lett. **70B**, 234 (1977).

18) Rubbia, C., McIntyre, P., and Cline, D., Proc. Int. Neutrino Conf., Aachen, 1976 (Vieweg, Braunschweig, 1977), p. 683.

19) Weinberg, S., Phys. Lett. **19**, 1264 (1967).
Salam, A., Proc. 8th Nobel Symposium, Aspenäsgården, 1968, ed. N. Svartholm (Almqvist and Wiksell, Stockholm, 1968), p. 367.

20) For a review see Bég, M.A.B., and Sirlin, A., Phys. Rep. **88**, 1 (1982).

21) Kobayashi, M., and Maskawa, K., Prog. Theor. Phys. **49**, 652 (1973).

22) Cabibbo, N., Phys. Rev. Lett. **10**, 531 (1963).

23) Ellis, J. et al., Ann. Rev. Nucl. Part. Sci. **32**, 443 (1982).

24) Glashow, S.L., Phys. Rev. **D2**, 1285 (1970).

25) Asano, Y. et al., Phys. Lett. **107B**, 159 (1981).

26) Drell, S.D., and Yan, T.M., Ann. Phys. (NY) **66**, 578 (1971).

27) Badier, J. et al., Phys. Lett. **87B**, 398 (1979).

28) Aurenche, P., and Lindfors, J., Nucl. Phys. **B185**, 274 (1981).
Humpert, B., and Van Neerven, W.L., Phys. Lett. **93B**, 456 (1980).

29) Jacob, M., Nuovo Cimento **9**, 826 (1958).

30) Arnison, G. et al. (UA1), preprint CERN-EP/85-108 (1985), submitted to Physics Letters B.

31) Paige, F.E., and Protopopescu, S.D., ISAJET program, BNL 29777 (1981).

32) Altarelli, G., Ellis, R.K., Greco, M., and Martinelli, G., Nucl. Phys. **B246**, 12 (1984).
Altarelli, G., Ellis, R.K., and Martinelli, G., Z. Phys. **C27**, 617 (1985).

33) Deshpande, N.G. et al., Phys. Rev. Lett. **54**, 1757 (1985).

34) Glück, M., Hoffman, E., and Reya, E., Z. Phys. **C13**, 119 (1982).

35) Geer, S., and Stirling, W.J., Phys. Lett. **152B**, 373 (1985).

36) Eichten, E. et al., Rev. Mod. Phys. **56**, 579 (1984).

37) Alpgård, K. et al. (UA5), Phys. Lett. **121B**, 209 (1983).
Banner, M. et al. (UA2), Phys. Lett. **122B**, 322 (1983).

38) These calculations were made using the EUROJET Monte Carlo program which contains all first-order (α_s^2) and second-order (α_s^3) QCD processes [see van Eijk, B., UA1 Technical Note UA1/TN/84-93 (1984), unpublished, and Ali, A., Pietarinen, A., and van Eijk, B., to be published]. The heavy-quark Q is fragmented using the parametrization of Peterson, C. et al., Phys. Rev. **D27**, 105 (1983), and the V-A matrix elements.

39) Fabjan, C.W., Acta Physica Austriaca Suppl. XIX, 621 (1978).

40) Arnison, G. et al. (UA1), preprint CERN-EP/85-19 (1985), submitted to Physics Letters B.

THEORETICAL EXPECTATIONS AT COLLIDER ENERGIES

E. Eichten

Fermi National Accelerator Laboratory

P. O. Box 500, Batavia, IL 60510

Abstract

This series of seven lectures is intended to provide an introduction to the physics of hadron-hadron colliders from the $S\bar{p}pS$ to the SSC. Applications in perturbative QCD ($SU(3)$) and electroweak theory ($SU(2) \otimes U(1)$) are reviewed. The theoretical motivations for expecting new physics at (or below) the TeV energy scale are presented. The basic theoretical ideas and their experimental implications are discussed for each of three possible types of new physics: (1) New strong interactions (e.g. Technicolor), (2) Composite models for quarks and/or leptons, and (3) Supersymmetry (SUSY).

1 INTRODUCTION TO COLLIDER PHYSICS

These lectures are intended to provide a introduction to the physics of hadron-hadron colliders present and planned. During the last twenty years, great theoretical advances have taken place. The situation in elementary particle physics has been transformed from the state (twenty years ago) of a wealth experiment results for which there was no satisfactory theory to the situation today in which essentially all experimental results fit comfortably into the framework of the **Standard Model**. The current generation of hadron-hadron colliders will allow detailed tests of the gauge theory of the strong interactions, QCD; while the hadron-hadron colliders which are being planned now will be powerful enough to probe the full dynamics of the electroweak interactions of the Weinberg-Salam model. The experiments performed at these colliders will confront this standard model and may show it inadequate for as we will discuss it is very likely incomplete.

After a brief review of the status of the standard model and experimental facilities present and planned, this introductory lecture will deal with the basics. The connenction between hadron-hadron collisions and the elementary subprocesses will be reviewed, along with a discussion of the parton distribution functions which play a central role in this connection.

The second lecture will concentrate of QCD phenomenology. The basic two to two parton subprocess will be reviewed and applications to jet physics discussed. The two to three processes and their relation (in leading logarithmic approximation) to the two to two processes is demonstrated. Finally the production of the top quark is discussed.

The third lecture will concentrate on the other half of the standard model gauge theory, the electroweak interactions. The Weinberg-Salam model is review. The main focus of this lecture is the fermions and gauge bosons of the electroweak model; the scalar sector is left to lecture four. The production and decay properties of single $W^\pm$'s and Z^0's are considered at present and future collider energies. Electroweak gauge boson pair production is also considered, with emphasis on what can be learned about the structure of the gauge interactions from measurement of pair production. Finally minimal extensions of the standard model are considered. In particular, the possiblities of a fourth generation of quarks and leptons and a W' or Z' are considered.

The fourth lecture will be devoted to the scalar sector of the electroweak theory. The limits on the Higgs mass (or self coupling) and fermion masses (Yukawa couplings) imposed by the condition of perturbative unitarity are presented. The prospects for discovery of the standard Higgs are discussed. Finally 't Hooft's naturalness condition is used to argue the unnaturalness (at the TeV energy scale) of the Weinberg-Salam model with elementary scalars. The possiblities for building a natural theory are discussed in the remaining three lectures.

In the fifth lecture the possibility of a new strong interaction at the one TeV scale will be examined. The basics of Technicolor, Extended Technicolor, and mass generation for technipions are reviewed. The phenomenological implications of both a miminal model and the more elaborate (and somewhat more realistic) Farhi-Susskind model are discussed.

The sixth lecture is devoted to the possibility that quarks and/or leptons are composite. Since no realistic models of compositeness have been proposed, the emphasis will be on the general theoretical requirements of a composite model, e.g. 't Hooft's constraint, and the model independent experimental signatures of compositeness.

In the last lecture the idea of a supersymmetric extension of the standard model is investigated. The basic idea of N=1 global supersymmetry and the present experimental constraints on the superpartners of known particles are reviewed. The production rates and detection prospects for superpartners in hadron-hadron collisions are presented.

There are many very good references to the various aspects of collider physics I will be discussing in these lectures and I will attempt to give some sources for each of the lectures as I discuss the material. It is however appropriate to mention one source before I begin, since I have drawn heavily on it and will refer frequently to it. This reference is "Supercollider Physics" by Eichten, Hinchliffe, Lane, and Quigg[1], hereafter denoted EHLQ. It contains a compendium of the physics opportunities for the next generation of hadron-hadron colliders, the so-called Super Colliders.

1.1 Status of the Standard Model

To begin, the present theory of elementary particles and and their interactions, the standard model, is a great success:

- The fundamental constituents of matter have been identified as leptons and quarks.

- A gauge theory encompassing the weak and electromagnetic interactions has been developed.

- Quark confinement has been explained by a asymptotically free gauge theory of colored quarks and gluons, QCD.

No known experimental results are inconsistent with the present theory. In fact, the basics of the standard model are in a number of recent textbooks[2].

1.1.1 The Fundamental Constituents

The elementary leptons and quarks are arranged into families, or generations.

For the leptons:

$$\begin{pmatrix} \nu_e \\ e \end{pmatrix}_L \quad \begin{pmatrix} \nu_\mu \\ \mu \end{pmatrix}_L \quad \begin{pmatrix} \nu_\tau \\ \tau \end{pmatrix}_L$$
$$e_R \quad\quad\quad \mu_R \quad\quad\quad \tau_R$$

and for the quarks:

$$\begin{pmatrix} u \\ d \end{pmatrix}_L \quad \begin{pmatrix} c \\ s \end{pmatrix}_L \quad \begin{pmatrix} t \\ b \end{pmatrix}_L$$
$$u_R, d_R \quad c_R, s_R \quad t_R, b_R$$

All the left-handed fermions appear in $SU(2)_L$ weak doublets and the right-handed fermions are singlets. The vertical columns form the elements of a single generation of quarks and leptons. This pattern is repeated three times, i.e. there are three known generations. The only missing constituent is the top quark, for which preliminary evidence has been reported by the UA1 Collaboration[3] at CERN.

The fundamental constituents have very simple basic properties:

- Pointlike and structureless down to the smallest distance scales we have probed ($\approx 10^{-16}$ cm)

- Spin 1/2

- Universal electroweak interactions

- Each quark comes in three colors

1.1.2 The Gauge Principle

The gauge principle has become the central building block of all dynamical models of elementary particles. As is well known, the gauge principle promotes a global symmetry of the Lagrangian, such as a phase invariance or invariance under a set of non-Abelian gauge charges to a dynamics associated with the associated local (space-time dependent) symmetry. If, for example, the Lagrangian for a set of free the Fermion fields

$$\mathcal{L} = i\overline{\psi}(x)\gamma^\mu \partial_\mu \psi(x) \tag{1.1}$$

is invariant under a set of global charges Q_a

$$\psi_\alpha(x) \to e^{i\alpha_a Q_a}\psi(x) \,, \tag{1.2}$$

then to preserve the symmetry under local gauge variations, $\alpha_a(x)$,

$$\psi_\alpha(x) \to e^{i\alpha_a(x)Q_a}\psi(x) \tag{1.3}$$

massless gauge fields $A_a^\mu(x)$ transforming according to

$$Q_a A_{a\alpha}^\mu(x) \to e^{i\alpha_a(x)Q_a}[Q_a A_a^\mu(x) - i\partial^\mu]e^{-i\alpha_a(x)Q_a} \tag{1.4}$$

must be introduced and the the Lagrangian must be modified to include an interaction between the fermions and these gauge bosons as well as gauge boson self interactions. The form of these interactions is dictated by the requirement of local gauge invariance. The Lagrangian becomes:

$$\mathcal{L} = i\overline{\psi}(x)\gamma^\mu \mathcal{D}_\mu \psi(x) + \frac{1}{2g^2}Tr([\mathcal{D}_\mu, \mathcal{D}_\nu]^2) \qquad (1.5)$$

where $\mathcal{D}_\mu^A$ is the gauge covariant derivative

$$\mathcal{D}_\mu^A \equiv \partial_\mu + igQ_a A_{a\mu}(x) . \qquad (1.6)$$

The Lagrangian (Eq. 1.5) describes a set of massless non Abelian gauge bosons interacting with massless fermions but the physical spectrum may realize the gauge symmetry in one of three different phases[4]:

- Confinement Phase - all physical states are singlets under the non-Abelian charges. This is the realization in the case of the color $SU(3)$ gauge interactions which describe the strong interactions.

- Higgs Phase - the symmetry is spontaneously broken only a subgroup of the original symmetry is manifest in the physical sprectrum the other symmetries are "hidden". In this case the gauge bosons associated with the broken symmetries acquire a mass. The $SU(2)_L \otimes U(1)$ electroweak interactions exhibits this behaviour.

- Coulomb Phase - This is the simpliest realization. The symmetry is manifest and the gauge bosons are massless physical degrees of freedom. Quantum Electrodynamics exhibits this phase.

Therefore, all three phases of a gauge theory are found in nature.

In addition to fermions and the gauge interactions, fundamental scalars are introduced in the standard model which interact via gauge interactions with the electroweak gauge bosons with the fermions and the electroweak vector bosons. The scalar self interactions (Higgs potential) are introduced to produce spontaneous symmetry breaking at the electroweak scale. There is as yet no direct experimental evidence for the scalar sector of the standard model. A detailed discussion of these scalar and their interactions is postponed until the third and fourth lectures.

The covariant derivative coupling of matter fields (fermions or scalars) with the carrier of the color interactions the gluon G is given by:

$$D_\mu^G = \partial_\mu + ig_s Q_s^a G_{a\mu}(x) \tag{1.7}$$

where Q_s^a is the color charge matrix of the matter field. While the covariant derivative coupling of matter fields to carrier of the $SU(2)_L$ electroweak interactions the W gauge triplet is giver by:

$$D_\mu^W = \partial_\mu + ig_2 Q_W^a W_{a\mu}(x) \tag{1.8}$$

where Q_W^a is the $SU(2)_L$ charge matrix of the matter field. The matter fields interact with the carrier of the $U(1)$ gauge interaction B by an Abelian gauge interaction (as in QED) with coupling strength g_1.

One can write the standard model including both the strong and electroweak interactions in a compact form using these covariant derivatives:

$$\mathcal{L} = \sum_{j=1,2,3} i\overline{\psi}_{q_j}\gamma^\mu D_\mu^G \psi_{q_j} + \frac{1}{2g_s^2}Tr([D_\mu^G, D_\nu^G]^2)$$

$$+ \sum_{j=1,2,3}\sum_{f=q,l} i\overline{\psi}_{f_j}\gamma^\mu \frac{(1-\gamma_5)}{2}D_\mu^W \psi_{f_j} + \frac{1}{2g_2^2}Tr([D_\mu^W, D_\nu^W]^2)$$

$$+ \sum_{j=1,2,3}\sum_{f=q,l} i\overline{\psi}_{f_j}\gamma^\mu [\partial_\mu + ig_1(\frac{1-\gamma_5}{2}L_{f_i} + \frac{1+\gamma_5}{2}R_{f_i})B_\mu]\psi_{f_j}$$

$$- \frac{1}{4}(\partial_\mu B_\nu - \partial_\nu B_\mu)^2$$

$$+|\mathcal{D}_\mu^W + i\frac{g_1}{2}B_\mu)\phi|^2 - [-\mu^2|\phi|^2 + \lambda(|\phi|^2)^2]$$

$$-[\sum_{i=1,2,3} \overline{\psi}_{L\ e_i}\Gamma_{e_i}\psi_{R\ e_i}\phi + h.c.$$

$$+ \sum_{i,j=1,2,3} \overline{\psi}_{L\ q_i}\Gamma_{ij}^{up}(\frac{1+\gamma_5}{2})\psi_{R\ q_j}\tilde{\phi} + h.c.$$

$$+ \sum_{i,j=1,2,3} \overline{\psi}_{L\ q_i}\Gamma_{ij}^{down}(\frac{1+\gamma_5}{2})\psi_{R\ q_j}\phi + h.c.] \tag{1.9}$$

with the notation

$$\psi_{q_i} \Rightarrow \begin{pmatrix} u_i \\ d_i \end{pmatrix}_L \text{ and } \begin{pmatrix} u_i \\ d_i \end{pmatrix}_R \tag{1.10}$$

and

$$\psi_{l_i} \Rightarrow \begin{pmatrix} \nu_i \\ e_i \end{pmatrix}_L \text{ and } e_{iR} \tag{1.11}$$

for the fermion fields, and

$$\phi = \begin{pmatrix} \phi_+ \\ \phi_0 \end{pmatrix}$$

$$\tilde{\phi} = \begin{pmatrix} \phi_0^t \\ \phi_- \end{pmatrix} \tag{1.12}$$

for the scalar fields. The indices i,j denote the generation. A possible CP violating strong interaction term (the so-called θ term) as well as gravitation interactions have not been included.

1.1.3 Unanswered Questions

In spite of the great success of the standard model, there are still many open questions. A partial list would include:

- What determines the pattern of quark and lepton masses and the mixing angles of the Kobayashi-Maskawa (K-M) matrix[5]?

- Why do the quark - lepton generations repeat? How many generations are there?

- Why are there so many arbitary parameters? In the standard model the arbitrary paramters are:

  3   coupling parameters α_s, α_{EM}, and $\sin\theta_w = g_1/\sqrt{g_1^2 + g_W^2}$

  6   quark masses

  3   generalized Cabibbo angles

  1   CP-violating phase in K-M matrix

  2   parameters of the Higgs potential

  3   charged lepton masses

  1   QCD vacuum phase angle

  A total of 19 arbitrary parameters. This number is not generally less in Grand Unified Models (GUTS) such as SU(5).

- Is the spontaneous symmetry breaking of the electroweak interactions due to the instability of the Higgs potential with elementary scalars as in the Weinberg-Salam model or does it have a dymanical origin? If the scalars are elementary what determines the mass of the Higgs scalar and is there more than one doublet of scalars?

- Why are all the interactions we know build on the gauge principle?

- What is the origin of CP-violation?

- How is gravity included in a unified way?

- Are the quarks and leptons of the standard model elementary of composite? The known fundamental fields in the standard model are:

18 quarks $\quad 3 \times (u \quad d \quad c \quad s \quad t \quad b)$

6 leptons $\quad (\nu_e \quad e \quad \nu_\mu \quad \mu \quad \nu_\tau \quad \tau)$

1 photon

3 intermediate bosons $\quad (W^+ \quad Z^0 \quad W^-)$

8 colored gluons

1 Higgs scalar $\quad$ (not yet observed)

1 graviton $\quad$ (not yet observed)

A total of 38 "elementary particles" - compare Air, Fire, Earth, and Water (A FEW). Is there a more economical substructure?

There are many theoretical speculations on these questions, but we are not likely to advance without new experimental observations.

1.2 Experimental Facilities

Further progress in understanding elementary particles and their interactions will depend on the study of phenomemna at higher energies/shorter distances. The experimental high energy facilities which exist or will exist by 1990 are listed below:

Date	Reaction	Location	Accelerator	Energy (CM)
Now	$\bar{p}p$ collisions	CERN	$S\bar{p}pS$	630 GeV
1986	$\bar{p}p$ collisions	Fermilab	TEV I	1,800 GeV
1987	e^+e^- collisions	Stanford	SLC	100 GeV
1989	e^+e^- collisions	CERN	LEP	100 GeV (phase 1)
				$\approx$ 200 GeV (phase 2)
1990	$e^{\pm}p$ collisions	DESY	HERA	314 GeV

Even though the center of mass energies of the hadron machines shown above are considerably higher than those of the lepton machines, the center of mass energies for the elementary constituent subprocesses are comparable. This is because the energy of a hadron is is shared among its quark constituents, so that energy carried by a given quark or gluon is typically only a small fraction of the total energy.

The conclusion drawn from a careful study of the physics potential of the facilities above is that elementary processes with center of mass energies up to a few hundred GeV will be thoroughly explored by these machines[6,7,8]

However, center of mass energy of 1 TeV is an important watershed in particle physics. For example:

- Unitarity limits on the standard model become relevant at about 1 TeV as will be shown in lecture 4.

- If electroweak symmetry breaking is dynamical, the Higgs scalars would be fermion-antifermion composite particles. As will be discussed in lecture 5, this internal structure, if it exists, should be observable at the one TeV scale.

- Low energy supersymmetry, which relates bosons and fermions, requires new particles whose masses are very likely below one TeV/c^2.

Therefore general arguments as well as specific speculations indicate that new phenomena should be observed at the energy scale at or below 1 TeV. Exploration of this energy scale is therefore the minimum requirement of the explored by the next generation of accelerators.

The two types of machines which are capable of this exploration are:

- A e^+e^- collider with a beam energy of 1-3 TeV. or

- A hadron collider (pp or $\bar{p}p$) with a beam energy of 10-20 TeV, so that numerous elementary constitutent collisions occur with center of mass energy of a few TeV.

At present there is under serious consideration in the United States a proposal[9] to build by 1994 a the Superconducting Super Collider (SSC) which would be a pp collider operating at a center of mass energy of 40 TeV with a collision rate for the protons (luminosity) of $10^{33} cm^{-2} sec^{-1}$. With present magnet technology($\approx$ 5 tesla magnets) this accerator would be about 20 miles in diameter. A smaller version of the SSC, the Large Hadron Collider[10] (LHC), could be built in the existing LEP tunnel. Field strengths of 6-10 tesla would give center of mass energies of 10-18 TeV. The luminosity would depend on the choice of a pp or $\bar{p}p$ option for the beams.

The present hadron-hadron colliders in conjuction with the future SSC provide a formidable array of experimental resources for advancing our knowledge.

Let me now begin the detailed discussion of the physics potential of these machines.

1.3 Preliminaries

In order to understand the strong interactions within QCD, we must be able to interpret the hadron collisions in terms of quarks and gluons[11]. The quarks carry flavor, have spin 1/2, and are in the fundamental (triplet) representation of SU(3) color, whereas the only internal quantum number of the gluons, which are spin one bosons, is color and the gluons are in the adjoint (octet) representation of the SU(3) color gauge group. Color is confined, which means that all physical states are singlets of color SU(3). Therefore, unlike lepton physics in which the elementary particles can be studied directly, in strong interactions the physical particles at are disposal are the hadrons, which are bound states of the elementary quarks and gluons.

The basic property of QCD at short distance (high energy) is asymptotic freedom; i.e. the coupling strength of QCD becomes weak at short distance[12]. This property of QCD allows us to calculate in perturbation theory at high energy. The final states associated with high energy interactions are quarks and gluons in perturbation theory but must be hadrons in reality since color is confined. However not all memory of the underlying quark and gluon final state is lost as the hadrons appear in a striking way - as jets. For our purposes a jet is simply a well collimated isolated spray of hadrons (we leave the precise criteria for a jet to the experimentalists). By observing these hadronic jets the underlying quark and gluon interactions can be inferred. For example, in e^+e^- scattering into hadronic final states, the lowest order Q.C.D. process is shown

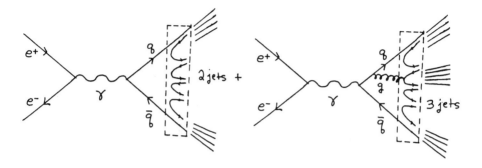

Figure 1: e^+e^- annihilation into (a) quark-antiquark pair and (b) quark-antiquark pair plus a gluon. Our ignorance of the hadronization process is contained within the dashed box in Figure 1.

The $q\bar{q}$ final state of QCD perturbation theory is not the physical final state. On a distance scale of the confinement scale ($\approx \Lambda_{QCD}$), the stong interactions will produce sufficient gluons and quark-antiquark pairs to locally neutralize the color and produce the color singlet hadrons, the physical final states of the process. This hadronization process, is nonperturbative and presently uncalculable. It can only be modeled phenomenologically[13] However, at high energies much of the information about the perturbative QCD interactions at short distance is remembered by the jets.[14] Crudely speaking the jets can be mapped one the one onto the quarks and gluons of the short distance (perturbative) process.

Figure 2 shows a e^+e^- event at $\sqrt{s} \approx 30$ GeV as seen in the JADE detector

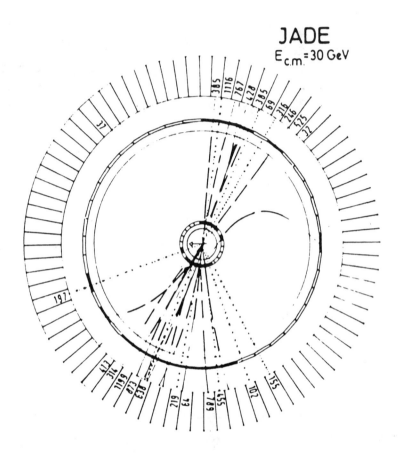

Figure 2: A two jet event in the JADE central detector[15]. The view is along the beam direction. Charged and neutral particles are denoted by solid and dotted lines respectively. The energy deposited into lead glass shower counters are given in MeV.

at PETRA[15]. This is a typical two jet event associated with the production of a quark-antiquark pair at high energy. The two hadronic jets are clearly visible in the event. The kinematic structure of two jet events retain knowledge of the production kinematics associated with the elementary process. For the production of two spin 1/2 fermions from the virtual photon the angular distribution is

$$\frac{d\sigma}{d\cos\theta} \sim 1 + \cos^2\theta \qquad (1.13)$$

where θ is the angle of the quark to the beam direction. To high accuracy[16], measured two jet events have this angular behaviour (characteristic of the production of two piontlike spin 1/2 fermions).

Sometimes in addition to a quark-antiquark a gluon is produced at short distance in e^+e^- collisions. The frequency of these events is dependent of the strength of the strong coupling α_s. These events should result in a three jet final state. Such three jet events are observed in e^+e^- collisions. An example is shown in Figure 3.

Unfortunately, PEP and PETRA energies are not sufficiently high to extract from the ratio of three to two jet events the value of the strong coupling without relying on the explicit modeling of the hadronization process[18]. Also no experimental procedure has yet been found which on a event by event basis can distinguish a jet associated with a light quark from one associated with a gluon. However all the qualitative features of these events agree well with expectations from QCD.[19]

For hadron-hadron collisions one would expect that it is much more difficult to expose the quark and gluon (partons) interactions, since the initial physical states (the hadrons) have a complicated structure in terms of the fundamental

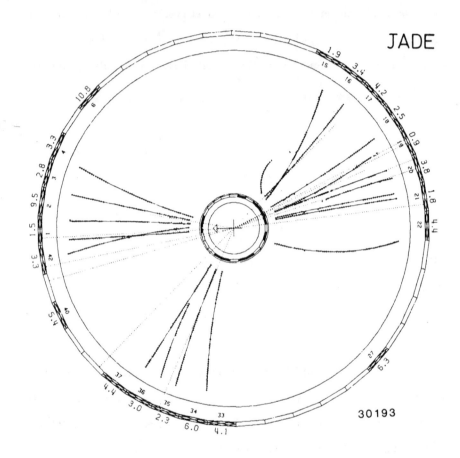

Figure 3: A three jet event in the JADE central detector[17].

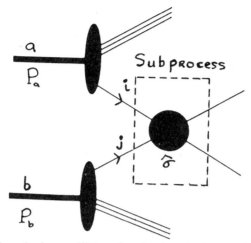

Figure 4: Hadron-hadron collision showing two to two parton subprocess.

constituents- the quarks and gluons. It is true in fact that, for many kinematic regions, hadron-hadron collisions can not be calculated using perturbative QCD. One simple example is the pp (or $\bar{p}p$) total cross section. This cross section grows as rapidly as the unitarity bound allows. For a detailed discussion of this "soft" interaction physics in hadron-hadron collisions see the excellent review of Block and Cahn[20]. However, the situation is not as bad for processes which involve a "hard" parton interaction. A hard parton interaction is one in which all the invariants (energy scales) of the process are large and thus QCD perturbation theory should apply. We will restrict our attention to these hard processes for the remainder of these lectures.

1.4 Parton Distributions

An example of a hadron-hadron collision process which involves a high energy subprocess is shown in Figure 4. The incident hadrons are composed of quarks and gluons and two of these partons, **i** and **j**, are assumed to interact at high energy. In such a case the final state will be recognizable as being composed of jets. However, to quantitatively understand the underlying parton interactions, it is necessary to separate out the effects of the physical hadrons. The inclusive cross-section for scattering of hadron **a** and hadron **b** to hadron **c** and anything **X** may be written as

$$d\sigma(a+b \to c+X) = \sum_{ij} \frac{1}{1+\delta_{ij}} \int dx_a dx_b [f_i^{(a)}(x_a, \hat{Q}^2) f_j^{(b)}(x_b, \hat{Q}^2)$$
$$+(i \leftrightarrow j)] d\hat{\sigma}(i+j \to c+X) \quad (1.14)$$

where $f_i^{(a)}(x_a, Q^2)$ is the probability that hadron **a** contains a parton **i** which carries a fraction x_a of the hadron's momentum. The cross-section for the subprocess $\hat{\sigma}$ involves only the elementary constitutents. The kinematic variables are:

- $s = (P_a + P_b)^2$ - The square of the total energy of the initial hadrons in their CM frame.

- $\hat{s} = (x_a P_a + x_b P_b)^2$ -The square of the total energy of the partons in the subprocess CM frame.

$$\hat{s} = x_a x_b s \equiv \tau \quad \text{for } (P_a^2, P_b^2 \ll s). \quad (1.15)$$

The parameter τ is used extensively in describing the physics of these collisions.

- $\hat{Q}^2$ - an invariant of the subprocess which characterizes the physical scales (e.g. $\hat{s}, \hat{t}$, or $\hat{u}$). The exact invariant depends on the process.

For SSC energies, we will be interested in Q^2 in the range :

$$(10 GeV)^2 \ll Q^2 \ll (10 TeV)^2 \quad (1.16)$$

Below 10 GeV we probably cannot analyze the subprocesses perturbatively; while above 10 TeV (even at SSC energies) the number of partons is insufficient to produce observable rates for known subprocesses. The typical x's are $\approx \sqrt{Q^2/s}$ so for $\sqrt{s} = 40$ TeV we must consider:

$$10^{-4} \leq x \leq 1 . \tag{1.17}$$

Clear experimental evidence for this jet structure had to wait for the UA1 and UA2 experiments at the CERN $S\bar{p}pS$ collider. Figure 5 shows a UA2 two jet event at $\sqrt{s} = 630$ GeV in the form of a "LEGO" plot[21]. This plot presents the energy deposition in the detector as a function of the solid angle measured from the interaction point. The horizontal axes are: ϕ, the azimuthal angle about the beam direction; and θ, the angle measured from the beam direction. The two jet structure of this event is obvious. Most of the events observed by UA1 or UA2 with total $E_T > 50$ GeV have this two jet energy deposition structure. The particular event shown in Fig. 5 is special in one way. This event has the highest transverse energy observed by UA2 in the 1984 run. The total observed transverse energy was 267 GeV in a proton-antiproton collision with a total energy of 630 GeV. The remaining energy in this event, can be accounted for by soft hadrons which did not deposit enough energy into a cell of the detector to pass a minimum energy cut or by hadrons which scattered into the far forward or far backward direction where the detector has poor effeciency. Clearly it is possible for the fundamental subprocess to have a significant fraction of the total available energy.

In order to quantitatively understand the quark/gluon subprocesses it is necessary to calculate the parton distribution functions $f_i^{(a)}(x_a, Q^2)$. The Q^2 dependence is due to QCD corrections to the Born approximation in the subprocess.

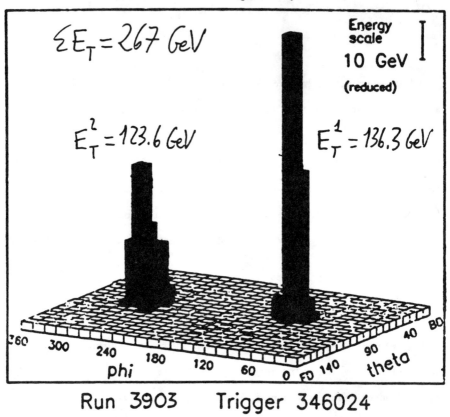

Figure 5: A LEGO plot of the event with the highest total transverse energy observed by UA2 in the 1984 run[21]. The height of each cell is proportional to the total energy deposition.

If the distribution function is known at some Q_0^2 which is high enough that QCD perturbation theory is valid, then the distribution function can be calculated in the leading logarithmic approximation (to all orders of perturbation theory) for values of $Q^2 > Q_0^2$ by use of the Altarelli-Parisi equations[22] (which are based on the renormalization group). This evolution gives the well-known Q^2 dependent scale violation of the parton distribution functions. Therefore the high Q^2 behaviour of these parton distribution functions is completely determined by measuring them at some sufficiently high Q_0^2 from which they are determined at all higher Q^2 by perturbative QCD.

The first step is to determine the parton distributions at some reference Q_0^2. In principle one should also be able to calculate these distributions in QCD, but this nonperturbative calculation is presently beyond our ability.

There are constraints on initial distribution functions which arise from valence quark counting for the proton (i.e. two up quarks and one down quark):

$$\int_0^1 dx[u(x,Q^2) - \bar{u}(x,Q^2)] = 2 \qquad (1.18)$$

$$\int_0^1 dx[d(x,Q^2) - \bar{d}(x,Q^2)] = 1 . \qquad (1.19)$$

Moreover flavor conservation of the strong interactions implies:

$$s(x,Q^2) = \bar{s}(x,Q^2)$$

$$c(x,Q^2) = \bar{c}(x,Q^2) \qquad (1.20)$$

$$\text{etc.}$$

Finally, from momentum conservation:

$$\int_0^1 dx\, x\, [\, g(x,Q^2) + u(x,Q^2) + \bar{u}(x,Q^2) + d(x,Q^2) + \bar{d}(x,Q^2) + 2s(x,Q^2)$$

$$+ 2c(x,Q^2) + 2b(x,Q^2) + 2t(x,Q^2) + \ldots] = 1 . \qquad (1.21)$$

Analysis of deep inelastic neutrino scattering data from the CDHS experiment[23] at CERN gives two sets of initial distributions corresponding to different values of the QCD scale parameter, Λ_{QCD}. The first set corresponds to $\Lambda_{QCD} = 200$ MeV for which the gluon distribution at the reference Q_0^2 is soft, i.e. it has a paucity of gluons at large x. The second set has $\Lambda_{QCD} = 290$ MeV and hard gluons, i.e. relatively more gluons at large x. Explicitly the CDHS analysis gives the following input parametrizations:

$$xu_v(x, Q_0^2) = 1.78 x^{0.5}(1 - x^{1.51})^{3.5}$$

$$xd_v(x, Q_0^2) = 0.67 x^{0.4}(1 - x^{1.51})^{4.5} \qquad (1.22)$$

and for **Set 1** with $\Lambda_{QCD} = 200$ MeV (the "soft gluons" distribution)

$$x\bar{u}(x, Q_0^2) = x\bar{d}(x, Q_0^2) = 0.182(1 - x)^{8.54}$$

$$x\bar{s}(x, Q_0^2) = 0.081(1 - x)^{8.54}$$

$$xG(x, Q_0^2) = (2.62 + 9.17x)(1 - x)^{5.90} \qquad (1.23)$$

while for **Set 2** with $\Lambda_{QCD} = 290$ MeV (the "hard gluons" distribution)

$$x\bar{u}(x, Q_0^2) = x\bar{d}(x, Q_0^2) = 0.185(1 - x)^{7.12}$$

$$x\bar{s}(x, Q_0^2) = 0.0795(1 - x)^{7.17}$$

$$xG(x, Q_0^2) = (1.75 + 15.575x)(1 - x)^{6.03} . \qquad (1.24)$$

For both distributions

$$xc(x, Q_0^2) = xb(x, Q_0^2) = xt(x, Q_0^2) = 0 . \qquad (1.25)$$

The CDHS fit to their measured structure functions F_2 and xF_3 is shown in Figure 6. The relation between these measured structure functions and the

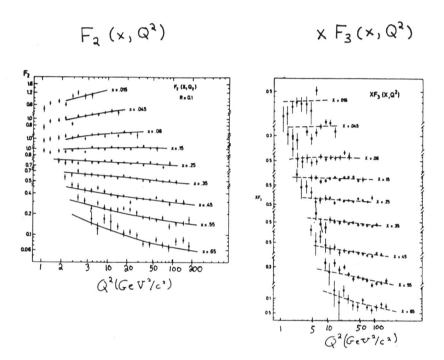

Figure 6: The structure functions F_2 and xF_3 versus Q^2 for different bins of x from CDHS[23]. The solid lines are the result of their fit **Set 1** to the data.

parton distribution functions is:

$$2xF_1 = x[(u + d + s + c + \ldots) + (\bar{u} + \bar{d} + \bar{s} + \bar{c} + \ldots)] \quad (1.26)$$

$$F_2 = 2xF_1 \frac{(1 + R(x, Q^2))}{1 + 4M_p^2 x^2/Q^2} \quad (1.27)$$

$$xF_3 = x[u - \bar{u} + d - \bar{d}] \quad (1.28)$$

where $R(x, Q^2)$ is the ratio of longitudinal to transverse cross section in deep-inelastic leptoproduction. R is predicted by QCD to go to zero at high Q^2 like $1/Q^2$, the data is not in disagreement with this behaviour however the measurements are not conclusive[24]. Different choices for R consistent with the data will affect the resulting distribution. The distribution Set 1 above uses $R = .1$ while Set 2 assumes the behaviour of R expected in QCD.

The up to down quark valence distributions can be separated using charged-current cross sections for hydrogen and deuterium targets. The parameterization use here is discussed by Eisele[25]. Once the valence distributions are known, the sea distribution may be determined from measurements of the structure function F_2 on isoscalar targets. It is also necessary to know the flavor dependence of the sea distribution. For this purpose, the strange quark distribution can be determined directly from antineutrino induced dimuon production[25]. Dileptons events arise mainly from production off the antistrange quarks in the proton hence the rate of opposite sign dilepton events gives information about the the ratio of strange to antiup quark distributions, assuming that both have the same x dependence. Also note that limits on same sign dimuon events put limits on the charm quark content of the proton[25].

Figure 7 shows a comparison of Set 2 of the distributions defined above with the results of the CHARM Collaboration[26]. We see that there is good agree-

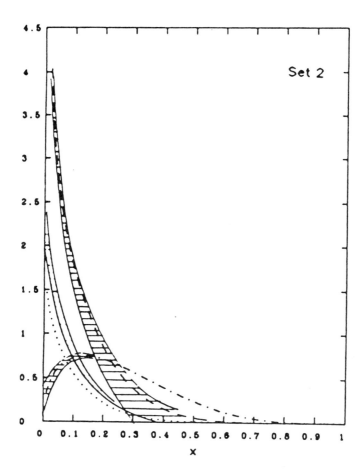

Figure 7: Comparison of the gluon distribution $xG(x,Q^2)$ (dashed line), valence quark distribution $x[u_v(x,Q^2)+d_v(x,Q^2)]$ (dot-dashed line), and the sea distribution $2x[u_s(x,Q^2)+d_s(x,Q^2)+s_s(x,Q^2)+c_s(x,Q^2)]$ (dotted line) of Set 2 with the determination (shaded bands) of the CHARM Collaboration[26].

ment with the results presented here except for the antiquark distributions. Also a second independent experiment measuring the structure fucntions CCFRR[27] finds that $F_2(x, Q_0^2)$ is more strongly peaked at small x that the CDHS measurements. This again suggests a larger sea distribution. Recently, the disagreement has been resolved, CDHS has made a new analysis[28] which disagrees with their old results and is in agreement with the CCFRR results. Thus the sea distributions used here are too small at Q_0^2. In general the effects of this error will be small since the Q^2 evolution washes out much of the dependence on the initial distribution, as we will see in the case of the gluon distributions shortly.

After the determination of these distribution functions has been carried out, it is necessary to extend them to higher values of Q^2 by means of renormalization group methods of Altarelli and Parisi. Although a detailed description of this procedure is beyond the scope of this lecture (see A. Mueller lectures[29] for a more extensive treatment), I will describe the basic idea of this evolution.

Let $q_v(x, Q^2)$ be the valence quark distribution for the proton $(u_v = u - \bar{u})$. If the quark is probed by a virtual photon of momentum Q^2 then this photon will be sensitive to fluctuations on the distance scale $\sqrt{1/Q^2}$. For example, if the quark has a fraction y of the proton's momentum, then it may virtually form a gluon and a quark which has a fraction $x < y$ of the initial proton momentum. Let $z = x/y < 1$. The probability of observing the quark with fraction z of the initial momentum of the parent quark is given by

$$\frac{\alpha(Q^2)}{\pi} P_{q_v \leftarrow q_v}(z) dln(Q^2) \tag{1.29}$$

in which the coupling strength α has been written explicitly. The splitting function $P(z)$ is calculable in QCD perturbation theory. Finally the renormalization group analysis of Altarelli and Parisi shows that

$$\frac{dq_v(x,Q^2)}{dln(Q^2)} = \frac{\alpha(Q^2)}{\pi} \int_x^1 \frac{dy}{y} q_v(y,Q^2) P_{q_v \leftarrow q_v}(z) \tag{1.30}$$

where the integral over z ($0 < z < 1$) has been replaced by integration over y ($x < y < 1$). This equation then determines the distribution functions for the valence quarks. Since the valence quark lines continues throughout the process, the evolution of the valence quark distributions is determined by the valence quarks alone while the distribution of non-valence quarks and gluons is determined by all of the various distribution functions.

The equation for the evolution in QCD of the valence quark distribution, $v(x,Q^2) = xu_v(x,Q^2)$ or $xd_v(x,Q^2)$, is

$$\frac{dv(x,Q^2)}{dln(Q^2)} = \frac{2\alpha_s(Q^2)}{3\pi} \int_x^1 dz \frac{(1+z^2)v(y,Q^2) - 2v(x,Q^2)}{1-z}$$

$$+ \frac{\alpha_s(Q^2)}{\pi}[1 + \frac{4ln(1-x)}{3}]v(x,Q^2) \tag{1.31}$$

where $y = x/z$. The result of numerical integration of these lowest order Altarelli-Parisi equations using the initial distributions of Set 2 (Eq. 1.24) is shown in Figure 8 for valence up quarks. As Q^2 increases from 10^2 to 10^6 the valence momentum distribution functions decrease at large x while increasing modestly at small x. This shift is caused by the fact that higher x quarks scatter into lower x quarks.

For the gluon distribution, $g(x,Q^2) \equiv xG(x,Q^2)$, the evolution is more complicated:

$$\frac{dg(x,Q^2)}{dln(Q^2)} = \frac{\alpha_s(Q^2)}{\pi} \int_x^1 dz \left[\frac{3[zg(y,Q^2) - g(x,Q^2)]}{1-z} + \frac{3(1-z)(1+z^2)}{z} g(y,Q^2) \right.$$

$$\left. + \frac{2}{3}\frac{1+(1-z)^2}{z} \sum_{\text{flavors } q} [yq_v(y,Q^2) + 2yq_s(y,Q^2)] \right]$$

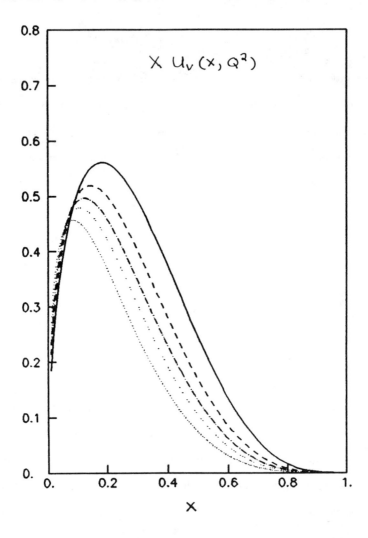

Figure 8: The valence up quark distribution of the proton, $xu_v(x,Q^2)$, as a function of x for various Q^2. The solid, dashed, dot-dashed, sparse dot, and dense dot lines correspond to $Q^2 = 10$, 10^2, 10^3, 10^4, and 10^6 $(GeV)^2$ respectively.

$$+\frac{\alpha_s(Q^2)}{\pi}[\frac{11}{4} - \frac{N_f}{6} + 3ln(1-x)]g(x,Q^2) \tag{1.32}$$

where N_f is the number of quark flavors. The evolution of the gluon distribution is feed by the valence (q_v) and sea (q_s) quark distributions as will as the gluon distribution (G) itself.

Figure 9 shows the evolution for the gluon distribution. The gluon distribtution is peaked at small x due to the high probability of emission of soft gluons from quarks(and other gluons).

The evolved gluon distribution functions at large Q^2 and small x (where they are peaked) are fairly insensitive to drastic modifications of their initial form at Q_0^2. This is because the gluon distributions are determined through the Altarelli-Parisi Equation (1.32) by the initial valence quark distributions at larger x. For instance, Figure 10 shows the result of modifying the initial gluon distribution of Set 1 (Eq. 1.23) for $x < .01$, values of x at which there is no existing data. The variations were:

$$xG(x,Q_0^2) = \begin{cases} 0.444x^{-.5} - 1.866 & (a) \\ 25.56x^{.5} & (b) \end{cases} \text{ for } x < .01. \tag{1.33}$$

These modifications match continuously at x=0.01 to Set 1 and are constrained to change the gluon momentum integral by no more than 10 percent. Fig 10 shows that a variation by a factor of 160 at $x = 10^{-4}$ for Q_0^2 yields only a factor of 2 difference at the same x for $Q^2 = 2000\ GeV^2$ This insensitivity at high Q^2 to the initial distribution is reassuring, for it implies that the gluon distribution at small x and large Q^2 is much better determined that our knowledge of the small x behaviour at Q_0^2 would lead one to expect.

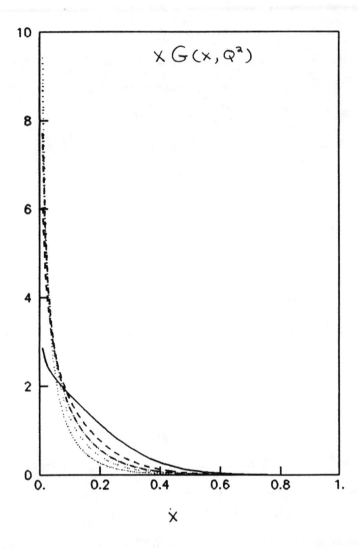

Figure 9: The gluon distribution of the proton, $xG(x,Q^2)$, as a function of x for various Q^2. The solid, dashed, dot-dashed, sparse dot, and dense dot lines correspond to $Q^2 = 10$, 10^2, 10^3, 10^4, and 10^6 $(GeV)^2$ respectively.

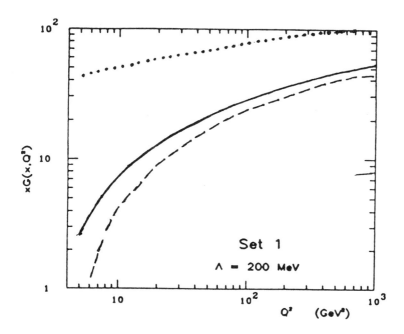

Figure 10: The Q^2 evolution of the gluon distribution $xG(x, Q^2)$ given in Set 1 (solid line) as compared to the two variations given in Eq. 1.33 for $x = 10^{-4}$. Distribution (a) is represented by a dotted line and distribution (b) is represented by a dashed line.

The light sea quarks, $l(x,Q^2) = xu_s(x,Q^2)$ or $xd_s(x,Q^2)$ or $xs_s(x,Q^2)$, evolve according to:

$$\frac{dl(x,Q^2)}{dln(Q^2)} = \frac{2\alpha_s(Q^2)}{3\pi}\int_x^1 dz[\frac{(1+z^2)l(y,Q^2) - 2l(x,Q^2)}{1-z} + \frac{3}{8}[z^2 + (1-z)^2]g(y,Q^2)]$$

$$+\frac{\alpha_s(Q^2)}{\pi}[1+\frac{4}{3}ln(1-x)]l(x,Q^2) \qquad (1.34)$$

The results of numerical evolution for the up antiquark distribution $(xu_s(x,Q^2))$ is shown in Figure 11. The total up quark distribution function is given by $xu_v(x,Q^2) + xu_s(x,Q^2)$.

The initial distribution at $Q^2 = Q_0^2$ was consistent with zero for the heavy quarks and antiquarks (xc_s, xb_s, xt_s). But the probably of finding a charm, bottom, or even top quark in the proton can become significant when the proton is probed at high Q^2. The evolution for heavy quarks is also dictated by the Altarelli-Parisi Equation but some method must be employed to treat the kinematic effects of the nonnegligible masses of the quarks and the associated production thresholds in perturbation theory. The method used was proposed by Gluck, Hoffman, and Reya[30]. For more details see EHLQ. The evolution equation in lowest order QCD for the heavy quark distribution, $h(x,Q^2) = xc_s(x,q^2)$ or $xb_s(x,Q^2$ or $xt_s(x,Q^2)$, in lowest order QCD is:

$$\frac{dh(x,Q^2)}{dln(Q^2)} = \frac{2\alpha_s(Q^2)}{3\pi}\int_x^1 dz[\frac{(1+z^2)h(y,Q^2) - 2h(y,Q^2)}{1-z}$$

$$+\frac{3}{4\beta}[\frac{1}{2} - z(1-z) + \frac{m_q^2(3-4z)z}{Q^2}\frac{1}{1-z} - \frac{16m_q^4 z^2}{Q^4}]g(y,Q^2)$$

$$-\frac{3m_q^2}{2Q^2}[z(1-3z) + \frac{4m_q^4 z^2}{Q^2}ln(\frac{1+\beta}{1-\beta})]g(y,Q^2)]\theta(\beta^2)$$

$$+\frac{\alpha_s(Q^2)}{\pi}[1+ln(1-x)]h(x,Q^2) \qquad (1.35)$$

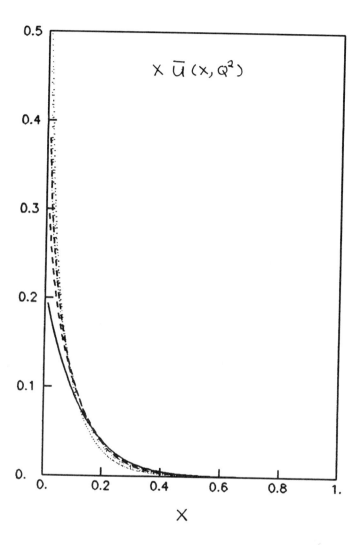

Figure 11: The up antiquark distribution of the proton, $xu_s(x,Q^2)$, as a fucntion of x for various Q^2. The solid, dashed, dot-dashed, sparse dot, and dense dot lines correspond to $Q^2 = 10$, 10^2, 10^3, 10^4, and 10^6 $(GeV)^2$ respectively.

wherethe velocity of the heavy quark is:

$$\beta = [1 - \frac{4m_q^2}{Q^2}(1-z)]^{\frac{1}{2}}, \qquad (1.36)$$

the strong coupling includes the heavy quark contribution

$$1/\alpha_s(Q^2) = \frac{25}{12\pi}ln(\frac{Q^2}{\Lambda^2}) - \frac{1}{6\pi}\sum_{i=b,t}\theta(Q^2 - 16m_q^2)ln(\frac{Q^2}{16m_q^2}), \qquad (1.37)$$

and $m_c = 1.8$ GeV/c^2, $m_b = 5.2$ GeV/c^2, and $m_t = 30$ GeV/c^2. The resulting distribution function for the bottom quark is shown in Figure 12.

As Q^2 increases the various quark distributions approach the asymptotic forms dictated by QCD. At infinite Q^2 the masses of the various quarks becomes unimportant and valence quark effects will be swamped by the virtual qaurk pair (i.e. the sea) ; hence there should be an $SU(6)$ flavor symmetry in this limit. Furthermore, QCD predicts[31] (at infinite Q^2) the the momentum fraction carried by any of these quark flavors to be 3/68 while that the momentum fraction carried by gluons should be 8/17. This approach to the asymptotic values is shown in Figure 13.

The effective parton-parton luminosity is:

$$\tau\frac{d\mathcal{L}_{ij}}{d\tau} \equiv \frac{\tau}{1+\delta_{ij}}\int_\tau^1 \frac{dx}{x}[f_i^{(P)}(x,\hat{s})f_j^{(P)}(\frac{\tau}{x},\hat{s}) + (i \leftrightarrow j)] \qquad (1.38)$$

This effective luminosity is the number of parton i - parton j collisions per unit τ with subprocess energy $\hat{s} = \tau s$. For a elementary cross section

$$\hat{\sigma}(\hat{s}) = \frac{\kappa}{\hat{s}} \qquad (1.39)$$

with coupling strength κ, the total number of events/sec,N is:

$$N(\text{events/sec}) = L_{\text{hadrons}}(\tau\frac{d\mathcal{L}}{d\tau})_{\text{partons}}\hat{\sigma}(\hat{s}) \qquad (1.40)$$

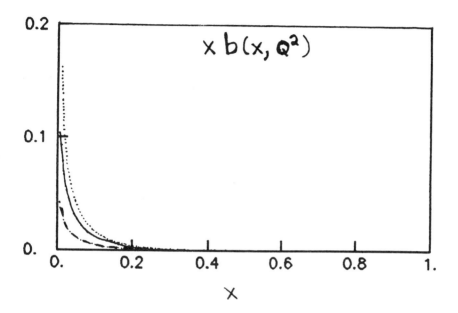

Figure 12: The bottom quark distribution, $xb(x, Q^2)$, as a function of x for various Q^2. The dot-dashed, solid, and dotted lines correspond to $Q^2 = 10^3$, 10^4, and 10^6 $(GeV)^2$ respectively.

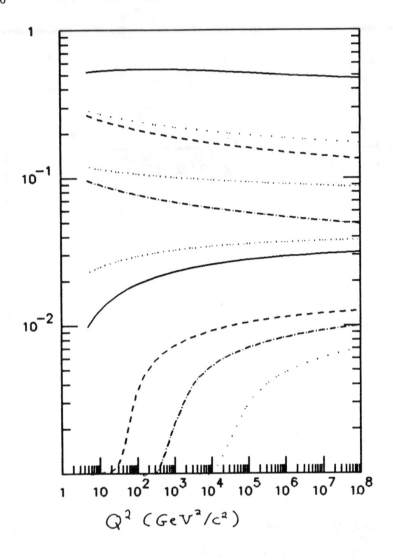

Figure 13: The fraction of the total momentum carried by each of the partons in the proton as a function of Q^2. From largest to smallest momentum fraction these partons are: gluon, up quark, up (valence only), down quark, down (valence only), antiup (or antidown) quark, strange quark, charm quark, bottom quark, and top quark.

where $\mathcal{L}_{\text{hadron}}$ is the hadron-hadron luminosity (measured in $cm^{-2}\ sec^{-1}$). Thus the combination

$$\frac{\tau}{\hat{s}}\frac{d\mathcal{L}}{d\tau} \qquad (1.41)$$

contains all the kinematic and parton distribution dependence of the rate. Hence this quantity can be used to make quick estimates of rates for various processes knowing only the coupling strength κ of the subprocess. This expression (Eq. 1.41) is shown for gg, $u\bar{u}$, $b\bar{b}$, and $t\bar{t}$ initial parton pairs in Figures 14-16 for the energies of the $S\bar{p}pS$ and Tevatron colliders. the corresponding figures for SSC energies are given in EHLQ (Figures 32-56).

Finally, it is possible at high enough Q^2, to have substantial distributions for any elementary particles which couple to either quarks or gluons: For example, the luminosities for top quark-antiquark interactions is shown in Figure 17. An even more exotic example is the luminosity for electroweak vector boson pairs.[32] The quantity $(\tau/\hat{s})d\mathcal{L}/d\tau$ is shown for transverse and longitudinal $W^{\pm}$ and Z^0 bosons at $\sqrt{s} = 40$ TeV in Figures 18(a) and 18(b) respectively. This property, that hadron collisions at high energies contain a broad spectrum of fundamental constituents as initial states in elementary subprocesses, is one of the most attractive features of using a hadron collider for the exploration of possible new physics at the TeV scale.

To summarize, the extraction of the elementary subprocesses from hadron-hadron collisions require knowledge of the parton distributions of the proton. By combining experimental data at low Q^2 and the evolution equations determined by perturbation theory in QCD we can obtain these distributions to sufficient accuracy at high energies to translate from the elementary subprocesses to estimates of experimental rates in hadron collisions. The evidence for

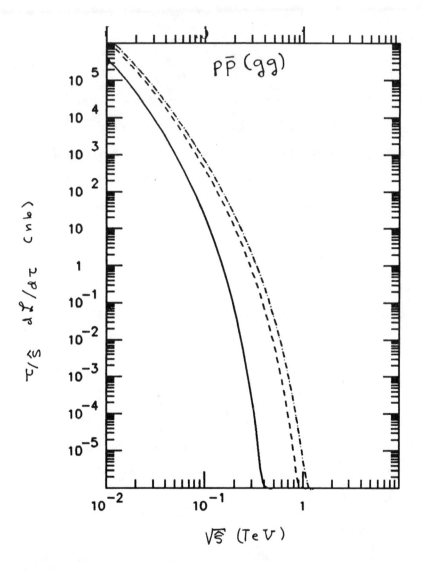

Figure 14: Quantity $(\tau/\hat{s})d\mathcal{L}/d\tau$ (in nb) for gg interactions in proton-antiproton collisions at energies: 630 GeV (solid line), 1.6 TeV (dashed line), and 2.0 TeV (dot-dashed line). $\sqrt{\hat{s}}$ is the subprocess energy (in TeV).

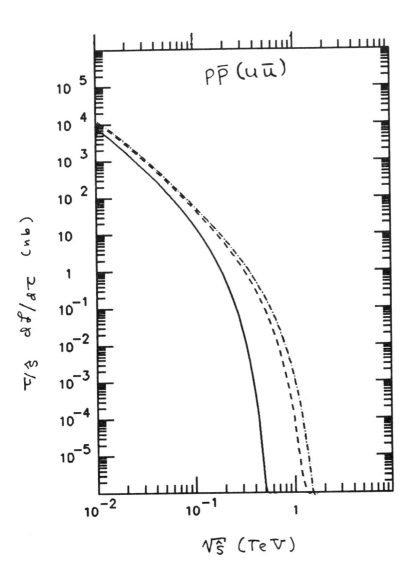

Figure 15: Quantity $(\tau/\hat{s})d\mathcal{L}/d\tau$ (in nb) for $u\bar{u}$ interactions in proton-antiproton collisions at energies: 630 GeV (solid line), 1.6 TeV (dashed line), and 2.0 TeV (dot-dashed line). $\sqrt{\hat{s}}$ is the subprocess energy (in TeV).

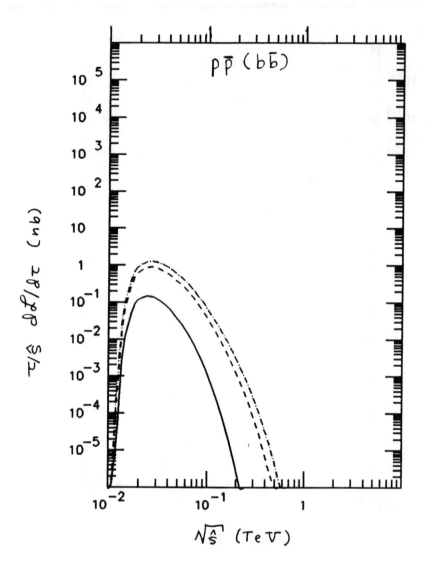

Figure 16: Quantity $(\tau/\hat{s})d\mathcal{L}/d\tau$ (in nb) for $b\bar{b}$ interactions in proton-antiproton collisions at energies: 630 GeV (solid line), 1.6 TeV (dashed line), and 2.0 TeV (dot-dashed line). $\sqrt{\hat{s}}$ is the subprocess energy (in TeV).

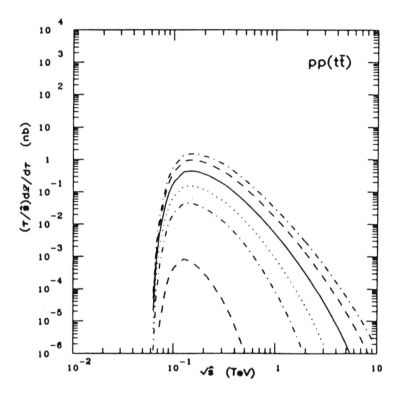

Figure 17: Quantity $(\tau/\hat{s})d\mathcal{L}/d\tau$ (in nb) for $t\bar{t}$ interactions in proton-proton collisions at energies: 2 TeV (dashed line), 10 TeV (dot-dashed line), 20 TeV (dotted line), and 40 TeV (solid line). $\sqrt{\hat{s}}$ is the subprocess energy (in TeV).

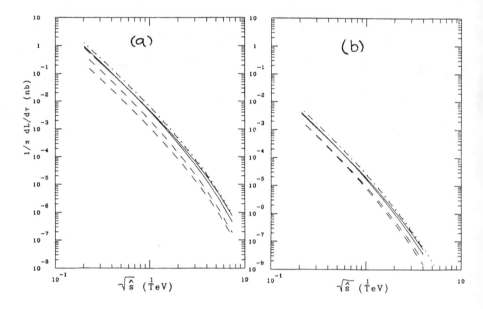

Figure 18: Quantity $(\tau/\hat{s})d\mathcal{L}/d\tau$ (in nb) for intermediate vector bosons interactions as a function of $\sqrt{\hat{s}}$ (in TeV) for proton-proton collisions at $\sqrt{s} = 40$ TeV. Transverse and longitudinal intermediate vector bosons are shown in Figures (a) and (b) respectively. In each figure, W^+W^-, W^+W^+, W^-Z^0, W^-W^-, and Z^0Z^0 pairs are denoted by dot-dashed, upper solid, lower solid, upper dashed, and lower dashed lines respectively. Figure from Ref. 32.

this conclusion is:

- Cross sections obtained using different parameterizations (Set 1 and Set 2 of Eqs. 1.22-1.25) generally differ by less than 20 percent at SSC energies[1].

- The evolved gluon distribution $G(x, Q^2)$ is very insensitive to drastic modifications of the small x $(x < 10^{-2})$ behaviour at $Q_0^2 = (5\ GeV)^2$ where it is unknown experimental[1].

- Corrections to the lowest order Altarelli-Parisi evolution equations for $f_i(x, Q^2)$ due to $ln(x)$ terms at small x and $ln(1-x)$ terms at large x do not give important contributions to the distrubutions functions in the range of x and Q^2 relevant to new physics at either the present colliders or the SSC[33].

2 THE STRONG INTERACTIONS

This lecture is devoted to understanding the jet physics in hadron-hadron collisions in terms of the underlying QCD processes.

2.1 Two Jet Physics

Consider, first, the two to two parton scattering subprocess as shown in Figure 19.

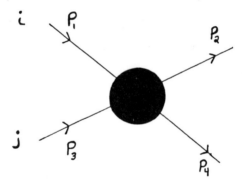

Figure 19: Two to two scattering process.

The invariants are:

$$\begin{aligned}\hat{s} &= (p_1 + p_3)^2 \\ \hat{t} &= (p_1 - p_2)^2 \\ \hat{u} &= (p_1 - p_4)^2\end{aligned} \quad (2.1)$$

When $\hat{s}$ and $\hat{t}$ are both large the physical final state will consist of two jets. Two variables that will be very useful in describing the jet kinematics are:

- $y \equiv \frac{1}{2} ln(\frac{E+p_{\|}}{E-p_{\|}})$, the jet rapidity. The relation between jet rapidity and angle of the jet relative to the beam direction is shown in Figure 20.

- $p_\perp$, the magnitude of the momentum of a jet perpendicular to the beam direction.

The differential cross section for incident hadrons a and b to produce a two jet final state with rapidities y_1 and y_2 and with given $p_\perp$ is

$$\frac{d^3\sigma}{dy_1 dy_2 dp_\perp} = \frac{2\pi\tau}{\hat{s}} p_\perp \sum_{ij} \frac{1}{(1+\delta_{ij})} [f_i^{(a)}(x_a, Q^2) f_j^{(b)}(x_b, Q^2) \hat{\sigma}_{ij}(\hat{s}, \hat{t}, \hat{u})$$
$$+ f_j^{(a)}(x_a, Q^2) f_i^{(b)}(x_b, Q^2) \hat{\sigma}_{ji}(\hat{s}, \hat{t}, \hat{u})] \qquad (2.2)$$

where $f_i^{(a)}$ is the probability distribution function for the ith parton in the hadron a as discussed in the previous lecture. The sum is over all initial state quarks and/or gluons which can contribute and the cross section is summed over all final states which are not distinguishable experimentally. A crossed term must be included because parton i may have come from either hadron a or hadron b; and a symmetry factor is included to avoid overcounting in the case of identical partons in the initial state. Also, because the scale (Q^2) dependence of the distibution functions, it is necessary to know the appropriate value of Q^2 for the given subprocess. To give a complete determination of this quantity requires analysis beyond the Born approximation. An partial estimate of Q^2 has been done[34] which suggests $Q^2 = p_\perp^2/4$.

The final ingredient needed to determine the differential cross section is the Born approximation for the elementary subprocesses. The differential cross section for two to two parton scattering can be expresses as:

$$\frac{d\hat{\sigma}}{d\hat{t}} = \frac{\pi \alpha_s^2}{\hat{s}^2} |M|^2 \qquad (2.3)$$

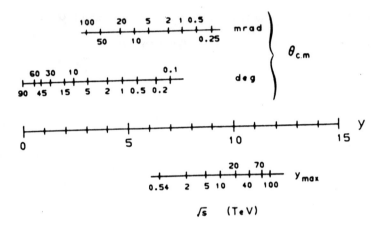

Figure 20: Correspondence of angles to the CM rapidity scale. Also shown is the maximum rapidity, $y_{max} = ln(\sqrt{s}/M_{proton})$ accessible for light secondaries.

and the invariant matrix element squared, $|M|^2$, are listed in Table 1 for all the two to two processes[34]. All partons have been assumed to be massless. In the subprocess CM frame the relationship between the scattering angle θ and $\hat{t}$ or $\hat{u}$ is

$$\hat{t} = -\frac{\hat{s}}{2}(1 - cos\theta)$$
$$\hat{u} = -\frac{\hat{s}}{2}(1 + cos\theta) \qquad (2.4)$$

The third column in Table 1 gives the value of $|M|^2$ at $90°$ in the CM frame. Two features of these cross sections will be particularly important. First, by far the largest cross-section is for the process $gg \to gg$. Second, reactions which initial parton type is perserved are considerably larger than those in which the final partons are different from the initial partons.

Using the structure functions of Set 2 determined in the last lecture from deep-inelastic leptoproduction and the subprocess cross section of Eq. 2.3, the single jet inclusive cross section at $\sqrt{s} = 540$ GeV is obtained from Eq. 2.1 simply by integrating over y_2. The single jet production rate can then be compared to the data from UA1[35,36] and UA2[37,38]. As shown in Figure 21, one obtains good agreement for $\Lambda = 290$ MeV and $Q^2 = p_\perp^2/4$ at rapidity $y = 0$ ($90°$ in hadron-hardon CM frame). Note that at low $p_\perp$ gluon-gluon scattering is dominant whereas at higher $p_\perp$ quark-gluon scattering dominates, and at the highest $p_\perp$ quark-quark scattering gives the leading contribution. Presently it is not possible to distinguish a light quark from a gluon jet experimentally; theoretical knowledge of which type of jet should be dominate at a given $p_\perp$ will be very helpful in finding their distinct experimental signature.

In the running at $\sqrt{s} = 540$ GeV there was a total integrated luminosity of

Table 1: Two to two parton subprocesses. $|M|^2$ is the invariant matrix element squared. The color and spin indices are averaged (summed) over initial (final) states. All partons are assumed massless. The scattering angle in the center of mass frame is denoted θ.

Process	$\|M\|^2$	$\theta = \pi/2$
$q\,q' \to q\,q'$	$\dfrac{4}{9} \dfrac{\hat{s}^2 + \hat{u}^2}{\hat{t}^2}$	2.22
$q\,q \to q\,q$	$\dfrac{4}{9} \left(\dfrac{\hat{s}^2 + \hat{u}^2}{\hat{t}^2} + \dfrac{\hat{s}^2 + \hat{t}^2}{\hat{u}^2} \right) - \dfrac{8}{27} \dfrac{\hat{s}^2}{\hat{u}\hat{t}}$	3.26
$q\,\bar{q} \to q'\,\bar{q}'$	$\dfrac{4}{9} \dfrac{\hat{t}^2 + \hat{u}^2}{\hat{s}^2}$	0.22
$q\,\bar{q} \to q\,\bar{q}$	$\dfrac{4}{9} \left(\dfrac{\hat{s}^2 + \hat{u}^2}{\hat{t}^2} + \dfrac{\hat{t}^2 + \hat{u}^2}{\hat{s}^2} \right) - \dfrac{8}{27} \dfrac{\hat{u}^2}{\hat{s}\hat{t}}$	2.59
$q\,\bar{q} \to g\,g$	$\dfrac{32}{27} \dfrac{\hat{t}^2 + \hat{u}^2}{\hat{t}\hat{u}} - \dfrac{8}{3} \dfrac{\hat{t}^2 + \hat{u}^2}{\hat{s}^2}$	1.04
$g\,g \to q\,\bar{q}$	$\dfrac{1}{6} \dfrac{\hat{t}^2 + \hat{u}^2}{\hat{t}\hat{u}} - \dfrac{3}{8} \dfrac{\hat{t}^2 + \hat{u}^2}{\hat{s}^2}$	0.15
$g\,q \to g\,q$	$-\dfrac{4}{9} \dfrac{\hat{s}^2 + \hat{u}^2}{\hat{s}\hat{u}} + \dfrac{\hat{u}^2 + \hat{s}^2}{\hat{t}^2}$	6.11
$g\,g \to g\,g$	$\dfrac{9}{2} \left(3 - \dfrac{\hat{t}\hat{u}}{\hat{s}^2} - \dfrac{\hat{s}\hat{u}}{\hat{t}^2} - \dfrac{\hat{s}\hat{t}}{\hat{u}^2} \right)$	30.4

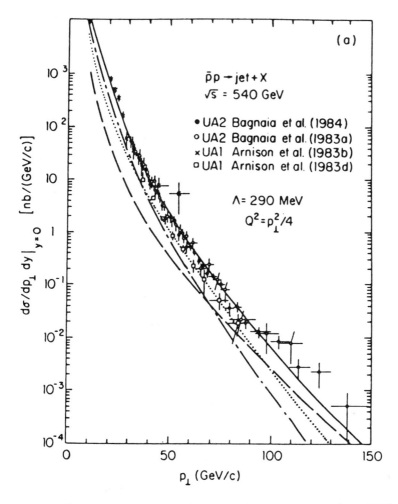

Figure 21: Differential cross section for jet production at $y = 0$ (90° CM frame) in $\bar{p}p$ collisions at 540 GeV according to the parton distributions of Set 2. The data are from Arnison et. al. (1983a is Ref. 35, 1983d is Ref. 36) and Bagnaia et.al. (1983b is Ref. 37, 1984 is Ref. 38).

about 100nb^{-1}. (One nanobarn (nb) is $10^{-33} cm^2$.) Thus if the minimum signal for jet study is 10 events/10GeV $p_\perp$ bin then the highest observable jet $p_\perp$ is about 100 GeV where the cross section becomes 10^{-2} nb.

In Figure 22 the data from UA2[39] is shown for both $\sqrt{s} = 540$ GeV and $\sqrt{s} = 630$ GeV along with our theoretical expectations.

Given the total running at $\sqrt{s} = 630$ GeV corresponds to an integrated luminosity of $\approx 400 nb^{-1}$, the maximum observable jet $p_\perp$ is ≈ 125 GeV/c.

If we extrapolate to SSC energies $\sqrt{s} = 40$ TeV, jets with very high $p_\perp$ will be observable. From EHLQ Fig.78, it is found that jets with $p_\perp \approx 4$ TeV are produced at the rate of 10 events per 10 GeV bin with an integrated luminosity 10^{40} cm^{-2}, about a year of running at the planned luminosity 10^{33} cm^{-2} sec^{-1}. The dominate two jet final states at various total transverse energy of the two jets $E_T \approx 2p_\perp$ is shown as a function of $\sqrt{s}$ for pp collisions in Figure 23. Also displayed are the values of E_T at which there be one jet event per bin of .01 $p_\perp$ for integrated luminosities of 10^{38} and 10^{40} cm^{-2} sec^{-1}. Notice that at $\sqrt{s} = 40$ TeV the quark-quark final states never dominates below these limiting E_T's even for integrated luminosity of 10^{40}.

One can investigate the angular distributions for various processes as a function of the subprocess CM scattering angle. The scattering processes for lowest order QCD given in Table 1 exhibit a forward and backward peak which is due to the exchange of a vector particle and is familiar from QED. In fact defining a variable

$$\chi = \frac{(1 + cos\theta)}{(1 - cos\theta)} \tag{2.5}$$

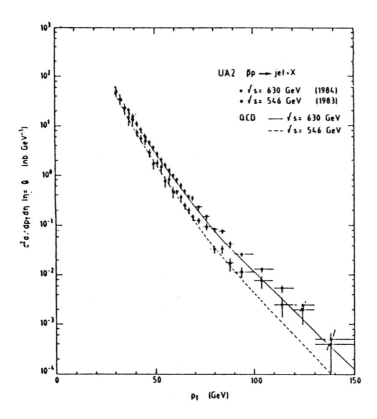

Figure 22: Inclusive jet production cross section (from UA2) at $y = 0$ as a function of the jet transverse momentum $p_\perp$. The data points for two collision energies 540 GeV (open circles) and 640 GeV (full circles) are compared to QCD predictions of Set 2 (solid lines). The additional systematic uncertainity in the data is ±45%.[39]

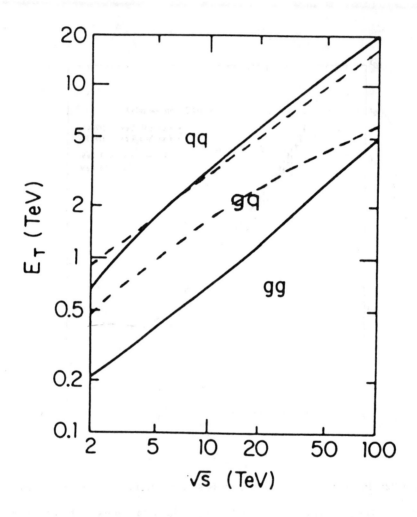

Figure 23: Parton composition of the two jet final states produced in pp collisions at 90° in the CM frame. The solid curves separate the regions in which gg, qg, and qq final states are dominant. The upper (lower) dashed line give maximum E_T for integrated luminosity of 10^{40} (10^{38}) cm^2.

a differential cross section behaving like

$$\frac{d\sigma}{d\cos\theta} \sim \frac{1}{(1-\cos\theta)^2} \qquad (2.6)$$

becomes

$$\frac{d\sigma}{d\chi} \sim \text{constant} . \qquad (2.7)$$

Therefore to a good approximation the behaviour of $d\sigma/d\chi$ in lowest order perturbation theory is a constant. The expected angular distribution agrees well with the UA1 data[40] as shown in Figure 24.

2.2 Multijet Events

Multijet events are also observed. Most of these events are composed of three jets, an example from the UA1 data[40] is shown in Figure 25. There are also some four jet events; one example of a four jet event from the UA2 data[41] is shown in Figure 26. In this event the four jets emerge at equal angles in a plane perpendicular to the beam direction. Four jet events will not be considered further here, since the theoretical calculations for the two to four parton QCD processes are still in progress[42].

The three jet events arise from the two to three parton scattering subprocess as shown in Figure 27. One invariant is:

$$\hat{s} = (p_4 + p_5)^2 \qquad (2.8)$$

In the subprocess CM frame $\vec{p_1} + \vec{p_2} + \vec{p_3} = 0$.

The kinematics is determined in terms of five variables in addition to the total energy of the subprocess $\sqrt{\hat{s}}$. Three of these variables, x_i for i=1,2,3, are the fraction of one half the CM energy taken by parton i in the final state; that is,

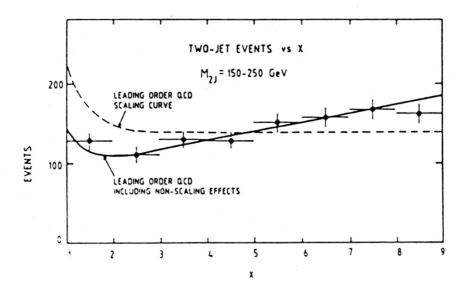

Figure 24: The two jet angular distribution plotted versus $\chi = (1+\cos\theta)/(1-\cos\theta)$. The broken curve shows the leading order QCD prediction, and the solid curve include scale breaking corrections. (From Ref. 40)

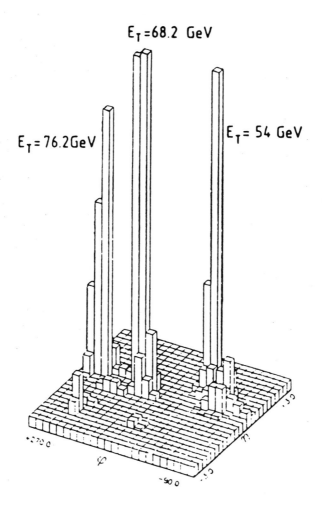

Figure 25: A typical LEGO plot for a three jet event from the UA1 data. (From Ref. 40)

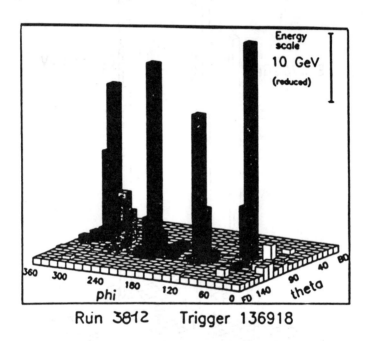

Figure 26: A LEGO plot for a four jet event observed in the UA2 data. (From Ref. 41)

$E_i = \hat{x}_i \sqrt{\hat{s}}/2$. Transverse momentum conservation ensures that $0 \leq \hat{x}_i \leq 1$, and overall energy conservation requires $\sum \hat{x}_i = 2$. The other variables are chosen to be θ, the angle of the plane formed by the final state partons with the beam direction; and ϕ, the azimuthal orientation of this plane with respect to the beam axis. Let

$$y_{boost} = (y_1 + y_2 + y_3)/3 \tag{2.9}$$

then the differential cross-section for this process can be written as

$$\frac{d\sigma}{d\hat{x}_1 d\hat{x}_2 dy_{boost} d\sqrt{\hat{s}} d\Omega} = \frac{\alpha_s^3(Q^2)\tau}{8\pi\sqrt{\hat{s}}} \sum_{ij} \frac{1}{1+\delta_{ij}} [f_i^{(a)}(x_a, Q^2) f_j^{(b)}(x_b, Q^2) |A_{ij}|^2 + (i \leftrightarrow j)] \tag{2.10}$$

where

$$\tau = \frac{\hat{s}}{s} \tag{2.11}$$

$$x_a = \sqrt{\tau} e^{y_{boost}} \tag{2.12}$$

$$x_b = \sqrt{\tau} e^{-y_{boost}} \tag{2.13}$$

$$Q^2 = s/4 . \tag{2.14}$$

and $|A_{ij}|^2$ is the absolute square of the invariant amplitudes computed by Berends, et. al.[43].

For the symmetric configuration, $x_i = 2/3$ for i = 1,2,3, $y_{boost} = 0$ and $\theta = 0$, the expected cross section is given in Figure 28 for $\sqrt{s} = 540$ GeV. Even at the highest $p_\perp$ shown the three quark jet final state does not dominate.

Instead of a detailed analysis of the kinematics for multi-jet processes it is more useful and instructive to do a simple theoretical calculation for a particular process. One of the most straightforward is the $gg \to ggg$ which has been computed by Berends et al.[43]. Defining a symmetric set of k_i for i=1,...,5 from the momenta of Figure 27 by:

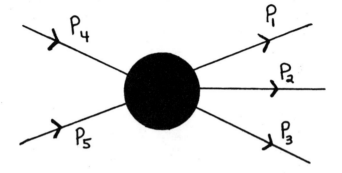

Figure 27: Two to three scattering process

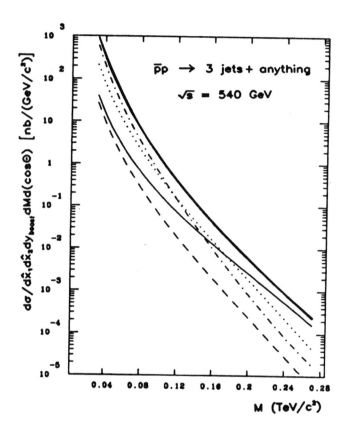

Figure 28: Differential cross section (thick line) for symmetric three jet production in $\bar{p}p$ collisions at 540 GeV, according to the distributions of Set 2. The ggg (dot-dashed line), ggq (dotted line), gqq (thin line), and qqq (dashed line) components are shown serarately.

$$k_3 = -p_1 \quad k_4 = -p_2 \quad k_5 = -p_3$$
$$k_1 = p_5 \quad k_2 = p_4$$

the invariant amplitude squared for the two to three process is:

$$|A|^2_{2\to 3} = \frac{27}{640} \frac{\sum_{m<n} k_{mn}^4}{\Pi_{m<n} k_{mn}} \sum_{\text{perms}} (12345) \qquad (2.15)$$

where:

$$k_{mn} = (k_m + k_n)^2/2 = k_m \cdot k_n \qquad (2.16)$$

for gluons on the mass shell, and

$$(12345) = k_{12} k_{23} k_{34} k_{45} k_{51} \qquad (2.17)$$

In the limit in which two of the final gluons become collinear (for example, 4 and 5), the amplitude simplifies considerably. In this case, k_4 and k_5 become parallel and may be written as

$$k_4 = (1-x)k \quad \text{and} \quad k_5 = xk \qquad (2.18)$$

where k is the total momentum of these two gluons and x is the fraction of this momentum which is carried by gluon 5. Then the leading behaviour of k_{45} is

$$k_{45} = x(1-x)k^2 \to 0 \,. \qquad (2.19)$$

The dominant contribution to the squared amplitude (Eq. 2.15) may now be calculated as follows. First the denominator is expanded to give the leading pole behaviour of $|A|^2_{2\to 3}$

$$\Pi_{m<n} k_{mn} \to k_{12} k_{13} k_{23} [x(1-x)]^4 k^2 (k \cdot k_1 \, k \cdot k_2 \, k \cdot k_3)^2 \qquad (2.20)$$

and the leading terms in the numerator are retained

$$\sum_{\text{perms}} (12345) \to \sum_{\text{perms}1,2,3} 10x^2(1-x)^2[(k \cdot k_1)(k \cdot k_3)^2(k \cdot k_2)(k_1 \cdot k_2)]$$

$$= 20x^2(1-x)^2(k \cdot k_1)(k \cdot k_2)(k \cdot k_3)[k_{12}(k \cdot k_3)$$

$$+ k_{23}(k \cdot k_1) + k_{31}(k \cdot k_2)] \qquad (2.21)$$

and

$$\sum_{m<n} k_{mn}^4 \to k_{12}^4 + k_{23}^4 + k_{31}^4 + (x^4 + (1-x)^4)[(k \cdot k_1)^4 + (k \cdot k_2)^4 + (k \cdot k_3)^4] \qquad (2.22)$$

where the expressions in Eqs. 2.18 and 2.19 have been used for k_4, k_5, and k_{45}. Equation 2.22 may be simplified to give:

$$(k_{12}^4 + k_{23}^4 + k_{31}^4)[1 + x^4 + (1-x)^4] \qquad (2.23)$$

where $k \cdot k_i = k_m \cdot k_n$ (i,m,n are cyclic permutations of 1,2,3), since $\sum_{i=1}^{3} k_i + k = 0$. Now the $2 \to 3$ gluon process is rewritten to give

$$|A|_{2\to 3}^2 \to \frac{27}{32} \frac{[1 + x^4 + (1-x)^4]}{x^2(1-x)^2 k^2} \frac{(k_{12}^4 + k_{23}^4 + k_{31}^4)}{k_{12}^2 k_{23}^2 k_{31}^2} [k_{12}^2 + k_{23}^2 + k_{31}^2] \qquad (2.24)$$

The above squared amplitude may be compared with the $2 \to 2$ gluon process (whose squared amplitude is given in Table 1) by reexpressing the momenta k_1, k_2, k_3, and k of Eq. 2.24 in terms of the invariants of the $2 \to 2$ process: $s = 2k_1 \cdot k_2$, $t = 2k_2 \cdot k_3$, and $u = 2k_1 \cdot k_3$. We have:

$$16(k_{12}^4 + k_{23}^4 + k_{31}^4) = s^4 + t^4 + u^4 = 2(s^2 t^2 + t^2 u^2 + u^2 s^2) \qquad (2.25)$$

and

$$4(k_{12}^2 + k_{23}^2 + k_{31}^2) = s^2 + t^2 + u^2 = -2(st + tu + us) \qquad (2.26)$$

Thus the factors containing the momenta k_{ij} in Eq. 2.24 becomes:

$$4[3 - \frac{tu}{s} - \frac{us}{t} - \frac{st}{u}] = \frac{8}{9}|A|_{2\to 2}^2 \qquad (2.27)$$

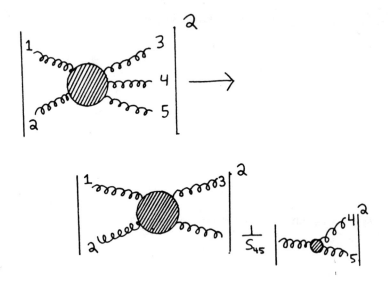

Figure 29: The leading pole behaviour of the $2 \to 3$ gluon process as two gluons become collinear. s_{45} is the propagator denominator for a gluon of momentum $k = k_4 + k_5$ and the squared vertex factor in exactly the Altarelli-Parisi splitting function for the gluon.

Therefore, the leading pole behaviour of the 2 → 3 squared amplitude as two gluons (4 and 5 here) become collinear is given in terms of the 2 → 2 process by:

$$|A|^2_{2\to 3} \to |A|^2_{2\to 2} \frac{1}{s_{45}} (3[\frac{x}{1-x} + \frac{(1-x)(1+x^2)}{x}]) \qquad (2.28)$$

where $s_{45} = 2x(1-x)k^2$. This equation is represented symbolically in Figure 29.

Thus the final result is that the leading behaviour of the two to three gluon scattering process as any two gluons become collinear is given by the product of: (1) The associated two to two process with the two collinear gluons being replaced by one on shell (massless) gluon with momentum, k , equal to the sum of their momenta, $k_4 + k_5$; (2) The propagator pole $1/s_{45}$, and (3) The Altarelli-Parisi (A-P) splitting function (see eq. 1.32) for $g \to gg$, i.e. the probability for a gluon to splitting into two carrying fractions x and 1-x respectively of the initial gluon momentum.

This result is a general feature of all the higher order processes in QCD, the leading behaviour of the N parton process as two parton momentum become collinear is given in terms of the associated N-1 parton process, a pole containing the singularity, and the relevant A-P splitting function. Since the amplitudes of QCD parton processes have only been calculated for $N < 4$ (even in tree approximation), Monte Carlo calculations of multijet processes in hadron collisions have taken advantage of the relation Eq. 2.28 to express the leading pole approximation to the N parton process in terms of the 2 → 2 process and the quark and gluon splitting functions. In fact, this approximation is then used everywhere even used outside of its range of validity; for example, for three jet events in which no two of the jets are collinear. The error made in this approximation varies with the values of the invariants (i.e. $\hat{s}$, $\hat{t}$, $\hat{u}$ etc.) but is generally less that a factor of two[44].

It is an easy exercise to show that if one starts with the $gg \to ggg$ amplitude squared and then lets one of the momenta of the final gluons approach zero, that the result is again proportional to the $gg \to gg$. The remaining factor is just the infrared correction to the $gg \to gg$ process and is again given by the gluon propagator in this limit times the appropriate splitting function.

2.3 Heavy Quark Production - The Top Quark

In addition to the jet structure of the strong interactions, there is another aspect of hadron physics of great current interest– heavy quark production and, in particular, top quark production. The top quark is the only missing constitutent of the three generations of the standard model. Heavy quarks can be produced through strong interactions either from the gg subprocesses or $q\bar{q}$ subprocesses shown in Figure 30 in lowest order QCD perturbation theory. The differential cross section for the gg production mechanism is given by:

$$\frac{d\hat{\sigma}}{d\hat{t}}(gg \to Q\overline{Q}) = \frac{\pi\alpha_s^2}{8\hat{s}^2}(\frac{6}{\hat{s}^2}(\hat{t}-m^2)(\hat{u}-m^2) + [(\frac{4}{3}\frac{(\hat{u}-m^2)}{(\hat{t}-m^2)}$$
$$\frac{4}{3}\frac{-2m^2(\hat{t}+m^2)}{(\hat{t}-m^2)^2} + 3\frac{(\hat{t}-m^2)(\hat{u}-m^2) + m^2(\hat{u}-\hat{t})}{\hat{s}(\hat{t}-m^2)}$$
$$+ (\hat{t} \leftrightarrow \hat{u})] - \frac{m^2(\hat{s}-4m^2)}{3(\hat{t}-m^2)(\hat{u}-m^2)}) \qquad (2.29)$$

where m is the heavy quark mass. The $q\bar{q}$ mechanism is less important since the quark-quark luminosities don't dominate at subenergies which give reasonable event rates (see Fig. 23) and because the color factors in the cross section are smaller than for gluon production. The differential cross section for $q\bar{q}$

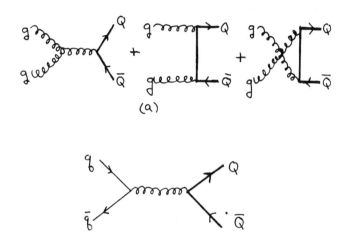

Figure 30: Lowest order Feynman diagrams for the process (a) $gg \to Q\overline{Q}$ and (b) $q\overline{q} \to Q\overline{Q}$

production is given by:

$$\frac{d\hat{\sigma}}{d\hat{t}}(q\bar{q} \to Q\bar{Q}) = \frac{4\pi\alpha_s^2}{9\hat{s}^2}[\frac{(\hat{t}-m^2)^2 + (\hat{u}-m^2)^2 + 2m^2\hat{s}}{\hat{s}^2}] \qquad (2.30)$$

Another mechanism for producing heavy quarks, one which is especially important at the $S\bar{p}pS$ and the Tevatron, is decay of the weak vector bosons $W^\pm$ and Z^0. A fuller discussion of electroweak processes will be given in the next lecture. The relevant lowest order diagrams are shown in Figure 31. The charged W's are produced by $u\bar{d}$ and $d\bar{u}$ initial states whereas the neutral boson Z is produced by $u\bar{u}$ and $d\bar{d}$ initial states.

For the top quark production at the $S\bar{p}pS$ or even the Tevatron, only real production of the weak bosons and subsequent decay to top quarks is significant. Away from the W or Z pole the addition power of α_{EM} makes the porduction rates negligible. To calculate the rate of top quark production associated with real weak boson production we must fold in the phase space factors for the final state containing a top quark, which given by

$$\frac{2p_W}{m_W} = \sqrt{[1 - \frac{(m_t + m_b)^2}{m_W^2}][1 - \frac{(m_t - m_b)^2}{m_W^2}]} \qquad (2.31)$$

for the $W^\pm$ and

$$\frac{2p_Z}{m_Z} = \sqrt{[1 - \frac{4m_Z^2}{m_W^2}]} \qquad (2.32)$$

for the Z^0.

The various contributions to the total cross section for heavy quark (top) production as a function of quark mass is shown in Figure 32 for $\sqrt{s} = 630$ GeV in $\bar{p}p$ collisions. The gluon production contribution drops rapidly as the top quark mass increases, while the W^+ contibution is flat up to the phase space factors above (Eqs. 2.31-32) for top quark masses low enough to be associated

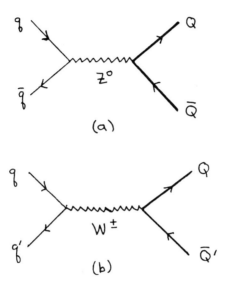

Figure 31: Lowest order Feynman diagrams for the production of (a) heavy quark pairs via a real or virtual Z^0 intermediate state, and (b) a heavy quark and a light quark via a real or virtual $W^\pm$ intermediate state.

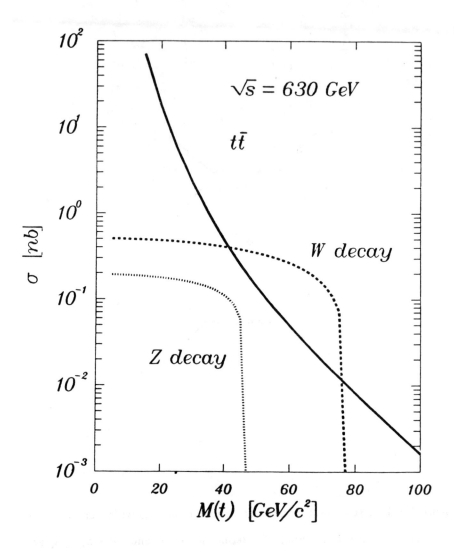

Figure 32: Cross sections for production of t or $\bar{t}$ quarks in $\bar{p}p$ collisions at $\sqrt{s} = 630$ GeV as a function of the mass of the heavy quark. The gluon production, W decay, and Z decay contributions are shown by solid, dashed, and dotted lines respectively.

with the decays of real $W^\pm$ and it drops precipitously when the W^+ mass is reached and only virtual W^+ production is possible. The Z boson contribution is always small compared to the W contribution. Gluon production dominates for $m_t < 40$ GeV, while at higher (observable) m_t W^+ production dominates.

2.4 A Higher Order QCD Process

As a final example of multijet processes, it is instructive to consider the next order QCD contribution to heavy quark production. For example, a typical reaction would be $gg \to g\overline{Q}Q$, which is suppressed by a factor of α_s^3 relative to α_s^2 of the lowest order process. For $\alpha_s = .1$, we would naively expect the higher order terms to be down by a factor of ten from the lowest order processes. Although this process has a different topology than the simple process $gg \to \overline{Q}Q$, it will contribute to inclusive top production. This process can be analyzed in the leading pole approximation as in the $gg \to ggg$ process discussed in Section 2.2. When the $\overline{Q}Q$ invariant mass, $\sqrt{s_{\overline{Q}Q}}$, although large, is assumed to be small compared with the total energy of the subprocess, $\sqrt{\hat{s}}$; then, we obtain the following expression for the subprocess cross section:

$$\sigma(a+b \to Q+\overline{Q}+c) \approx \sigma(a+b \to g+c) \int \frac{ds_{\overline{Q}Q}}{s_{\overline{Q}Q}} dx \frac{\alpha_s}{2\pi} P_{g \to Q\overline{Q}}(x) \quad (2.33)$$

which is shown schematically in Figure 33. Here the splitting function for a gluon into quark-antiquark is given by:

$$P_{g \to Q\overline{Q}}(x) = \frac{1}{2}[x^2 + (1-x)^2] \quad (2.34)$$

In most of the relavant kinemtic range, this approximation to the exact $gg \to g\overline{Q}Q$ process is good to within 20 percent.

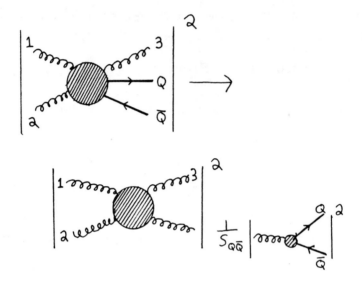

Figure 33: The $gg \to g\overline{Q}Q$ process in leading logarithmic approximation. As Q and $\overline{Q}$ become collinear we can express the process in terms of the on shell $gg \to gg$ process times the A-P splitting function for $g \to \overline{Q}Q$ and a singular propagator $1/s_{\overline{Q}Q}$.

To see the relative contribution to heavy quark production of this $2 \to 3$ one can simple calculate the ratio of this cross section to that for the $gg \to Q\overline{Q}$ process. Roughly for production at $90°$ in the CM frame one obtains:

$$\frac{\sigma(gg \to gQ\overline{Q})}{\sigma(gg \to Q\overline{Q})} = \frac{\sigma(gg \to gg)}{\sigma(gg \to Q\overline{Q})} \frac{\alpha_s}{3\pi} log(\frac{\hat{s}}{4m_Q^2}) \qquad (2.35)$$

Estimating each the terms we find that, even though there is the $\frac{\alpha_s}{3\pi} \approx .03$ suppression, the $gg \to gQ\overline{Q}$ cross section is considerably larger than the lower order $gg \to Q\overline{Q}$ process because the basic $gg \to gg$ cross section is larger than the $gg \to q\overline{q}$ cross section by a factor of **104**. Aside from a larger QCD color factor for two gluon final state, there some dynamic cancellations between terms for the quark-antiquark final state. The logarithmic term in Eq. 2.35 adds a factor of 2-4, leading to a total factor of 5-10 for the ratio of cross sections for production of heavy quark-antiquark plus energetic gluon to heavy quark-antiquark alone. The first to point out this fact was Z. Kunszt, et.al.[45]. It is important to point out that this is not a breakdown of QCD perturbation theory at high energy. These are two physically distinct processes. This exercise does however illustrate that naive α_s power counting is not always sufficient to determine the relative importance of various subprocesses in a given experimental situation.

The relative strengths of these two contributions to inclusive production of charm and bottom quarks at the $S\overline{p}pS$ collider has been analysed by Halzen and Hoyer[46]. The results are shown in Figure 34. Experimentally, these two process have very different topological structure and therefore can be differentiated. The $2 \to 2$ contribution produces relatively large heavy quark pair mass, $s_{\overline{Q}Q}/s$ near 1, and the Q and $\overline{Q}$ move in approximately opposite directions to each other; whereas, the $2 \to 3$ contribution tends to produce relatively small pair mass and both the Q and $\overline{Q}$ will tend to move in roughly the same direction. In Fig.

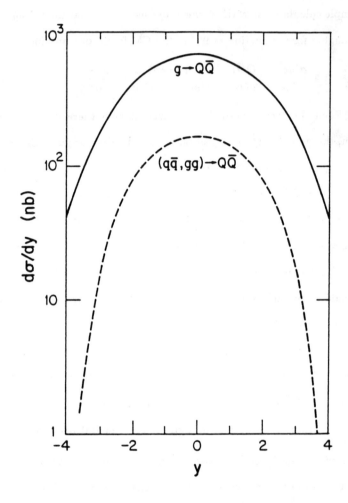

Figure 34: Rapidity distribution for heavy quark production (charm and bottom) by the 2 → 2 (dashed line) and 2 → 3 (solid line) processes. (From Ref. 46)

34 a cut on the minimum $p_\perp$ of 7 GeV/c has been imposed on the jet opposite the heavy quark to avoid a divergence in the 2 → 3 amplitude as the gluon momentum becomes soft. Moreover, folding in the experimental jet cuts will enhance the 2 → 2 contribution to the point that it dominates the observed events.

2.5 Jets - Present and Future

To summarize, jets have emerged at $S\bar{p}pS$ collider energies as clear and distinct tags of the underlying quarks and gluons. The predictions of QCD have been verified within the accuracy of present experimental data. Futhermore, the one to one association of jets with the underlying quarks and gluons will provide an important tool for studying QCD at all higher energy hadron colliders.

Future tests of QCD in hadron-hadron colliders will depend on detailed analysis of rare events and on precision measurements of basic jet physics. Jets will be used to identify the specific parent quark or gluon. But to do this methods are needed:

- To distinguish reliably light quark from gluon jets. We know that at the lowest $p_\perp$'s gluon jets dominate and as the $p_\perp$ increases the fraction of quark jets also increase. This should be of some aid in finding the distinguishing characteristics of light quark vis-a-vis gluon jets.

- To find a signature of heavy quark jets, t and b, which distinguishes them from ordinary quark jets. Heavy quark are produced copiously at very high energies (eg. at the SSC) and if the jets associated with heavy quarks could be tagged many properties of the quarks could be studied. Also heavy quark jets are important signatures of many of the new physics possiblities to be discussed in the last three lectures.

Both of these problems are presently being vigorously investigated by both theorists and experimentalists[47,48].

3 ELECTROWEAK PHYSICS

The electroweak interactions, which provide the remaining gauge structure in the standard model, is the topic of this lecture. Since both leptons and quarks participate in these interactions, the electroweak force is probed both in lepton and in hadron colliders.

The Lagrangian for the electroweak interactions has the gauge structure $SU(2)_L \otimes U(1)_Y$ with massless gauge bosons $W^\pm, W^3$ and B respectively. The charged $SU(2)_L$ bosons acquire mass from symmetry breakdown as well as one combination of the neutral W and B bosons, the Z^0 boson, while the other neutral boson, the photon, remains massless. The fermions are grouped into three generations (k), each generation repeating the $SU(2)_L$ and $U(1)_Y$ classifications of the quarks and leptons. For the first generation the fermionic structure is

$$\begin{array}{ccccc} & SU(2)_L & & U(1)_Y & \\ \begin{pmatrix} u \\ d \end{pmatrix}_L & u_R & d_R & 1/3 \quad 4/3 & -2/3 \\ \begin{pmatrix} \nu_e \\ e \end{pmatrix}_L & & e_R & -1 & -2 \end{array}$$

The full Lagrangian for the Weinberg-Salam standard model is given by:

$$\begin{aligned} \mathcal{L} = & \sum_k \sum_{f=u,d,\nu,e} [i\bar{f}_k \gamma^\mu (\partial_\mu - ig\frac{\tau^a}{2}(\frac{1-\gamma_5}{2})W_{\mu a} - ig'\frac{Q_Y}{2}B_\mu) f_k] \\ & - \frac{1}{4}\sum_a (\partial_\mu W_{\nu a} - \partial_\nu W_{\mu a} + g\epsilon_{abc}W_{\mu b}W_{\nu c})^2 - \frac{1}{4}(\partial_\mu B_\nu - \partial_\nu B_\mu)^2 \\ & + (\mathcal{D}_\mu \phi)^* (\mathcal{D}^\mu \phi) - [-\mu^2 \phi^* \phi + \lambda (\phi^* \phi)^2] \\ & + \sum_k \sum_{f=u,d,\nu,e} \Gamma_{kf} \bar{f}_L^{(k)} f_R^{(k)} \bar{\phi} + h.c. \end{aligned} \quad (3.1)$$

In addition to the gauge self interactions and the fermions with minimal couplings to the gauge fields, there are additional elementary scalar fields, sometimes called Higgs fields after the role they play in electroweak symmetry breaking. Here the Higgs particles are a complex doublet under $SU(2)_L$ with $Q_Y = 1$, so their gauge interactions are through the covariant derivative:

$$\mathcal{D}_\mu = \partial_\mu - ig\frac{\tau^a}{2}W_{\mu a} - ig'\frac{B_\mu}{2} \tag{3.2}$$

and their self interactions (the Higgs potential) are responsible for the electroweak symmetry breakdown. Finally, the Yukawa couplings of the Higgs particles to(the fermions are responsible for fermion mass generation.

3.1 Electroweak Symmetry Breaking

It is worthwhile examining in some detail the structure of the symmetry breakdown mechanism. The Higgs scalars are written in terms of a complex doublet field:

$$\phi \equiv \frac{1}{\sqrt{2}}\begin{pmatrix} \phi_1 + i\phi_2 \\ \phi_3 + i\phi_0 \end{pmatrix} \quad \text{and} \quad \overline{\phi} = i\sigma_2\phi^* \tag{3.3}$$

The Higgs potential consists of a mass term of the wrong sign with coefficient μ^2, and a quartic term of strength $\lambda > 0$:

$$V = -\mu^2(\phi^*\phi) + \lambda(\phi^*\phi)^2 \tag{3.4}$$

This potential is shown in Figure 35.

The symmetry of the Higgs potential is $SU(2)_L \otimes SU(2)_R \otimes U(1)$, a larger symmetry than the gauged $SU(2)_L \otimes U(1)_Y$ symmetry. To exposed this additional symmetry the Higgs potential can be rewritten in terms of

$$\Phi \equiv \frac{1}{\sqrt{2}}(\tau_0\phi_0 + i\tau_a\phi^a) \tag{3.5}$$

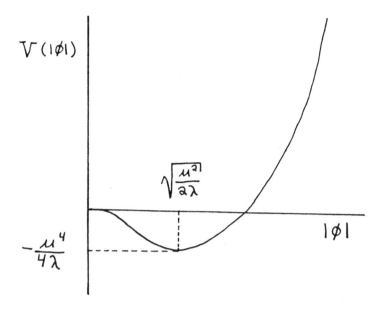

Figure 35: Higgs potential for a complex scalar field, ϕ.

where τ_a are the Pauli matrices and τ_0 is the 2x2 unit matrix. In terms of Φ, the scalar self interactions are given by:

$$\mathcal{L}_S = Tr(\partial_\mu \Phi^\dagger \partial^\mu \Phi) - \lambda [Tr(\Phi^\dagger \Phi) - \frac{\mu^2}{2\lambda}]^2 + \frac{\mu^4}{4\lambda}. \tag{3.6}$$

Now there is a manifest symmetry of this part of the Lagrangian under the transformations:

$$\Phi \to U_L \Phi U_R^\dagger \quad \text{where} \quad \begin{array}{l} U_L = e^{-i\Lambda_L^a \tau_a} \\ U_R = e^{-i\Lambda_R^a \tau_a} \end{array} \tag{3.7}$$

This symmetry is not valid for the full Lagrangian since:

- The Yukawa couplings, distinguishes members of the $SU(2)_R$ doublets. That is, $\Gamma_u \neq \Gamma_d$ in Eq. 3.1.

- The electroweak gauge interactions of the scalar fields also break the $SU(2)_R$ global symmetry of the Higgs potential.

Nonetheless, this extra invariance will be important to our discussions in lectures 4 and 5.

Now minimizing the Higgs potential in terms of Φ gives

$$<\Phi> = (v/\sqrt{2})\tau_0 \text{ where } v = \sqrt{\frac{\mu^2}{2\lambda}} \qquad (3.8)$$

Shifting the Higgs fields by this vacuum expectation value

$$\tilde{\Phi} \equiv \Phi - <\Phi> = \Phi - (v/\sqrt{2})\tau_0 \qquad (3.9)$$

replaces $Tr(\Phi^\dagger \Phi)$ by

$$Tr(\tilde{\Phi}^\dagger \tilde{\Phi}) - \frac{v}{\sqrt{2}}Tr[(\tilde{\Phi}^\dagger + \tilde{\Phi}) \cdot \tau_0] + v^2 \qquad (3.10)$$

The new expressions (Eqs. 3.9 and 3.10) are invariant under the vector subgroup of transformations $U_R = U_L$ so there is a residual $SU(2)_V \otimes U(1)_V$ symmetry. Rewriting the scalar potential in terms of a physical Higgs scalar $H = \frac{1}{\sqrt{2}}Tr(\tau_0 \tilde{\Phi})$ and $\tilde{\Phi}$:

$$V = -\frac{\mu^4}{4\lambda} + \frac{1}{2}m_H^2 H^2 + 2\sqrt{\lambda}m_H Tr(\tilde{\Phi}^\dagger \tilde{\Phi})H + \lambda[Tr(\tilde{\Phi}^\dagger \tilde{\Phi})]^2 \qquad (3.11)$$

where the Higgs scalar mass is $m_H^2 = 2\mu^2$ and the remaining three scalar fields are massless Goldstone bosons, ϕ^+, ϕ^0, and ϕ^-, corresponding to the three broken symmetries. These Goldstone bosons then provide masses for the W^+, Z^0, and W^- bosons by the usual Higgs mechanism[49]. The mass terms for the gauge bosons can be seen explicitly by writing the kinetic energy term for the scalar particles in terms of the field $\tilde{\Phi}$:

$$(D_\mu \Phi)^\dagger (D^\mu \Phi) = (D_\mu \tilde{\Phi})^\dagger (D^\mu \tilde{\Phi}) + g^2 v^2/2 \ W_\mu^+ W^{-\mu} + v^2/2 \ (gW_3^\mu - g'B^\mu)^2$$
$$- (v/\sqrt{2})[Tr([\tau_0(ig\frac{\tau_a}{2}W_{a\mu} + ig'\frac{B_\mu}{2})\partial_\mu \tilde{\Phi}] + h.c.) \qquad (3.12)$$

One linear combination of the two neutral gauge bosons, the Z^0, has acquired a mass while the other linear combination, the photon, remains massless. The mass eigenstates can be seen directly in Eq. 3.12 are:

$$Z^0 = \frac{gW_3 - g'B}{\sqrt{g^2 + g'^2}} \qquad m_Z = \sqrt{g^2 + g'^2}v/\sqrt{2} \qquad (3.13)$$

$$A = \frac{g'W_3 + gB}{\sqrt{g^2 + g'^2}} \qquad m_A = 0 \qquad (3.14)$$

$$W^\pm \qquad m_W = gv/\sqrt{2} \qquad (3.15)$$

The Weinberg angle, θ_w is defined by:

$$x_w \equiv sin^2\theta_w = \frac{g'^2}{g^2 + g'^2} \qquad (3.16)$$

Furthermore, we can define a parameter ρ, by:

$$\rho = \frac{m_W^2}{m_Z^2 cos^2\theta_w} \qquad (3.17)$$

Because of the custodial $SU(2)_V$ symmetry of the Higgs sector after symmetry breaking

$$\rho = 1 \qquad (3.18)$$

if quantum loop corrections are ignored.

Rewriting the interaction of the longitudinal degrees of freedom for the electroweak bosons in terms of mass eigenstates gives:

$$-\frac{v}{\sqrt{2}}Tr(\tau_0[ig\frac{\tau_a}{2}W_{\mu a} + ig'\frac{B_\mu}{2}]\partial^\mu \tilde{\Phi}) + h.c. \rightarrow gv(W_{1\mu}\partial^\mu \phi_1 + W_{2\mu}\partial^\mu \phi_2 + Z_\mu\partial^\mu \phi_3) \ . \qquad (3.19)$$

We see explicitly that the Goldstone bosons, ϕ_1, ϕ_2, ϕ_3, mix with the $W^\pm$ and Z^0 to become the longitudinal degrees of freedom for the corresponding gauge boson.

The charged weak currents have been described by the Fermi constant, G_F, long before the W-S model was proposed. The $W^\pm$ mass may be expressed in terms of this constant by:

$$m_W^2 = g^2 v^2/2 = g^2/(4\sqrt{2} G_F) \tag{3.20}$$

so that the vacuum expectaton value v is determined in terms of the Fermi constant as

$$tr(\Phi) = \sqrt{2} v = (2\sqrt{2} G_F)^{-\frac{1}{2}} = 246 GeV \tag{3.21}$$

This sets the scale of the weak interactions.

3.2 The $W^\pm$ and Z^0 Gauge Bosons

The gauge structure of the Weinberg-Salam model has been confirmed experimentally by the observation of W and Z bosons at the $S\bar{p}pS$ collider[50,51]. The Z^0 was observed in the decays into high energy $e^+ e^-$ and $\mu^+ \mu^-$ pairs. These events have essentially no backgrounds making identifying the Z^0 purely a matter of event rate. The $W^\pm$ decays are more numerous but their study is more complicated since only the charged lepton is observed directly. The neutrino escapes the dectector, hence its signature is large missing transverse energy E_T in the event. There is no actual resonance peak at the $W^\pm$ mass, although there is a Jacobian (phase space) peak in the charged lepton spectrum at a somewhat lower energy[52]. The measured masses are:

Experiment	m_W (GeV)	m_Z (GeV)
UA1 Ref. 50	$83.1 \, {}^{+1.3}_{-0.8} \pm 3$	$93.0 \pm 1.6 \pm 3$
UA2 Ref. 51	$81.2 \pm 1.1 \pm 1.3$	$92.5 \pm 1.3 \pm 1.5$

The first error quoted is the statistical error and the second is the systematic error. The width for the Z^0 measured by UA2[51] is

$$\Gamma_{\exp}(Z^0) = 2.19 \begin{array}{c} +1.7 \\ -0.5 \end{array} \pm 0.22 \text{GeV} \qquad (3.22)$$

The theoretical values[53] for the masses and other properties of the $W^\pm$ and Z^0 are collected in Table 2.

The value of x_w of $.217 \pm .014$ determined from other experiments[53] is used in this comparison. The theoretical calculations include one-loop corrections to the masses and widths. The calculations of their widths which assume a top quark mass of 40GeV. Theory and experiment agree within the present accuracy.

The background of ordinary two jet events with invariant pair mass equal to the W mass (within the experimental mass resolution) is of the same order of magnitude as the signal of hadronic W decays. Experimentalist are seeking addition cuts on such events which would establish a clear signal for hadronic decays of the W and Z, but as yet none have been identified. Experimental determination of the total cross sections for $W^\pm$ and Z^0 production times the leptonic branching ratio for $\bar{p}p$ collisions give:

$\sqrt{s}$ (GeV)	Experiment	$\sigma \cdot \text{BR}(e^+e^-)$ (in picobarns)	
		$W^+ + W^-$	Z^0
546	UA1 Ref. 50	$550 \pm 80 \pm 90$	$40 \pm 20 \pm 13$
	UA2 Ref. 51	$500 \pm 90 \pm 50$	$110 \pm 39 \pm 9$
630	UA1 Ref. 50	$630 \pm 50 \pm 90$	$79 \pm 21 \pm 12$
	UA2 Ref. 51	$530 \pm 60 \pm 50$	$52 \pm 19 \pm 4$

Table 2: Selected properties of the $W^{\pm}$ and Z^0 electroweak gauge bosons. Here $sin^2\theta_w \equiv x_w$ is assumed to be $.217 \pm .014$. Primes on down quarks denote electroweak eigenstates. The factor $f_s = 1 + \alpha_s$ includes the leading order QCD correction for decays.

Process	$W^{\pm}$	Z^0
Mass (GeV/c^2)	83.0 ± 2.8	93.8 ± 2.3
Branching Fractions Leptons ($l^+ = e^+, \mu^+,$ or τ^+) Light Quarks Top Quarks (mass = m_t)	$(l^+\nu) = 1$ $\begin{pmatrix} \bar{d'}u \\ \bar{s'}c \end{pmatrix} = 3f_s$ $(\bar{b'}t) = 3f_s \left(1 - \frac{m_t^2}{M_W^2}\right)^2$ $\left(1 + \frac{m_t^2}{2M_W^2}\right)$	$(\bar{\nu}\nu) = 2$ $(\bar{l}l) = 1 + (1 - 4x_w)^2$ $\begin{pmatrix} \bar{u}u \\ \bar{c}c \end{pmatrix} = 3f_s(1 + (1 - \frac{8}{3}x_w)^2)$ $\begin{pmatrix} \bar{d}d \\ \bar{s}s \\ \bar{b}b \end{pmatrix} = 3f_s(1 + (1 - \frac{4}{3}x_w)^2)$ $(\bar{t}t) = 3f_s\sqrt{1 - 4\frac{m_t^2}{M_Z^2}}$ $\left(1 - \frac{m_t^2}{M_Z^2} \frac{1 + \frac{16}{3}x_w - \frac{64}{9}x_w^2}{1 - \frac{8}{3}x_w + \frac{32}{9}x_w^2}\right)$
Total Width (GeV) ($m_t = 40 GeV/c^2$)	2.8	2.9

The experimental errors (statistical plus systematic) are sufficiently large that the growth of the cross sections with energy is not apparent. The theoretical predictions for the total cross sections at 630GeV, based on the analysis of Altarelli et. al.[54,55] are:

$$\sigma(\bar{p}p \to W^+ \text{ or } W^-) = 5.3 \, {}^{+\,1.6}_{-\,0.9}$$

$$\sigma(\bar{p}p \to Z^0) = 1.6 \, {}^{+\,0.5}_{-\,0.3} \qquad (3.23)$$

There are a number of sources for the theoretical uncertainties given above. The lower theoretical error takes into account the uncertainty in the determination of the parton distribution functions and the QCD Λ parameter. Altarelli et al.[54,55] considered a variety of different parameterizations for the parton distributions. The upper theoretical error also includes the uncertainty in what value to choose for the momentum scale which determines the scale violations in the distribution functions. This scale is determined in higher order in QCD perturbation theory, but these calculations have not yet been done. This ambiguity in this scale factor leads to an uncertainty in the total cross section. However, since the cross sections are being evaluated at high Q^2, a factor of two change in this momentum scale only results in small corrections to the cross section. The usual estimates are obtained by using the intermediate boson mass as this momentum scale. The upper error uses the transverse momentum of the final lepton as the momentum scale.

The ratio of the cross sections should be less sensitive to these theoretical ambiguities, and in fact theory and experiment are in good agreement for the ratio of W to Z total cross sections.

The relative branching ratios for various decays of the $W^\pm$ and Z^0 are also shown in Table 2 normalized to $W^+ \to l + \nu = 1$. Now using the theoretical branching ratio (assuming $m_t = 40\ GeV/c^2$) one predicts for the cross section, σ, times leptonic branching ratio, B, at $\sqrt{s} = 630$ GeV:

$$\sigma \cdot B(W^+ + W^-) = 460 \begin{array}{c} +140 \\ -80 \end{array} pb$$

$$\sigma \cdot B(Z^0) = 51 \begin{array}{c} +16 \\ -9 \end{array} pb \qquad (3.24)$$

The theoretical and experimental cross sections agree within the rather large errors although they do not coincide.

Table 3 shows the theoretical predictions for the total cross sections for single $W^\pm$ and Z^0 production at present and future hadron colliders. The structure functions of Set 2 (Eq. 1.24) are used for these cross sections.

A cross section of 10 (nb) corresponds to an expectation of 10^5 W^+ events/year for a luminosity of $10^{30} cm^{-2} sec^{-1}$. Hadron colliders provide a copious source of $W^\pm$ and Z^0 bosons. With such statistics:

- It is possible to study rare decays such as those expected in supersymmetry models (lecture 7), in extensions of the standard model with additional doublets of Higgs scalars, or in technicolor models (lecture 5).

- Precision (one loop) tests of the electroweak interactions will be possible. However most of these tests are better suited to e^+e^- colliders such as the SLC or LEPI, which provide a clean and copious source of Z^0 bosons.

- The total width of the Z^0 is sensitive to the number of generations, since there is a contribution of 186 MeV to the width of the Z^0 for every neutrino

Table 3: Total cross sections for production of single electroweak gauge bosons. All cross sections are in nanobarns.

Collider	$\sqrt{s}$ (TeV/c^2)	Gauge Boson		
		W^+	W^-	Z^0
$\bar{p}p$	.63	3.4	3.4	1.2
$\bar{p}p$	1.8	10.2	10.2	3.9
$\bar{p}p$	2.0	11.2	11.2	4.9
pp	10	41	28	22
pp	20	73	54	41
pp	40	122	95	72

type. Hence the measurment of the Z^0 width to an accuracy of 100 MeV will determine the number of standard generations.

At SSC energies and luminosities the rates for production of $W^\pm$ and Z^0 bosons are even more impressive. However since much of the production will be at sizable rapidities the events will not be as clean as at $S\bar{p}pS$ collider energies where the electroweak bosons are produced essentially at rest. Thus some ingenuity will be required to take advantage of these rates[56].

Next we will consider some of the details of $W^\pm$ production. For example the cross section for $p\bar{p} \to W^\pm + X$ is shown in Figure 36. This cross-section rises steeply near threshold because of both the threshold kinematics of the elementary process and the steep decrease in the the parton-parton luminosities as x approaches one. The production cross section in pp collisions (also shown in Fig. 36) is smaller than $\bar{p}p$ at the same $\sqrt{s}$ because of the lack of valence

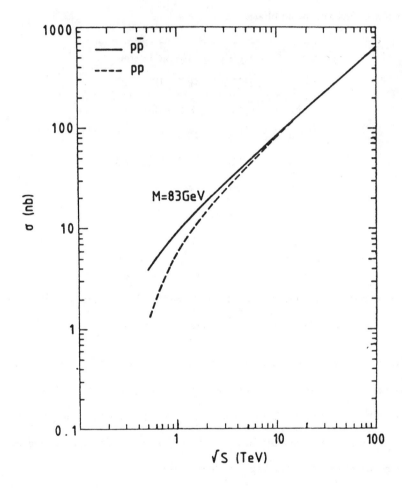

Figure 36: Total cross section for the production of W^+ and W^- (for $M_W = 83$ GeV/c^2) versus center of mass energy. The solid line is for $\bar{p}p$ collisions and the dashed line is for pp collisions. Adapted from Ref. 55

antiquarks in pp collisions. There are small differences between the W^+ and W^- production in pp collisions because the valence quarks contribute more to W^+ than to W^- production.

The rapidity distribution of W^+ production is relatively flat at SSC energies. The net helicity of W^+ inclusive production in pp and $\bar{p}p$ interactions can be calculated straightforwardly and is shown as a function of the rapidity for $\sqrt{s} = 40$ TeV in Figure 37. To understand qualitatively the behaviour of these helicities consider the two production modes:

- A W^+ can be produced from a u_L quark from the "beam" p or $\bar{p}$ carrying fraction x_1 of the beam momentum (defined to be in the $+z$ direction) and a $\bar{d}_R$ antiquark carrying fraction x_2 of the "target" proton. The resulting W^+ will have momentum along the beam direction

$$p_\| = (\frac{x_1 - x_2}{2})\sqrt{s} \qquad (3.25)$$

and spin $J_z = -1$, since the u_L has spin $J_z = -1/2$ and the $\bar{d}_R$ also has spin $J_z = -1/2$. Hence the helicity of the resulting W^+ is opposite to the sign of the longitudinal momentum $p_\|$.

- A W^+ can also be produced from a $\bar{d}_R$ antiquark from the beam p or $\bar{p}$ carrying fraction x_1 of the beam momentum and a u_L quark carrying fraction x_2 of the target proton. In this case, the resulting W^+ will now have spin $J_z = +1$, since the $\bar{d}_R$ has spin $J_z = 1/2$ and the u_L also has spin $J_z = 1/2$. Hence the helicity of the resulting W^+ is the same as the sign of the longitudinal momentum $p_\|$.

The net helicity of the W^+ results from the sum of these two production processes. For pp collisions the quark distributions are of course identical for

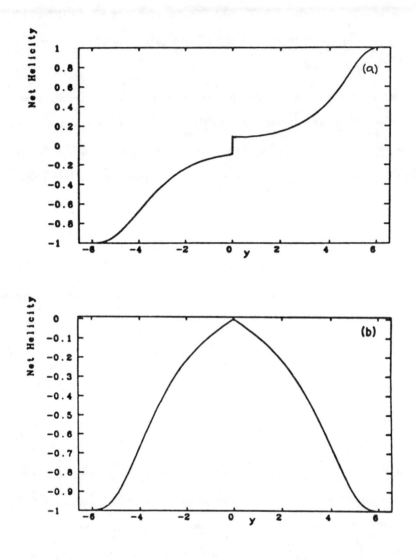

Figure 37: The net helicity of the W^+ as a function of the rapidity y. The W^+ production is shown both for $\bar{p}p$ (a) and for pp (b) collisions at $\sqrt{s} = 40$ TeV. Parton distributions of Set 2. (From EHLQ)

the "beam" and "target" particles. The contribution to W^+ production of the first process above for W^+ rapidity y equals the contribution of the second process for rapidity $-y$. Thus the net helicity $h_w(y)$ is symmetric about $x_1 = x_2$, (i.e. $y = 0$); thus $h_w(-y) = h_w(y)$. For $x_1 > x_2$, the valence quarks dominates so that for $y > 0$ the helicity is negative. For $\bar{p}p$ collisions the second process dominates since there both the quark and antiquark are valence quarks. Therefore the helicity is antisymmetric $h_w(-y) = -h_w(y)$; positive for $y > 0$; and is discontinuous at $y = 0$ since the net helicity does not vanish there.

The net helicity of the produced W^+ is a result of the W^+'s chiral coupling and leads to a measurable front-back asymmetry in the decay lepton spectrum. Measuring this helicity as a function of rapidity will distinguish chiral couplings (L,R) from nonchiral couplings (V,A) for the W or any new gauge boson of the electroweak type (i.e. coupling to both leptons and quarks).

3.3 Associated Production

In addition to the production of W's and Z's in the lowest order of QCD perturbation theory, there is the next order processes in which the W or Z are produced in association with a gluon or a gluon jet. These processes are shown in Figure 38.

Since the transverse momenta of the incoming partons is negligible at high energy, the gauge bosons produced by the lowest order subprocess have small transverse momentum whereas in associated production the gauge bosons may have large transverse momentum. One consequence of this associated production is the production of monojet events; which occur when the associated gluon or quark produces a jet with transverse momentum and the Z decays into an

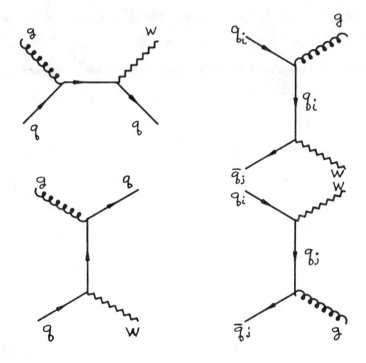

Figure 38: Lowest-order Feynman diagrams for the reactions $\bar{q} + q \to W + g$ and $g + q \to W + q$.

undetected $\nu\bar{\nu}$ pair. A few such events have been seen at UA1[57]. Calculations of the transverse momentum distribution of W's and Z's has been carried out by Altarelli et al.[54,55] for $\sqrt{s}$ = 630 GeV. In these calculations the leading log terms terms have been summed to all orders of perturbation theory. Their result is shown in Figure 39 for $y = 0$, i.e. at 90°. For example, at 630 GeV 1% of all W's associatively produced have transverse momementum greater than 45 GeV. At higher energies the resummation becomes less important at least at high p_T. The lowest order associated production of W's is shown in Figure 40. To get a feeling for event rates remember that a cross section of 10^{-5} (nb/GeV) corresponds to 100 events/yr/GeV for a luminosity of $10^{33} cm^{-2} sec^{-1}$. Therefore, very high transverse momentum W's are produced at SSC energies.

3.4 Electroweak Pair Production

The present experimental data show that the gauge bosons of the electroweak interactions exist and have approximately the properties required of them in the Weinberg-Salam model. However, the crucial property of the electroweak gauge theory, the non-Abelian self-couplings of the W's, Z's, and γ's was not mentioned. This coupling can be tested in hadron colliders in electroweak boson pair production.

An elementary calculation will show the importance of the non-linear gauge boson coupling and how the theory differs from an Abelian gauge model. The tree approximation to W^+W^- production from the $\bar{u}u$ initial state is given by the three Feynman diagrams shown in Figure 41. In an Abelian theory only the t-channel graph would exists. The kinematic variables are given in the figure along with the appropriate polarization tensors ($\epsilon_\pm$). Evaluating the t-channel

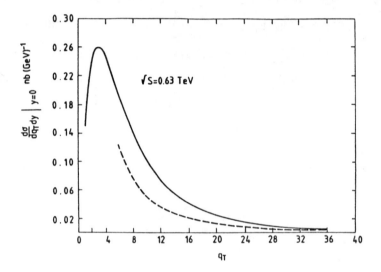

Figure 39: Comparison of the resummed expression for $d\sigma/dp_\perp dy|_{y=0}$ (solid line) with the first order perturbative expression (dashed line) at $\sqrt{s} = 630$ GeV. (From Ref. 55)

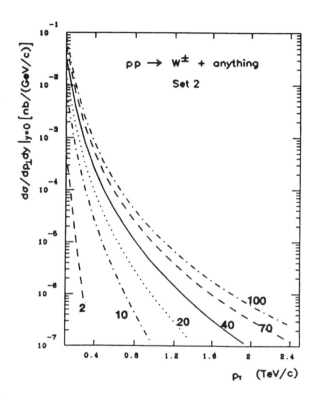

Figure 40: Differential cross section $d\sigma/dp_\perp dy|_{y=0}$ for the production of a W^+ as a function of the W^+ transverse momentum $p_\perp$ at $\sqrt{s} = 2$, 10, 20, 40, 70, and 100 TeV (from bottom to top curve). Parton distributions of Set 2 were used. (From EHLQ)

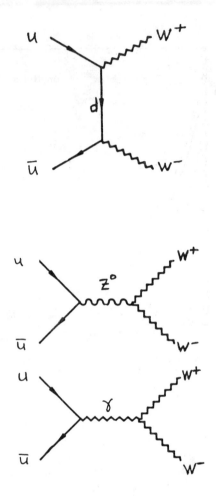

Figure 41: Lowest-order Feynman diagrams for the reaction $u + \bar{u} \to W^+ + W^-$. A direct channel Higgs boson diagram vanishes because the quarks are idealized as massless.

graph gives:
$$\mathcal{M}_1 = -i\frac{g^2}{2}\overline{v}(p_2)\not{\epsilon}_{-}\frac{\not{Q}}{Q^2}\not{\epsilon}_{+}(\frac{1-\gamma_5}{2})u(p_1) \tag{3.26}$$

where in the CM frame the momenta simplify:

$$p_1 = (P, 0, 0, P)$$

$$p_2 = (P, 0, 0, -P)$$

$$k_+ = (P, K\sin\theta\cos\phi, iK\sin\theta\sin\phi, K\cos\theta)$$

$$k_- = (P, K\sin\theta\cos\phi, -iK\sin\theta\sin\phi, K\cos\theta) \tag{3.27}$$

where $P^2 - K^2 = m_W^2$ and quark masses have been ignored. The polarization tensors are

$$\epsilon_\pm = (\vec{k}_\pm \cdot \hat{\epsilon}_\pm / m_W; \ \hat{\epsilon}_\pm + \vec{k}_\pm(\vec{k}_\mp \cdot \hat{\epsilon}_\pm)/[m_W(P+m_W)]) \tag{3.28}$$

in terms of the polarization states in the $W^\pm$ rest frame:

$$\hat{\epsilon}_\pm = \begin{pmatrix} (1, i, 0) \\ (1, -i, 0) \end{pmatrix}. \tag{3.29}$$

At high energies ($K \to \infty$) these tensors simplify to:

$$\epsilon_\pm \to k_\pm / m_W \tag{3.30}$$

Now inserting the above formulas into the expression for $\mathcal{M}_1$ in Eq. 3.26 and using the equation of motions for the $W^\pm$ fields give

$$\begin{aligned}\mathcal{M}_1 &= \frac{ig^2}{2m_W^2}\overline{v}(p_2)\frac{\not{Q}\not{Q}\not{Q}}{Q^2}(\frac{1-\gamma_5}{2})u(p_1) \\ &= iG_F 2\sqrt{2}\overline{v}(p_2)\not{Q}(\frac{1-\gamma_5}{2})u(p_1)\end{aligned} \tag{3.31}$$

for the amplitude. If this were the only contribution, then the invariant matrix element squared for this production process would be

$$|\mathcal{M}|^2 = 2G_F^2 s(s - 4m_W^2)\sin^2\theta \tag{3.32}$$

so that the total cross-section would be

$$\sigma(W^+W^-) \sim \frac{G_F^2 s}{3\pi} \qquad (3.33)$$

which grows linearly with s and violates unitarity at high energies (see lecture 4). Of course, including the gauge self interactions in the remaining Feynman diagrams of Fig. 41 restore unitarity. In the present case both the photon and Z^0 contributions must be included to recover unitarity.

We will explicitly show the cancellation between the t-channel and the s-channel exchange diagrams for left-handed initial quarks. The contributions for right-handed initial quarks must satisfy unitarity including only the s-channel photon and Z^0 exchanges, since the t-channel graph only exists for left-handed initial quarks. This behaviour for right-handed initial quarks can also be easily checked.

The three gauge boson vertices are:

$$ie \begin{pmatrix} 1 \\ \cot\theta_w \end{pmatrix} [g_{\alpha\beta}(k_+ - k_-)_\lambda]$$

$$g_{\beta\lambda}(k_- - k_3)_\alpha + g_{\lambda\alpha}(k_3 - k_+)_\beta \qquad (3.34)$$

and the quark-antiquark-gauge boson vertices are:

$$-ieQ_q\gamma_\lambda \qquad (3.35)$$

$$-i\frac{e}{sin2\theta_w}\gamma_\lambda[L_q(\frac{1-\gamma_5}{2})+R_q(\frac{1-\gamma_5}{2})] \qquad (3.36)$$

where

$$L_q = \tau_3 - 2Q_q sin^2\theta_w \qquad (3.37)$$

$$R_q = -2Q_q sin^2\theta_w \qquad (3.38)$$

and

$$\sqrt{2}G_F m_Z^2{}^{\frac{1}{2}} = \frac{e}{sin2\theta_w} = \frac{g'}{2sin\theta_w} = \frac{g}{2\theta_w} \qquad (3.39)$$

The amplitude for the two s-channel graphs for the initial state of a left handed up quark-antiquark is

$$\mathcal{M}_2 = i\frac{g}{2}\bar{v}(p_2)\gamma_\mu(\frac{1-\gamma_5}{2})u(p_1)[\frac{2Q_q sin^2\theta_w}{s}+\frac{1-2Q_q sin^2\theta_w}{s-m_W^2}]$$

$$[\epsilon_+\cdot\epsilon_-(k_+-k_-)^\mu+k_-\cdot\epsilon_+\epsilon_+^\mu-k_+\cdot\epsilon_-\epsilon_+^\mu] \qquad (3.40)$$

As $s \to \infty$ the amplitude simplifies as for $\mathcal{M}_2$ and in addition one has the relation $k_+ \cdot k_3 = k_- \cdot k_3 = s/2$ so that for large s the amplitude becomes:

$$\mathcal{M}_2 \rightarrow \frac{ig^2}{16m_W^2}\bar{v}(p_2)(\not{k}_+ - \not{k}_-)(\frac{1-\gamma_5}{2})u(p_1)$$

$$= -iG_F 2\sqrt{2}\bar{v}(p_2)\not{Q}(\frac{1-\gamma_5}{2})u(p_1) \qquad (3.41)$$

where again the equation of motion has been used. To leading order in s the sum of $\mathcal{M}_1$ and $\mathcal{M}_2$ (Eqs. 3.31 and 3.40) cancel so that the elementary cross section goes as:

$$\sigma \rightarrow \text{constant}/s \qquad (3.42)$$

as $s \rightarrow \infty$. Hence unitarity is explicitly maintained.

The cross section for $pp \rightarrow W^+W^-$ pair production is shown in Figure 42. The slow rise with collision energy of the total cross section is the result of the combined effects of the $1/\hat{s}$ behaviour of the elementary cross section and growth (at fixed $\hat{s}$) of the $\bar{q}q$ luminosities with s. The top curve gives the total production cross section without any rapidity cuts. However large rapidities are associated with production of W's near the beam direction (see Fig. 20) where measurements are very difficult; hence more realistic rates are obtained when rapidity cuts are included.

Similar gauge cancellations occur in the $W^\pm\gamma$ and $W^\pm Z^0$ total cross sections. The Z^0Z^0 and $Z^0\gamma$ cross sections are uninteresting in the present context, since the only graphs which appear are present in the Abelian theory and therefore the non-Abelian gauge couplings are not probed. The rates of electroweak pair production are interesting only at supercollider energiees. For an integrated luminosity of 10^{40} cm^{-2} at $\sqrt{s} = 40$ TeV there are $\approx 2 \times 10^6$ W^+W^- pairs produced.

Some other tests of the non-Abelian gauge couplings are the following:

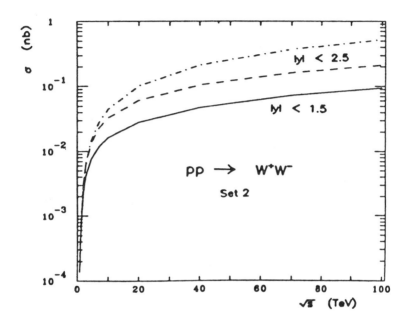

Figure 42: Yield of W^+W^- pairs in pp collisions, according to the parton distributions of Set 2. Both W's must satisfy the rapidity cuts indicated. (From EHLQ)

Table 4: Total cross sections for pair production of electroweak gauge bosons. The invariant mass of the $W^\pm\gamma$ (or $Z^0\gamma$) pair is more than 200 GeV/c^2. All cross sections are in *picobarns*.

Collider	$\sqrt{s}$ (TeV/c^2)	Process W^+W^-	$W^\pm Z^0$	Z^0Z^0	$W^+\gamma$	$Z^0\gamma$
$\bar{p}p$	.63	.037	.006	.003	.001	.003
$\bar{p}p$	1.8	2.4	.69	.28	.16	.41
$\bar{p}p$	2.0	3.1	.90	.37	.21	.55
pp	10	45	16.5	6.5	3.6	10
pp	20	102	38	15.3	8.2	23
pp	40	214	73	33	18	50

- If the $W^\pm$ were just a massive spin one boson, then the W kinetic interaction

$$-\frac{1}{2}(\partial_\mu W_\nu^+ - \partial_\nu W_\mu^+)(\partial_\mu W_\nu^- - \partial_\nu W_\mu^-) \quad (3.43)$$

would generate the minimal QED coupling with the photon given by

$$(D_\mu^{EM} W_\nu^+ - D_\nu^{EM} W_\mu^+)(D^{\mu EM} W^{\nu-} - D^{\nu EM} W^{\mu-}) = -\frac{1}{2}(F_{\mu\nu}^+ F^{-\mu\nu})_{\text{EMpart}} \quad (3.44)$$

Therefore the non-Abelian term

$$-ieF_{\mu\nu}^{EM} W^{+\mu} W^{-\nu} \quad (3.45)$$

of the W-S model is a nonminimal coupling from the point of view of QED - a Pauli term which generates an anomolous magnetic moment for the W. However, without this additional term the high energy behaviour of the $W^\pm\gamma$ production cross section will violate unitarity at sufficiently high energy[58].

- The lowest order production cross-section for $W^\pm\gamma$ has a zero in the Born amplitude at

$$\hat{t} = 2\hat{u} \qquad (3.46)$$

or equivalently at CM angle

$$\cos\theta_{\rm CM} = -\frac{1}{3} \qquad (3.47)$$

due to specific form of the non-Abelian couplings[59]. There is a dip in the elementary cross section which is still visible when the parton distributions have been folded in to give the hadron-hadron production cross section. (See EHLQ Fig. 137)

3.5 Minimal Extensions

The simplest and most natural generalization of the standard model is the posibility of an fourth generation of fermions. This possibility requires no modification of our basic ideas; in fact, we have no explanation why there are three generations in the first place. So it is natural to consider a new quarks and/or leptons within the context of our discussion of the standard model.

In general, consistency of the $SU(2)_L \otimes U(1)_Y$ gauge interactions requires that any additional quarks and leptons satisfy the anomaly cancellation conditions[60]:

$$\sum_f Q_Y^3(f) = 0 \qquad (3.48)$$

and

$$\sum_f Q_Y(f) = 0 \qquad (3.49)$$

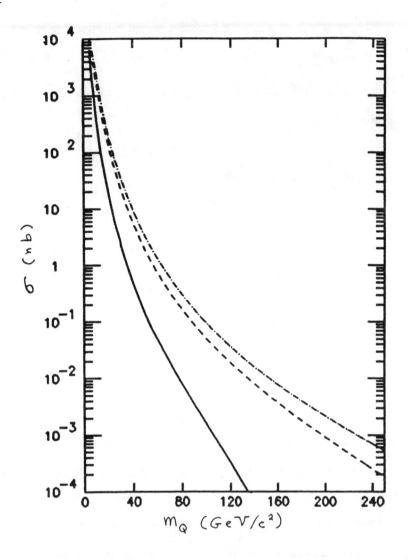

Figure 43: The total cross section for heavy quark production as a function of heavy quark mass, m_Q, for $\bar{p}p$ collisions at $\sqrt{s} = 630$ GeV (solid line), 1.8 TeV (dashed line), and 2.0 TeV (dot-dashed line). The parton distributions of Set 2 used.

where $Q_Y(f)$ is the weak hypercharge of the new fermion f. Hence a new quark doublet with standard weak charge assignments would require new leptons as well to avoid gauge anomalies. Of course a fourth generation satisfies these conditions in exactly the same way as each of the three ordinary generations.

3.5.1 New Fermions

The production of new heavy quarks in hadron colliders occurs via the same mechanisms as already discussed for top quark production (Section 2.3): gluon production and production via the decays of real (or virtual) $W^\pm$ and Z^0 bosons. For new quark masses above $\approx m_W$ the main mechanism is gluon production. Figure 43 shows the cross section for heavy quark production as a function of m_Q for $\bar{p}p$ collisions at $S\bar{p}pS$ and Tevatron energies. The corresponding cross sections at SSC energies are given in EHLQ (p. 646)

New sequential leptons will be pair produced via real and virtual electroweak gauge bosons in the generalized Drell-Yan mechanism[61]. For the $S\bar{p}pS$ and Tevatron collider energies, only decays of real $W^\pm$ and Z^0 can be significant. Hence the discovery limit for a new charged lepton, $L^\pm$, is ≈ 45 GeV in Z^0 decays: while if the associated neutral massless (neutrino-like) lepton, N^0, the discovery limit for the $L^\pm$ is extended to ≈ 80 GeV in $W^\pm$ decays.

At Supercollider energies, higher mass charged leptons can be produced through virtual electroweak gauge bosons. The pair production of charged heavy leptons proceeds via virtual γ and Z^0 states. The cross section at various energies is shown in Figure 44 for pp collisions.

Neutral lepton pairs, $N^0 N^0$, can be produced by virtual Z^0 states however in the most conventional case in which N^0 is effectively stable these events are

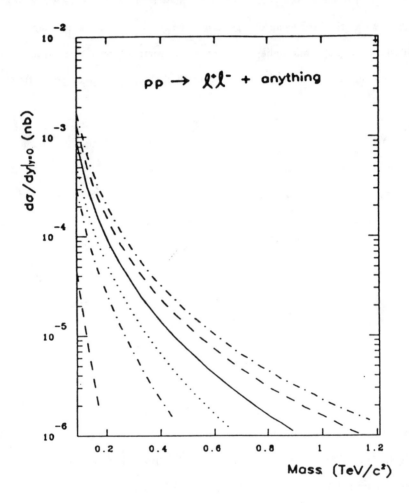

Figure 44: Cross section $d\sigma/dy|_{y=0}$ for the production of L^+L^- pairs in pp collisions. The contributions of both γ and Z^0 intermediate states aree included, and the calculation is carried out using distribution Set 2. The energies $\sqrt{s}$ are 2, 10, 20, 40, 70, 100 TeV for the bottom to top curve. (From EHLQ)

undetectable. Also, heavy leptons can be be produced by the mechanism:

$$pp \to W^{\pm}_{\text{virtual}} \to L^{\pm} N^0 \ . \tag{3.50}$$

If the neutral lepton is essentially massless as in the most convential cases, then significantly higher chagred lepton masses are accessible at a given luminosity and $\sqrt{s}$. the cross section for this process at Supercollider energies is shown in Figure 45. The principal decays of very heavy fermions will involve the emmision of a real W. If, for example, $Q_u > Q_d$ then Q_u will decay into a real W^+ and a light charge $-1/3$ quark or Q_d (if kinematically allowed). Q_d will decay into a W^- and a charge $2/3$ quark. While for a new lepton, $L^{\pm}$, the decay will give a real $W^{\pm}$ and its neutral partner, N^0. These signals should be relatively easy to identify experimentally, so it is likely that 100 produced events will be enought to discover a new quark or lepton. The discovery limits using this criterion is given in Table 5 for both present and future colliders.

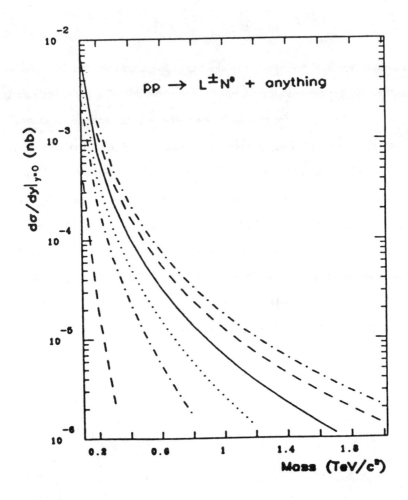

Figure 45: Cross section $d\sigma/dy|_{y=0}$ for the production of $L^\pm N^0$ pairs in pp collisions. The N^0 is assumed to be massless, and the parton distributions are those of Set 2. The energies are the same as in Fig. 44. (From EHLQ)

Table 5: Expected discovery limits for new generation of quarks and leptons at present and planned hadron colliders. Basic discovery condition is 100 produced events. A more detailed discussion can be found in EHLQ.

Collider		$\sqrt{s}$ (TeV)	$\int dt \mathcal{L}$ (cm)$^{-2}$	Mass limit (Gev/c^2)		
				New Quark Q	New Lepton	
					$\mathbf{L}^{\pm}$ or $\mathbf{L}^0$ $m(\mathbf{L}^{\pm}) = m(\mathbf{L}^0)$	$\mathbf{L}^{\pm}$ $m(\mathbf{L}^0) = 0$
$S\bar{p}pS$	$\bar{p}p$	.63	3×10^{36}	70	40	60
upgrade			3×10^{37}	90	45	70
$TEVI$	$\bar{p}p$	1.8	10^{37}	135	48	75
upgrade		2	10^{38}	220	55	95
SSC	pp	40	10^{38}	1,250	100	340
			10^{39}	1,900	250	610
			10^{40}	2,700	550	1,400

There are interesting constraints on the masses of new fermions which arise from the requirement that partial wave unitarity be respected perturbative in the standard model. I will leave the discussion of these limits until the next lecture.

3.5.2 New Electroweak Bosons

A number of proposals have been advanced for enlarging the electroweak gauge group beyond the $SU(2)_L \otimes U(1)_Y$ of the standard model. One class contains the "left-right symmetric" models[62] based on gauge groups containing

$$SU(2)_L \otimes SU(2)_R \otimes U(1)_Y \qquad (3.51)$$

which restores parity invariance at high energies. Other models, notably the electroweak sector derived from $SO(10)$ or E_6 unified theories, exhibit additional $U(1)$ invariances[63]. These will contain extra neutral gauge bosons.

All these models have new gauge coupling constants which are of the order of the $SU(2)_L$ coupling of the standard model. They imply new gauge bosons with masses of a few GeV/c^2 or more.

Assuming a new charged gauge boson, W', with the same coupling strengths as the ordinary W we obtain the cross section for production in $\bar{p}p$ collisions cross section shown in Figure 46 for present collider energies, and in EHLQ (p.648) for supercollider energies.

For a new neutral gauge boson, Z', with the same coulping strengths as the ordinary z we obtain the production cross sections shown in Figure 47 for present collider energies, and in EHLQ (p.649) for supercollider energies.

Requiring 100 produced events for discovery, the mass limits for discovering

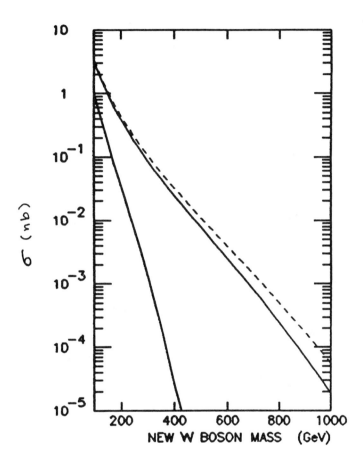

Figure 46: Total cross section, σ (nb), for production of a new charged gauge boson, $W'^{\pm}$ in $\bar{p}p$ collisions at $\sqrt{s} = 630$ GeV (lower solid line), 1.8 TeV (upper solid line), and 2.0 TeV (dashed line). The parton distributions of Set 2 used. The same couplings as the standard $W^{\pm}$ assumed.

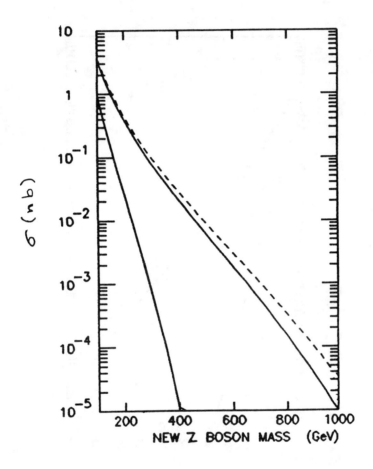

Figure 47: Total cross section, σ (nb), for production of a new neutral gauge boson, Z'^0 in $\bar{p}p$ collisions at $\sqrt{s} = 630$ GeV (lower solid line), 1.8 TeV (upper solid line), and 2.0 TeV (dashed line). The parton distributions of Set 2 used. The same couplings as the standard Z^0 assumed.

Table 6: Expected discovery limits for new intermediate gauge bosons $W'^{\pm}$ and Z'^0 at present and planned hadron colliders. Standard model couplings are assumed and 100 produced events are required for discovery.

				Mass limit (Gev/c^2)	
Collider		$\sqrt{s}$	$\int dt \mathcal{L}$	Intermediate Boson	
		(TeV)	(cm)$^{-2}$	$\mathbf{W'^{\pm}}$	$\mathbf{Z'^0}$
$S\bar{p}pS$	$\bar{p}p$	.63	3×10^{36}	200	180
upgrade			3×10^{37}	280	240
$TEVI$	$\bar{p}p$	1.8	10^{37}	520	500
upgrade		2	10^{38}	730	700
SSC	pp	40	10^{38}	2,200	1,800
			10^{39}	4,000	3,300
			10^{40}	6,700	5,200

a new W' or Z' in present and future hadron colliders is given in Table 6.

4 THE SCALAR SECTOR

4.1 The Higgs Scalar

4.1.1 Lower Bound on the Higgs Mass

A lower bound of the Higgs mass (m_H) arises from requiring that the symmetry breaking miminum of the potential $V(\phi)$ be stable with respect to quantum corrections[64]. If m_H is too small there could be tunneling to a symmetry preserving vacuum.

To illustrate this, we do a simple one loop calculation using the standard symmetry breaking potential for a Higgs doublet[65]:

$$V(\phi^+\phi) = \mu_0^2 \phi^+\phi + |\lambda|(\phi^+\phi)^2 \,. \tag{4.1}$$

It is sufficient to consider an external scalar field with its only non-zero component along the direction of symmetry breaking. This amounts to taking only the real neutral component so that $<\phi^+\phi> = <\phi>^2$. This field couples to those particles that aquire mass as a result of the symmetry breaking, $W^\pm$ and Z^0 and the fermions ψ.

$$<\phi^+\phi>[\frac{g^2}{2}W_\mu^+ W^{-\mu} + \frac{(g^2+g'^2)}{4}Z_\mu^0 Z^{-\mu}] + \sum_{i=1}^{3}\phi[\Gamma_{u_i}\overline{\psi}_{u_i}\psi_{u_i} + \Gamma_{d_i}\overline{\psi}_{d_i}\psi_{d_i}] \tag{4.2}$$

Because the Yukawa couplings are small we shall ignore the fermions and only consider the contribution to the effective potential from vector particle loops, with ϕ insertions:

The form of the integral for these processes is:

$$-i\int \frac{d^4k}{2\pi^4} \frac{k^2}{k^2 - \frac{g^2}{4}<\phi>^2} \quad (4.3)$$

which may be regulated to give :

$$A_0 + A_1 <\phi>^2 + \frac{g^4}{128\pi^2} <\phi>^4 ln(\frac{<\phi>^2}{A_2}) \quad (4.4)$$

That is, a sum of a quartically, quadratically and logarithmically divergent term.

When the effective potential is renormalized we can ignore A_0, and absorb A_1 into the scalar mass renormalization. The term A_2 is absorbed into the scalar coupling renormalization, while the finite part appears with a renormalization scheme dependent scale parameter M in the resulting one-loop effective potential.

$$V_{1loop}(<\phi>^2) = -\mu^2 <\phi>^2 + C <\phi>^4 ln(\frac{<\phi>^2}{M^2}) \quad (4.5)$$

This is the form of the general answer. A careful calculation taking into account fermions and scalars as well was performed by E. Gildner and S. Weinberg[66]. They obtained

$$C = \frac{1}{64\pi^2} \frac{1}{<\phi>_0^4}(3(2M_W^4 + M_Z^4) - 4\sum_F m_F^4 + m_H^4) \quad (4.6)$$

where $<\phi>_0^2 = 1/(2\sqrt{2}G_F)$ and the Yukawa and gauge couplings are reexpressed in terms of particle masses.

In models with a non minimal Higgs sector, m_H^4 would be replaced with $\sum m_H^4$. Note that $C > 0$ as is required for overall stability at large values of $<\phi>^2$.

This potential has a local minimum at $<\phi>^2 = <\phi>_0^2$ where $\frac{\partial V}{\partial <\phi>^2}|_{<\phi>_0^2} = 0$ so

$$<\phi>_0^2 (ln(\frac{<\phi>_0^2}{M^2}) + \frac{1}{2}) = \frac{\mu^2}{2C} . \quad (4.7)$$

To check that this is an absolute minimum, because in general there is another local minimum at $<\phi>^2 = 0$, we must ensure that $V(<\phi>_0^2) < V(0)$, which requires

$$ln(\frac{<\phi>_0^2}{M^2}) > -1 \,. \tag{4.8}$$

This condition, that the symmetry breaking minimum is more stable that the symmetry preserving one can be expressed as a limit on m_H, by using the definition $m_H^2 = \frac{\partial V}{\partial \phi^2}|_{<\phi>_0^2}$. This implies

$$m_H^2 > 2C <\phi>_0^2 \simeq \frac{3G_F\sqrt{2}}{16\pi^2}(2M_W^4 + M_Z^4) = 7.1 GeV/c^2 \,. \tag{4.9}$$

In the context of the minimal Higgs model, this represents a strict lower bound for m_H consistent with symmetry breaking. A slightly simpler calculation[67] can be done for the case $\mu = 0$, leading to $m_H > 10 GeV/c^2$, however this assumption has no theoretical basis.

4.1.2 Unitarity Bounds

The simplest upper bound on m_H arises from the requirement of preturbative unitarity. That is, on the assumption that the couplings are sufficiently weak to make perturbation theory valid, we require that all processes obey the constraint of unitarity. Of course, it is possible that perturbation theory is not valid, in that case there is likely to be other physics associated with the interactions becoming strong, at a similar scale. We postpone that discussion until the next lecture.

Unitarity in general requires:

$$S^\dagger S = (1 + iT^\dagger)(1 - iT) = 1 \tag{4.10}$$

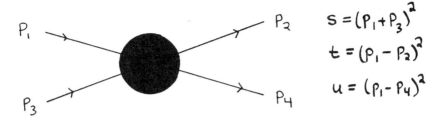

Figure 48: Kinematics of the 2 → 2 scattering amplitude for equal mass scalar particles.

or

$$i(T - T^\dagger) = -ImT = T^\dagger T \tag{4.11}$$

To set up the unitarity argument in its simplest form we only consider two particle quasielastic scattering for equal mass scalar particles (i.e. internal quantum numbers but not masses can change from the initial to final state). The scattering process in the center of mass frame is shown in Fig. 48 In this simple case, the T matrix is:

$$T_{fi} = (2\pi)^4 \delta^4(p_1 + p_3 - p_2 - p_4) \frac{1}{(2\pi)^6} \frac{1}{s} M_{fi}(s,t) \tag{4.12}$$

where $s = (p_1 + p_3)^2 = (p_2 + p_4)^2$ and $t = (p_1 - p_2)^2$ and the scattering angle in the CM Frame, θ, is given by

$$t = -\frac{(s - 4m^2)}{2}(1 - cos\theta) . \tag{4.13}$$

The invariant amplitude can be expanded into partial waves:

$$M = 16\pi \sum_{J=0}^{\infty} (2J + 1) A_J(s) P_J(cos\theta) . \tag{4.14}$$

For s below the inelastic threshold, so that there is only two particle intermediate states, the unitarity condition may be written entirely in terms of the two particle amplitude $\mathcal{M}$ as:

$$-Im\mathcal{M}(s,cos\theta_{fi}) = \frac{1}{64\pi^2}sqrt\frac{s-4m^2}{s}\int d\Omega_k \mathcal{M}^*(s,cos\theta_{fk})\mathcal{M}(s.cos\theta_{ki}) \quad (4.15)$$

where the momentum integration has been done to obtain the phase space factor $\sqrt{(s-4m^2)/s}$.

Now, using the partial wave expansion for $\mathcal{M}$ and performing the angular integration we find that each partial wave satisfies:

$$|A_J(s)| \geq \sum_{n=1}^{\infty} A_J^{(n)} = \sum_{n=1}^{\infty} c_{Jn}g^{2n} \quad . \quad (4.16)$$

The Born contribution (first order) corresponds to $A_J^{(1)} = c_{J1}g^2$. The criterion for validity of perturbation theory is that succesive terms in the expansion are smaller, i.e. $|c_1g^2| \gg |c_2g^4| \gg \ldots$ etc.. Thus we will consider only the lowest order terms in the following.

The $J = 0$, s-wave scattering condition is

$$-ImA_0 \geq |A_0|^2 \quad (4.17)$$

so

$$-Im(c_1g^2 + c_2g^4 + \ldots) \geq |c_1g^2 + c_2g^4 + \ldots|^2 \quad . \quad (4.18)$$

It is a property of the Born amplitude that at high energy $(s \rightarrow \infty)$ it is essentially real. Thus the imaginary term of c_1 can be dropped to obtain:

$$|c_1g^2| > -Im(c_2g^4) \geq |c_1g^2|^2 \quad . \quad (4.19)$$

Thus the perturbative unitarity constraint on the Born amplitude is[68]

$$1 > |c_1g^2| = |A_0^{(1)}| \quad . \quad (4.20)$$

We proceed to apply this constraint to the scalar sector of the standard (W-S) electroweak theory.

4.1.3 Upper Bound on the Higgs Mass[68]

Now we apply the general unitarity arguments specifically to the W-S Model. We start by showing that at high energy we need only consider an effective scalar theory, so that the simple bound (Eq. 4.21) just derived can be applied even in this more complicated theory[69].

As discussed in Section 2, massive vector particle (V_μ) scattering has potentially bad behaviour at high energy. This is apparent from the form of the polarization sum:

$$-\sum_\lambda \xi^\mu(k,\lambda)\xi^\nu(k,\lambda) = g^{\mu\nu} - k^\mu k^\nu/M_V^2 \ . \tag{4.21}$$

The dominant term here is the $k^\mu k^\nu/M_V^2$ piece which comes from the longitudinal polarization (λ_L)

$$\xi^\mu(k,\lambda_L) = \frac{1}{M_V}(|\vec{k}|, k^0 \frac{\vec{k}}{|\vec{k}|}) \tag{4.22}$$

and has the potential to violate the unitarity bounds.

It has been shown that the only renormalizable theories with massive non-Abelian vector bosons are thoses in which the masses arise from a spontaneously broken gauge symmetry[70]. In such cases one can replace the longitudinal components of the vector fields by scalar fields and get an effective Lagrangian that is valid at high energy.

The most appropriate gauge for showing this is the t'Hooft-Feynman gauge[71],

$$\partial_\mu V^\mu + M_V \phi = 0 \tag{4.23}$$

where ϕ is the Goldstone Boson associated with giving mass to the vector boson. In momentum space, the longitudinal component of the vector field is

$$\tilde{V}_L(k) = \xi_L^\mu \tilde{V}_\mu(k) = \frac{1}{M_V}(|\vec{k}|\tilde{V}_0 - k^0 \frac{\vec{k}}{|\vec{k}|}\cdot\vec{\tilde{V}}) \ ; \tag{4.24}$$

which, using the gauge condition (Eq. 4.24), just becomes the scalar field in the high energy limit:

$$\tilde{V}_L(k) \to -i\tilde{\phi} + O(\frac{M_V}{|\vec{k}|}) \quad for \ |\vec{k}| \gg M_V \ . \tag{4.25}$$

In discussing the high energy unitarity constraints we may therefore ignore the transverse degrees of freedom of the vector fields and only consider an effective Lagrangian describing scalars interacting with fermions. The scalars include both the Higgs (h) and the 'eaten' Goldstone Bosons (w^+, w^-, z) that describe the longitudinal degrees of freedom of the gauge bosons. The notation is

$$\phi = \frac{1}{\sqrt{2}} \begin{pmatrix} \sqrt{2}\ iw^+ \\ v + h - iz \end{pmatrix} \tag{4.26}$$

with $m_H^2 = 2\mu^2$, $v^2 = 1/(\sqrt{2}G_F)$, and $\lambda = G_F m_H^2/\sqrt{2}$.

The full effective Lagrangian is given by:

$$\begin{aligned}
\mathcal{L}_{eff} =\ & (\partial_\mu w^+)(\partial^\mu w^-) - m_W^2 w^+ w^- + \frac{1}{2}(\partial_\mu z)(\partial^\mu z) - \frac{1}{2}m_Z^2 z^2 \\
& + \frac{1}{2}(\partial_\mu h)(\partial^\mu h) - \frac{1}{2}m_H^2 h^2 \\
& - \lambda v h(2w^+ w^- + z^2 + h^2) - \frac{1}{4}\lambda(2w^+ w^- + z^2 + h^2)^2 \\
& + \bar{u}\gamma^\mu D_\mu^G u - m_u \bar{u}u + \bar{d}\gamma^\mu D_\mu^G d - m_d \bar{d}d \\
& + \bar{\nu}\gamma^\mu D_\mu^G \nu + \bar{e}\gamma^\mu D_\mu^G e - m_e \bar{e}e \\
& - (\Gamma_u[i\ \bar{d}\frac{(1+\gamma_5)}{2}u\ w^- + \bar{u}\frac{(1+\gamma_5)}{2}u\ \frac{(h+iz)}{\sqrt{2}}] \\
& + \Gamma_d[i\ \bar{u}\frac{(1+\gamma_5)}{2}d\ w^+ + \bar{d}\frac{(1+\gamma_5)}{2}d\ \frac{(h-iz)}{\sqrt{2}}] \\
& + \Gamma_e[i\ \bar{\nu}\frac{(1+\gamma_5)}{2}e\ w^+ + \bar{e}\frac{(1+\gamma_5)}{2}e\ \frac{(h-iz)}{\sqrt{2}}] + h.c.) \tag{4.27}
\end{aligned}$$

Using this Lagrangian, all Born amplitudes for neutral channels can be easily calculated. The results are summarized in Fig. 49.

$w^+ w^- \to w^+ w^-$:

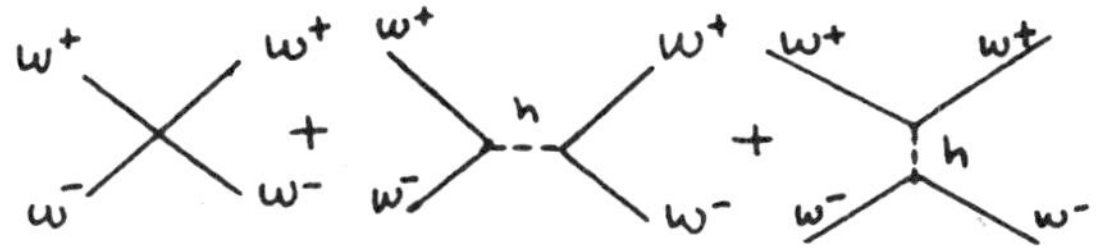

$$-2i\lambda[2 + \frac{m_H^2}{s - m_H^2} + \frac{m_H^2}{t - m_H^2}]$$

$zz \to zz$:

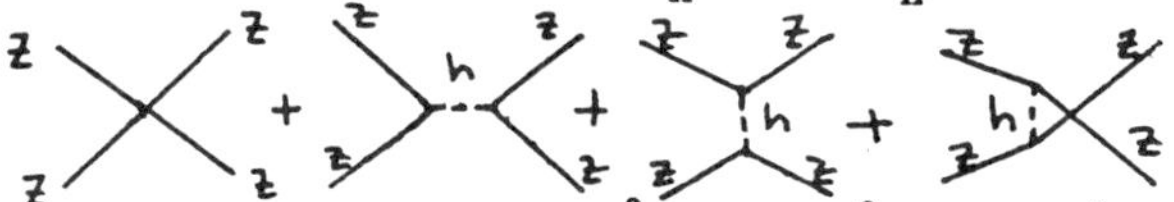

$$-2i\lambda[3 + \frac{m_H^2}{s - m_H^2} + \frac{m_H^2}{t - m_H^2} + \frac{m_H^2}{u - m_H^2}]$$

$hh \to hh$:

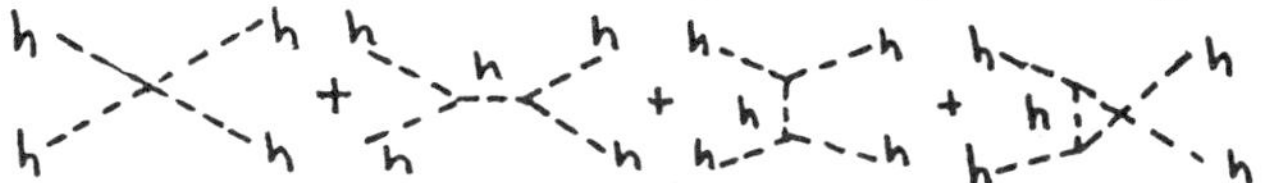

$$-6i\lambda[1 + 3\frac{m_H^2}{s - m_H^2} + 3\frac{m_H^2}{t - m_H^2} + 3\frac{m_H^2}{u - m_H^2}]$$

$hz \to hz$:

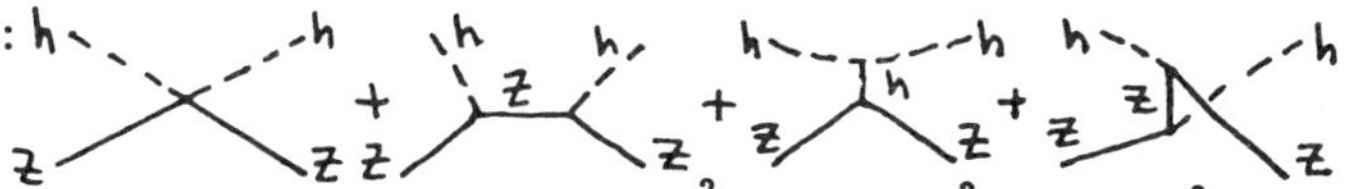

$$-2i\lambda[1 + \frac{m_H^2}{s - m_H^2} + 3\frac{m_H^2}{t - m_H^2} + \frac{m_H^2}{u - m_H^2}]$$

$w^+ w^- \to zz$:

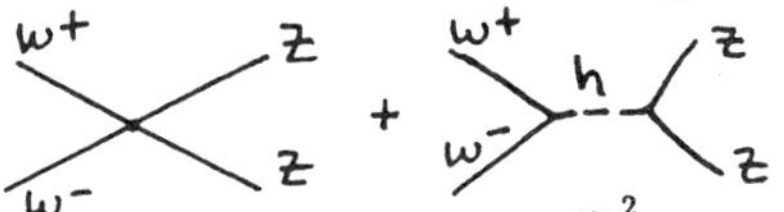

$$-i\lambda[1 + \frac{m_H^2}{s - m_H^2}]$$

$w^+ w^- \to hh$:

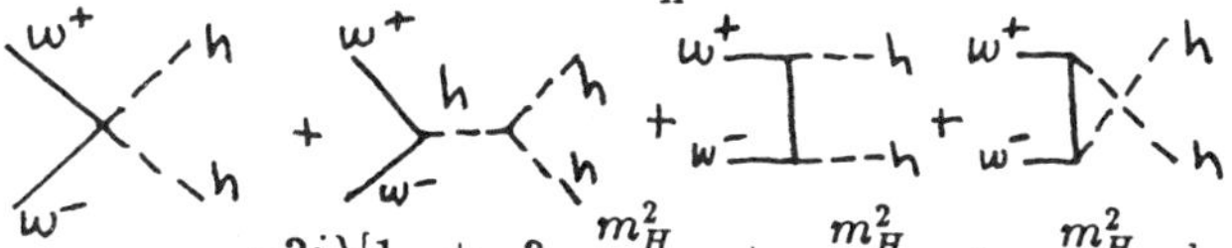

$$-2i\lambda[1 + 3\frac{m_H^2}{s - m_H^2} + \frac{m_H^2}{t - m_H^2} + \frac{m_H^2}{u - m_H^2}]$$

$zz \to hh$:

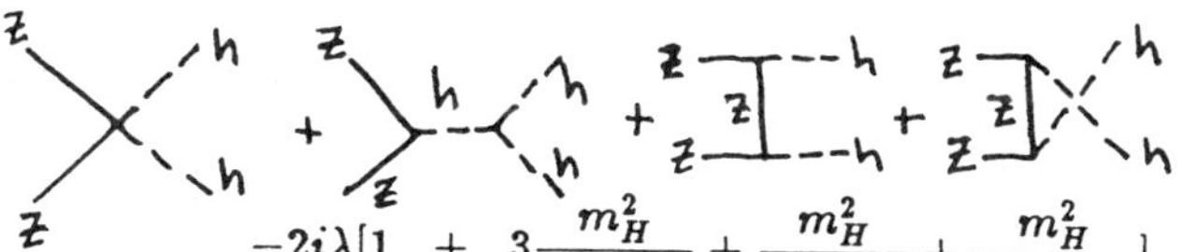

$$-2i\lambda[1 + 3\frac{m_H^2}{s - m_H^2} + \frac{m_H^2}{t - m_H^2} + \frac{m_H^2}{u - m_H^2}]$$

Figure 49: Born amplitudes for neutral channels.

The limiting behaviour of these processes at high energy $(s \gg m_H^2 > m_W^2, m_Z^2)$ is collected in matrix form

$$\mathcal{M} = -2\sqrt{2}G_F m_H^2 \begin{pmatrix} 1 & \frac{1}{\sqrt{8}} & \frac{1}{\sqrt{8}} & 0 \\ \frac{1}{\sqrt{8}} & \frac{3}{4} & \frac{1}{4} & 0 \\ \frac{1}{\sqrt{8}} & \frac{1}{4} & \frac{3}{4} & 0 \\ 0 & 0 & 0 & \frac{1}{2} \end{pmatrix} \begin{matrix} w^+w^- \\ z^2/\sqrt{2} \\ h^2/\sqrt{2} \\ hz \end{matrix} \quad (4.28)$$

As in the previous section we expand in partial waves and identify the s-wave Born term as

$$A_0^{(1)} = \frac{\mathcal{M}}{16\pi} \equiv \frac{G_F m_H^2}{4\pi\sqrt{2}} t_0 . \quad (4.29)$$

To obtain the best bound we diagonalize the matrix t_0 (defined above). The largest eigenvalue is $3/2$, for the combination of channels above which correspond to the isoscalar channel $(2w^+w^- + zz + hh)$.

Substituting this into the perturbative unitarity condition $|A_0^{(1)}| \leq 1$ we find an upper bound on m_H:

$$m_H \leq \sqrt{\frac{8\pi\sqrt{2}}{3G_F}} = .98 TeV . \quad (4.30)$$

We close this section with a comment on the nature a perturbative unitarity bound. If such a bound is violated then perturbative expansion must be invalid since the Lagrangian is unitary. That is, the interactions are strong and perturbation theory is therefore unreliable. An up to date analysis of the physics of a strongly interacting scalar sector has been given by Chanowitz and Gaillard[72].

Whether the scalar sector of the W-S Model is, in fact, strongly interacting is presently unknown. Because the scalar sector is protected by an order of α_{em} from showing up in low energy electroweak measurements (e.g. in the ρ parameter[73]) no experiment to date rules out the possibility of a strongly interacting Higgs sector. Only direct obsevation of the Higgs scalar or strong interactions at (or below) the TeV scale will settle this question experimentally.

4.2 Constraints on Fermion Masses

4.2.1 Perturbative Unitarity Bounds

We can use the same W-S effective Lagrangian (Eq. 4.27) and perturbative unitarity for the Yukawa couplings to put upper bounds on fermion masses. In general because of spin the perturbative unitarity condition will be more complicated than the one we derived (Eq. 4.20), futhermore the neutral fermion-antifermion channels $(\overline{F}F)$ will couple to the channels w^+w^-, zz, hh, and zh already discussed.

The general case is disscused fully by M. Chanowitz, M. Furman, and I. Hinchliffe[74]. However in the $J=0$ partial wave things are simpler and if we further assume that m_H is small relative to the fermion masses to be bounded we can aviod having a coupled problem. In this case the helicity amplitudes in the CM Frame for the $\overline{F}F$ channels are defined by:

$$\frac{\vec{\Sigma}\cdot\vec{p}}{|\vec{p}|}u^{(\lambda)}(p) = \lambda u^{(\lambda)}(p)$$
$$\frac{\vec{\Sigma}\cdot\vec{p}}{|\vec{p}|}v^{(\lambda)}(p) = -\lambda v^{(\lambda)}(p) \qquad (4.31)$$

If $F = \begin{pmatrix} F_1 \\ F_2 \end{pmatrix}$ is a quark (or lepton) doublet then the relevant Born amplitudes are shown in Figure 50.

For the amplitudes in Fig. 50 we can construct a matrix of the $J=0$ partial wave amplitudes for the various channels just as in the scalar case (Eq. 4.28). The only complication is that we must consider each helicity channel as well.

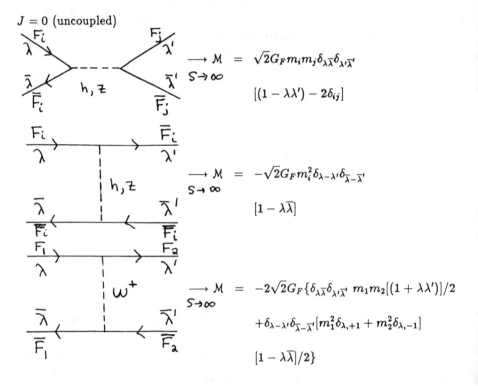

Figure 50: Born graphs for the $\overline{F}F$ amplitudes in the uncoupled limit ($m_H \ll m_i$). $\mathcal{M}$ is the amplitude for the $J = 0$ partial wave in the high energy limit.

The non zero helicity amplitudes are:

$$\begin{pmatrix} + & + & \to & + & + \\ - & - & \to & - & - \\ + & - & \to & - & + \\ - & + & \to & + & - \end{pmatrix}. \qquad (4.32)$$

The unitarity condition is simply $|\mathcal{M}_0^{(1)}| \leq 1$ as before. Applying this condition to the largest eigenvalue of $|\mathcal{M}|$ in the fermion case leads to the following upper bounds for a quark doublet[74]:

$$\frac{G_F}{8\sqrt{2}\pi}[3(m_1^2 + m_2^2) + \sqrt{9(m_1^2 - m_2^2) + 8m_1^2 m_2^2}] \leq 1 \qquad (4.33)$$

which for equal mass quarks $(m = m_1 = m_2)$ becomes:

$$m \leq (\frac{4\sqrt{2}\pi}{5G_F})^{1/2} = 530 Gev/c^2. \qquad (4.34)$$

For a lepton doublet, the bound is:

$$\frac{G_F}{8\sqrt{2}\pi}[m_1^2 + m_2^2 + |m_1^2 - m_2^2|] \leq 1. \qquad (4.35)$$

and for the case $m_1 = 0$, $m_2 = m$ the limit becomes:

$$m \leq (\frac{4\sqrt{2}\pi}{G_F})^{1/2} = 1.2 TeV/c^2. \qquad (4.36)$$

A slightly better bound for leptons of $\approx 1 TeV/c^2$ comes from considering the more complicated case of the $J = 1$ partial wave[74].

Although only one generation of quarks and/or leptons has been considered it is possible to interpret the bounds as being on the sum over generations of masses with the other quantum numbers the same. Ofcourse in practice this sum is dominated by the heaviest fermion in any case.

It is interesting to compare these unitarity bounds on fermion masses within the standard model with the discovery limits of the various hadron colliders present and planned. These limits are shown in Table 5. We see that the SSC will be able to discover any new fermion with mass satisfying the bounds given above.

4.2.2 Experimental Bounds

In addition to the lower bounds on the masses of new quark or leptons arising from discovery limits summarized in Table 5 there is also the possibility of upper bounds on fermion masses arising from experimental measurements. This was first was realized by M. Veltman[75]. The basic point is that the Higgs sector of the W-S theory has an $SU(2)_L \otimes SU(2)_R$ symmetry as we discussed in Section 3.2. This symmetry is spontaneously broken down to an $SU(2)_V$ symmetry when the scalar field acquires it's vacuum expectation value. It is the residual $SU(2)_V$ symmetry that ensures:

$$\frac{M_W^2}{M_Z^2 \cos^2\theta_w} \equiv \rho = 1 . \tag{4.37}$$

The Yukawa couplings and electroweak gauge interactions break the $SU(2)_V$ symmetry explicitly. In particular, for $\Gamma_u \neq \Gamma_d$, the fermion one loop corrections to the $W^\pm$ and Z^0 masses will change the value of the ρ parameter from 1.

For a heavy fermion doublet the correction is[75]

$$\rho = 1 + \xi \frac{G_F}{8\sqrt{2}\pi^2} [\frac{2m_1^2 m_2^2}{m_1^2 - m_2^2} \log(\frac{m_2^2}{m_1^2}) + m_1^2 + m_2^2] \tag{4.38}$$

where ξ is 1 for leptons and 3 for quarks. For example, in the case of the leptons, with $m_1 = 0, m_2 = m$:

$$\rho = 1 + \frac{G_F}{8\sqrt{2}\pi^2} m^2 . \tag{4.39}$$

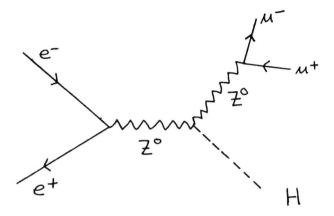

Figure 51: Associated Production Mechanism for a Low Mass Higgs Boson.

A compilation of the present data yields a measured value for ρ [76]

$$\rho = 1.02 \pm 0.02 \tag{4.40}$$

which does not leads to the bounds on new lepton and new quark masses of

$$m_L \leq 620 GeV/c^2 \tag{4.41}$$

and

$$|m_u^2 - m_d^2|^{1/2} \leq 350 GeV/c^2 \tag{4.42}$$

respectively.

4.3 Finding the Higgs

4.3.1 Higgs Mass Below $2M_W$

Finding a Higgs boson with a low mass $m_H \leq M_Z$ is possible through real or slightly virtual Z^0 production by the mechanism shown in Figure 51

Although hadron colliders will produce 10^4 to 10^5 Z^0's a year, the best place to find the Higgs boson in this mass range is an e^+e^- collider where the energy can be tuned the the region of the $Z,^0$ pole to yield a clean, high statistics sample of Z^0 decays. In particular LEP should have approximately 10^7/yr Z^0 decays

For the intermediate mass range ($M_Z \leq m_H \leq 2M_W$) no convincing signal for detecting a Higgs boson is presently known. The production rate (by the mechanism in Fig. 51) is small even in a e^+e^- collider with $\sqrt{s} = 200$ GeV. On the other hand, in hadron colliders additional production mechanisms exist and the total rate of Higgs boson production in the mass range can be substantial. Thus hadron colliders provide the best hope for finding a Higgs boson with a mass in this intermediate range.

The first and most obvious additional Higgs production mechanism in a hadron collider is direct production by a quark pair (shown in Figure 52a). Because the Higgs coupling is proportional the mass of the fermion, we might expect the heaviest pair, namely the top quarks, be the dominant subprocess. Indeed,

$$\sigma(\bar{p}p \to H_0 + X) = \frac{G_F \pi}{3\sqrt{3}} \sum_i \frac{m_i^2}{m_H^2} \tau \frac{d\mathcal{L}_{i\bar{i}}}{d\tau}$$

$$= 3.36 nb \sum_i \frac{m_i^2}{m_H^2} \tau \frac{d\mathcal{L}_{i\bar{i}}}{d\tau} \quad (4.43)$$

where m_i is the mass of the ith quark flavor. However, referring back to Fig. 17 we see that the $\bar{t}t$ luminosity is small even at supercollider energies. For example at $\sqrt{s} = 40$ TeV assuming a 30 GeV/c^2 top quark the Higgs production cross section

$$\sigma(\bar{t}t \to H^0) = 9 \ pb \ . \quad (4.44)$$

For lighter quarks, where the luminosity is greater, the mass porportional coupling suppresses production.

There is however a second production mechanism which gives large production cross sections. This is the gluon fusion process shown in Figure 52b. This one loop coupling of gluons to the Higgs through a quark loop takes advantage of both the large number of gluons in a proton at these subenergies and the large coupling of the Higgs to heavy quarks in the loop. The cross section is[77]:

$$\sigma(\bar{p}p \to H^0 + X) = \frac{G_F \pi}{32\sqrt{2}} \left(\frac{\alpha_s}{\pi}\right)^2 |\eta| \tau \frac{d\mathcal{L}_{gg}}{d\tau} \quad (4.45)$$

where $\eta = \sum_i \eta_i$ and

$$\eta_i = \frac{\epsilon_i}{2}[1 + (\epsilon_i - 1)\phi(\epsilon_i)] \quad (4.46)$$

and $\epsilon_i = \frac{m_i^2}{m_H^2}$ and

$$\phi(\epsilon) = \begin{cases} -[sin^{-1}(1/\sqrt{\epsilon})]^2 & \epsilon > 1 \\ \frac{1}{4}[ln(\frac{1+\sqrt{1-\epsilon^2}}{1-\sqrt{1-\epsilon^2}}) + i\pi]^2 & \epsilon < 1 \end{cases} \quad (4.47)$$

For small ϵ_i, η can be approximated by $0.7 m_i^2/m_H^2$.

This gluon fusion mechanism leads to large cross sections for Higgs production:

m_H	$\sigma(\bar{p}p \to H^0 + X)$	
(GeV/c^2)	via gluon fusion	
100	3 pb	300 pb
200	.1 pb	25 pb

In this mass range the principal decay mode of the Higgs is the heaviest fermion pair avaiable, presumably top. Hence a top jet pair with the invariant mass m_H is the signal of the Higgs. However, this signal is buried in the background of QCD jet pairs. Even if a perfectly efficient means of tagging top

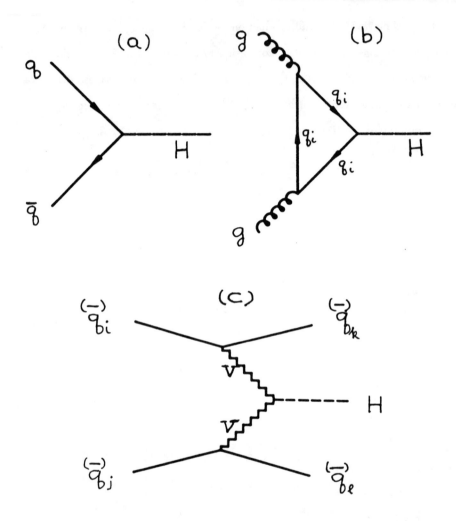

Figure 52: Higgs production mechanisms in hardon colliders: (a) direct production from quark-antiquark annihilation, (b) gluon fusion, and (c) intermediate vector boson fusion.

quark jets existed, the signal/background ratio is hopeless small. For example, at $\sqrt{s} = 40$ TeV with a 30 GeV/c^2 top quark

M_{pair}	$\frac{d\sigma}{dM_{pair}}(\bar{p}p \to t + \bar{t} + X)$
100 GeV	7 nb
200 GeV	.7 nb

which swamps the gluon fusion cross sections given above.

At SSC energies it may be possible the find a Higgs in this intermediate mass range by associated production with $W^\pm$ or Z from a $q\bar{q}$ initial state. This is basically the same mechanism used for seeing a low mass Higgs in e^+e^- shown in Fig. 51. Although the production rate is low even for SSC energies, the signal/background ratio is much better than in the gluon fusion mechanism because the associated $W^\pm$ or Z^0 can be identified through its leptonic decays. The rate is marginal and the success of the method depends on the effieiency of detecting top jets. For a detailed discussion of these issues see Ref. 78.

4.3.2 Higgs Mass Above 2 M_W

For high mass Higgs, there is a new production mechanism, in addition to direct production (Fig. 52a) and gluon fusion (Fig 52b), intermediate vector boson (IVB) fusion[79] shown in Figure 52c. This mechanism becomes significant because (as we saw in Fig. 18) the proton contains a substantial number of electroweak gauge bosons constituents at high energies.

The total width (along with the principal partial widths) is shown in Figure 53 for the Higgs bosons with masses above the threshold for decay in W^+W^- and Z^0Z^0 pairs. The decays into W^+W^- and Z^0Z^0 pairs dominate for Higgs masses above 250 GeV/c^2; hence the detection signal ofor a Higgs in the high mass range is a resonance in electroweak gauge boson pair production. The width of this resonance grows rapidly with the Higgs mass. For a Higgs as massive as the unitarity bound (1 TeV/c^2) the width is approximately 500 GeV/c^2, making the resonance difficult to observe.

The cross section for the production and decay

$$\bar{p}p \rightarrow H^0 + \text{anything}$$
$$\phantom{\bar{p}p \rightarrow} \hookrightarrow Z^0 Z^0 \qquad (4.48)$$

at $\sqrt{s} = 40$ TeV is shown in Figure 54. The rapidity of each Z^0 is restricted so that $|y_Z| < 2.5$ and m_t is assumed to be 30 GeV/c^2. The cut ensures that the decay products of the Z^0 will not be confused with the forward-going beam fragments. The contributions from gluon fusion and IVB fusion are shown separately.

The background from ordinary $Z^0 Z^0$ pairs is given by

$$\Gamma \frac{d\sigma(pp \rightarrow ZZ + X)}{dM} \qquad (4.49)$$

where $M = m_H$ and $\Gamma = \max(\Gamma_H, 10 \text{ GeV})$. As can be seen from Fig. 54, the background of standard $Z^0 Z^0$ pairs is small.

To compare the reach of various machines the following criterion to establish the existence of a Higgs boson have been adopted in EHLQ. There must be at least 5000 events, and the signal must stand above background by five standard deviations. The 5000 events should be adequate even if we are restricted to observing the leptonic decay modes of the Z^0 (or $W^\pm$). In particular, 18 detected events would remain from a sample of 5000 $Z^0 Z^0$ pairs where both z's decay into e^+e^- or $\mu^+\mu^-$. Figure 55 shows the maximum detectable Higgs mass in the

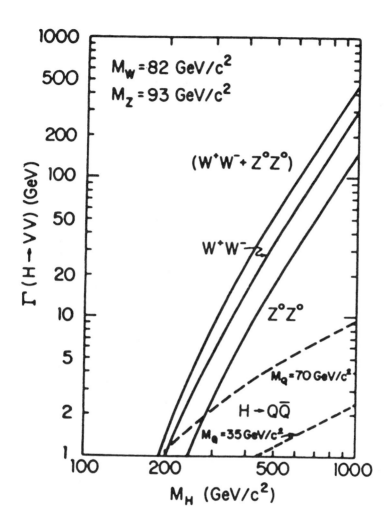

Figure 53: Partial decay widths of the Higgs boson into intermediate boson pairs as a function of the Higgs mass. For this illustration $M_W = 82$ GeV/c^2 and $M_Z = 93$ GeV/c^2. (From EHLQ)

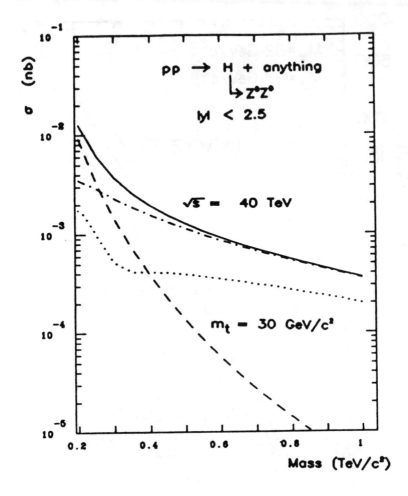

Figure 54: Cross section for the reaction $pp \to (H \to ZZ)+$ anything according to EHLQ parton distribution with $\Lambda = .29$GeV. The contribution of gluon fusion (dashed line) and IVB fusion (dotted-dashed line) are shown separately. Also shown (dotted line) is ZZ pair background.

Z^0Z^0 final state, with $|y_W| < 2.5$, and $m_t=30$ GeV/c^2 as a function of $\sqrt{s}$ for various integrated luminosities. Similar limits apply for the W^+W^- final state. More details of this analysis can be found in EHLQ.

The assumptions made in the analysis resulting in the discovery limits of Fig. 55 are conservative. It was assumed that $m_t=30$ Gev/c^2 and that there are no additional generations of quarks. If m_t is heavier or there are additional generations then the Higgs production rate will increase considerably. Hence we can safely conclude that at the SSC with $\sqrt{s} = 40$ TeV and $\mathcal{L} = 10^{33}cm^{-2}sec^{-1}$ the existence of a Higgs with mass $m_H > 2M_W$ can be established. If at least one Z^0 can be detected in a hadronic mode then $\mathcal{L} = 10^{32}cm^{-2}sec^{-1}$ would be sufficient.

4.4 Unnaturalness of the Scalar Sector

Presently there is no experimental evidence that requires the modification or extension of the standard model. The motivations for doing so are generally based upon aesthetic principles of theoretical simplicity and elegance. Perhaps the most compelling argument that the standard model is incomplete is due to t'Hooft[80].

In general the Langrangian $\mathcal{L}(\Lambda)$ provides a description of the physics at energy scales at and below Λ in terms of fields (degrees of freedom) appropriate to the scale Λ. In this sense any Lagrangian should be considered as an effective Lagrangian describing physics in terms of the fields appropriate to the highest energy scale probed experimentally. One can never be sure that at some higher energy Λ' $\mathcal{L}(\Lambda')$ may not involve different degrees of freedom. This in fact has happened many times before in the history of physic; the most recent being the

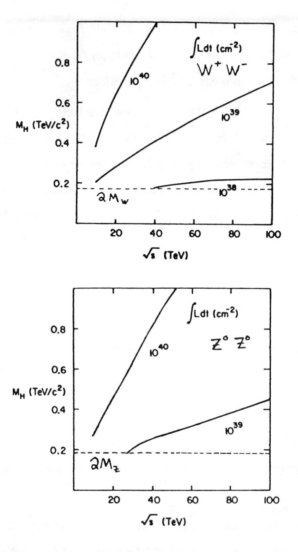

Figure 55: Discovery limit of m_H as a function of $\sqrt{s}$ in $pp \to H \to W^+W^-$ and $pp \to Z^0 Z^0$ for integrated luminosities of 10^{40}, 10^{39}, and (for the W^+W^- final state) $10^{38} cm^{-2}$, according to the criteria explained in the text. The dashed line is the kinematic threshold for the appropriate Higgs decay.

replacement of hadrons with quarks at energy scales above a GeV.

It is a sensible to ask which type of effective Lagrangian can consistently represent the low energy effective interactions of some unknown dymanics at some higher energy scale. This type of question is in a sense metaphysical since it concerns the theory of theories, however much can be learned from studying the classes of possible theories. In this respect one very important property of a Lagrangian is whether it is "natural" or not. There are many different properties of a theory which have been called naturalness[81] . Here I am discussing only the specific definition of t' Hooft[80] :

A Lagrangian $\mathcal{L}(\Lambda)$ is natural at the energy scale Λ if and only if each small parameter ξ(in units of the appropriate power of Λ) of the Lagrangian is associated with an approximate symmetry of $\mathcal{L}(\Lambda)$ which in the limit $\xi \to 0$ becomes an exact symmetry.

Within the context of an effective Lagrangian this definition of naturalness is simply a statement that it would require a dynamical accident to obtain small ξ except as defined above. This definition of Naturalness has two important properties: First to determine whether a theory is nature at some energy scale Λ does not require any knowledge of physics above Λ; and Second, If a Lagrangian becomes unnatural at some energy scale Λ_0 then it will be unnatural at all higher scales Λ. Hence if naturalness is to be a property of the ultimate theory of interactions at very high energy scales, then the effective Lagrangian at all lower energy scales must have the property of naturalness. The W-S theory will elementary scalars becomes unnatural at or below the electroweak scale as we shall see below; therefore if we demand that the final theory of everything is natural, then the standard model must be modified at or below the electroweak scale!

The problem with naturalness in the W-S Model comes from the scalar sector. To see the essential difficulty, we consider a simple ϕ^4 theory:

$$\mathcal{L} = \frac{1}{2}(\partial_\mu \phi)^2 - \frac{1}{2}m^2 \phi^2 - \frac{\lambda}{4!}\phi^4 \qquad (4.50)$$

λ can be a small parameter naturally because in the limit $\lambda = 0$ the theory becomes free and hence there is an additional symmmetry, ϕ number conservation. For the parameter, m^2, the limit $m^2 = 0$ apparently enhances the symmetry by giving a conformally invariant Lagrangian; however this symmetry is broken by quantum corrections[82] and thus can not be used to argue that a small m^2 is natural.

Finally, if both λ and m^2 are taken to zero simultaneously, we obtain a symmetry $\phi(x) \to \phi(x) + c$. Hence we can have an approximate symmetry at energy Λ_0 where:

$$\lambda \sim O(\epsilon) \quad \text{and} \quad m^2 \sim O(\epsilon \Lambda_0^2) \qquad (4.51)$$

Therefore

$$\Lambda_0 = O(\frac{m}{\sqrt{\lambda}}) \qquad (4.52)$$

ignoring factors of order one. Thus naturalness breaks down for $\Lambda \geq \Lambda_0$.

Returning to the W-S Lagrangian of Eq. 3.1, we can ask if there is any approximate symmetry which can allow for a small scalar mass consistent with naturalness? We have seen that the only possibility is the symmetry $\phi \to \phi + c$. But this symmetry is broken by both the gauge interactions and the scalar self interactions; hence

$$\frac{m_H^2}{\Lambda^2} \geq O(\lambda) \geq O(\alpha_{em}) \qquad (4.53)$$

and remembering that $m_H^2 = \bigcirc(\lambda v^2)$ Eq 4.53 implies

$$\Lambda \leq \bigcirc(v) = 246 GeV \tag{4.54}$$

the electroweak scale. The W-S model becomes unnatural at approximately the electroweak scale. Also

$$m_H = \bigcirc(\frac{\sqrt{\lambda}}{g} M_W) \geq \bigcirc(M_W) \tag{4.55}$$

Hence values of m_H much below M_W are unnatural.

To summarize, the W-S model is unnatural at energy scales $\Lambda > G_F^{-\frac{1}{2}}$ because m_H^2/Λ^2 is a small parameter which does not has any associated approximate symmetry of the Lagrangian. This unnaturalness is not cured in GUTS models (e.g. SU(5)). The theory must be modified at the electroweak scale in order to remain natural.

Two solutions have been proposed to retain naturalness of the Lagrangian above the electroweak scale:

- Eliminate the scalars as fundamental degrees of freedom in the Lagrangian for $\Lambda \gg G_F^{-\frac{1}{2}}$. We will consider this possibility in the next two lectures on Technicolor and Compositeness.

- Associate an approximate symetry with the scalars being light. The only posible symmetry known is Supersymmetry, which we will discuss in the last lecture. Since supersymmetry relates boson and fermion masses, and chiral symmetry protects zero values for fermion masses; by combining these two symmetries we can associate a symmetry with masses of scalar fields being zero. However to be effective in protecting scalar masses at the electroweak scale the scale of supersymmetry breaking must be of the order of a TeV or less.

Hence both alternatives for removing the unnaturalness of the standard model require new physics at or below the TeV scale. We will consider the possible physics in detail in the remaining lectures.

5 A NEW STRONG INTERACTION ?

As we discussed in the last section, the Weinberg-Salam Lagrangian is unnatural for $\Lambda \gg G_F^{-\frac{1}{2}}$. One remedy is to make the scalar doublet of the standard model composite. Then the usual Lagrangian is only the appropriate effective Lagrangian for energies below the scale Λ_T of the new strong interaction which binds the constituents of the electroweak scalar doublet. Clearly this new scale Λ_T cannot be much above the electroweak scale if it is to provide a solution to the naturalness problem.

It should be noted that the standard model itself will be strongly interacting for m_H near the unitarity bound of Eq 4.30 since $m_H^2 = 2\lambda v^2$. So many results presented here will be applicable to that case as well. See M.K. Gaillard's lectures in this school for a detailed discussion of this possibliity[83].

5.1 Minimal Technicolor

5.1.1 The Model

The simplest model for a new strong interaction is called technicolor and was first proposed by S. Weinberg[84] and L. Susskind[85]. This model is build upon our knowledge of the ordinary strong interactions (QCD).

The minimal technicolor model introduces a new set of fermions (technifermions) interacting via a new non-Abelian gauge interaction (technicolor). Specifically the technicolor gauge group is assumed to be $SU(N)$ and the technifermions are assumed to be massless fermions transforming as the $\mathbf{N} + \overline{\mathbf{N}}$ representation. None of the ordinary fermions carry technicolor charges.

The technifermions will be denoted by U and D. In the minimal model the technifermions have no color and transform under the $SU(2) \otimes U(1)$ as:

$$\begin{array}{ccc} & SU(2)_L & U(1)_Y \\ \begin{pmatrix} U \\ D \end{pmatrix}_L & 2 & 0 \\ U_R & & 1 \\ & 1 & \\ D_R & & -1 \end{array}$$

The values of the weak hypercharge Y of the technifermions is consistent with the requirement of an anomaly free weak hypercharge gauge interaction. With these assignments the technifermion charges are:

$$Q = I_3 + Y/2 \quad \text{thus} \quad Q_U = +1/2 \text{ and } Q_D = -1/2 \tag{5.1}$$

The usual choice for N is $N = 4$.

Technicolor becomes strong at the scale Λ_T at which $\alpha_T(\Lambda_T) \approx 1$. As with the ordinary strong interactions, the chiral symmetries of the technifermions

$$SU(2)_L \otimes SU(2)_R \tag{5.2}$$

are spontaneously broken to the vector subgroup[86]

$$SU(2)_V \tag{5.3}$$

by the condensate $< \overline{\Psi}\Psi > \neq 0$. The $SU(2)_L \otimes SU(2)_R$ symmetry of the technifermions accounts for the $SU(2)_L \otimes SU(2)_R$ symmetry of the effective Higgs potential. Associated with each of the three broken symmetries is a Goldstone boson. These are $J^{PC} = 0^{-+}$ isovector massless states:

$$\begin{aligned} \Pi_T^+ &\sim \overline{D}\gamma_5 U \\ \Pi_T^0 &\sim \frac{1}{\sqrt{2}}(\overline{U}\gamma_5 U - \overline{D}\gamma_5 D) \\ \Pi_T^- &\sim \overline{U}\gamma_5 D \end{aligned} \tag{5.4}$$

Goldstone bosons associated with the spontaneous breakdown of global chiral symmetries of the technifermions are commonly called technipions.

The couplings of the three Goldstone bosons to the EW currents are given by current algebra:

$$< 0|J_a^\mu(0)|\Pi_b(q) > \; = \; iq^\mu F_\pi \delta_{ab} g/2$$

$$< 0|J_Y^\mu(0)|\Pi_b(q) > \; = \; iq^\mu F_\pi \delta_{a3} g'/2. \quad (5.5)$$

These couplings determine the couplings of the Goldstone bosons to the $W^\pm$ and Z^0. To see how Higgs mechanism works here, consider the contribution of the Goldstone bosons to the polarization tensor of an electroweak boson:

$$\int d^4x e^{ik\cdot x} < 0|T J_\alpha^\mu(x) J_\beta^\nu(0)|0 > = -i\Pi_{\alpha\beta}^{\mu\nu}(k) = -i(g^{\mu\nu}k^2 - k^\mu k^\nu)\Pi(k) \quad (5.6)$$

Using the coupings of the Goldstone bosons to the currents given in Eq. 5.4, we see that the Goldstone bosons contribute to give a pole to $\Pi(k)$ as $k^2 \to 0$ so

$$\Pi(k) = \frac{M_{\alpha\beta}^2}{k^2} \quad (5.7)$$

where

$$\mathcal{M}^2 = \frac{F_\pi^2}{4} \begin{pmatrix} g^2 & 0 & 0 & 0 \\ 0 & g^2 & 0 & 0 \\ 0 & 0 & g^2 & -gg' \\ 0 & 0 & -gg' & g'^2 \end{pmatrix} \begin{matrix} + \\ - \\ 3 \\ Y \end{matrix} \quad (5.8)$$

This is simply the standard Higgs mechanism with the scalars being composite bosons. The mass matrix $\mathcal{M}$ gives a massive $W^\pm$ and Z^0 with

$$M_W/M_Z = cos(\theta_w) = \frac{g}{\sqrt{g^2 + g'^2}} \quad (5.9)$$

and a massless photon. To obtain the proper strength of the weak interactions we require

$$F_\pi = 246 GeV . \quad (5.10)$$

The usual theory of the spontaneously broken symmetries of the $SU(2)_L \otimes U(1)_Y$ model is complelely reproduced. The custodial $SU(2)_V$ symmetry of the technicolor interactions (Eq. 5.2) guarantee the correct W to Z mass ratio.

Technicolor provides an elegant solution to the naturalness problem of the standard model; however it has one major deficiency. The chiral symmetries of ordinary quarks and leptons remain unbroken when the technicolor interactions become strong. Hence no quark or lepton masses are generated at the electroweak scale. Another way of saying the same thing is that the interactions generated by the technicolor do not generate effective Yukawa couplings between the ordinary quarks and leptons and the composite scalars. We return to discuss attempts to remedy this problem later.

5.1.2 Technicolor Signatures

Knowing the spectrum of ordinary hadrons, and attributing its character to QCD, we may infer the spectrum of the massive technihadrons. The spectrum should minic the QCD spectrum with two quark flavors. It will include:

- An isotopic triplet of $J^{PC} = 1^{--}$ technirhos

$$\begin{aligned} \rho_T^+ &= \overline{D}\gamma^\mu U \\ \rho_T^0 &= \frac{1}{\sqrt{2}}(\overline{U}\gamma^\mu U - \overline{D}\gamma^\mu D) \\ \rho_T^- &= \overline{U}\gamma^\mu D \end{aligned} \quad (5.11)$$

The masses and widths of the technirho mesons can be estimated using the QCD analogs and large N arguments[87]. We obtain

$$m_{\rho_T} = m_\rho \frac{F_\Pi}{f_\pi}\sqrt{\frac{3}{N}} \approx 2TeV\sqrt{\frac{3}{N}} \quad (5.12)$$

$$\Gamma(\rho_T \to \Pi_T \Pi_T) = \Gamma(\rho \to \pi\pi)(\frac{3}{N})[\frac{m_{\rho_T}}{m_\rho}][1 - 4m_\pi]^{-\frac{3}{2}} \quad (5.13)$$

For the choice $N = 4$, $M_{\rho_T} = 1.77$ TeV and $\Gamma_{\rho_T} = 325$ GeV.

- An isoscalar $J^{PC} = 1^{--}$ techniomega

$$\omega_T^0 = \frac{1}{\sqrt{2}}(\overline{U}\gamma^\mu U + \overline{D}\gamma^\mu D) \quad (5.14)$$

with a mass approximately degenerate with the technirho and which decays principally into three technipions.

- An isoscalar $J^{PC} = 0^{-+}$ technieta

$$\eta_T = \frac{1}{\sqrt{2}}(\overline{U}\gamma_5 U + \overline{D}\gamma_5 D) \quad (5.15)$$

with a mass ≈ 1 TeV/c^2.

- An isoscalar $J^{PC} = 0^{++}$ technisigma,

$$H_0 = \frac{1}{\sqrt{2}}(\overline{U}U + \overline{D}D) \quad (5.16)$$

with a mass expected to be $\approx 2\Lambda_T$ and ordinary technicolor strong decays. The technisigma is the analogy of the physical Higgs scalar in the Weinberg-Salam model. Here the dynamics determines the mass of the Higg-like scalar; it is not a free parameter as it is in the standard model; and in particular, it cannot be light.

In addition there are other more massive scalars, axial vectors, and tensors. There will also be a rich spectrum of (T^N) technibaryons. Some of these might well be stable against decay, within technicolor.

In hadron-hadron collisions, technifermions of the minimal model will be pair produced by electroweak processes. One possible experimantal signature is the creation of stable technibaryons, which for all odd values of N would carry half-integer charges. The production rate cannot exceed the overall rate of technifermion pair production, which even at the SSC will be minuscule - on the order of the Drell-Yan cross section.

The signature of the minimal technicolor scheme is the expected modifications to the electroweak processes in the 1-TeV regime. Thus only a supercollider will have sufficient energy to observe these signals. The most prominent of these are the contributions of the s-channel technirho to the pair production of gauge bosons. Because of the weak hypercharge assignments of the technifermions the techniomega (unlike the omega in QCD) does not mix with the photon or Z^0 to produce a s-channel resonance.

Because of the strong coupling of the technirhos to pairs of longitudinal W's or Z's (the erstwhile technipions), the processes[88]

$$q_i \bar{q}_i \to (\gamma \text{ or } Z^0) \to W_L^+ W_L^- \qquad (5.17)$$

and

$$q_i \bar{q}_i \to W^\pm \to W_L^\pm Z_L^0 \qquad (5.18)$$

where the subscript L denotes longitudinal polarization, will lead to significant enhancements in the pair production cross sections.

Including the s-channel technirho enhancement, the differential cross section for production of W^+W^- is given by

$$\begin{aligned}\frac{d\sigma}{dt}(u\bar{u} \to W^+W^-) =& \frac{\pi\alpha^2}{12s^2}\Big\{2(1 + \frac{M_Z^2}{s-M_Z^2}L_u)(\frac{ut-M_W^4}{st} - \frac{2M_W^2}{t}) + \frac{ut-M_W^4}{t^2} \\ & -4(\frac{M_Z^2}{s-M_Z^2})L_u + \frac{M_Z^2 s\beta_W^2}{(s-M_Z^2)^2}\frac{(L_u^2+R_u^2)}{1-x_w} \\ & +\frac{(ut-M_W^4)}{s^2}[2+X+(\frac{6M_W^2-sX}{s-M_Z^2})(\frac{L_u}{1-x_w})]\end{aligned}$$

$$+(\frac{s}{s-M_Z^2})^2[\beta_w^2 - 1 + X + \frac{12M_W^4}{s^2}]\frac{(L_u^2 + R_u^2)}{4(1-x_w)^2}\} \quad (5.19)$$

where $\beta_W \equiv \sqrt{1 - 4M_W^2/s}$, $L_u = 1 - 4x_w/3$, $R_u = 4x_w/3$, and

$$X = \frac{M_{\rho_T}^4}{(s - M_{\rho_T}^2)^2 + M_{\rho_T}^2 \Gamma_{\rho_T}^2} \quad (5.20)$$

All the effects of the technirho are contained in the factor X, setting $X = 1$ corresponds to the standard model expression. The corresponding expresions for the contribution of technirhos to $d\sigma/dt$ for $d\bar{d} \to W^+W^-$ and $u\bar{d} \to W^+Z^0$ are given in EHLQ(Eqs. 6.22-6.23 respectively). There is no ρ_T^0 enhancement in the Z^0Z^0 final state since ρ_T^0 has $I_T = 1$ and $I_{T3} = 0$ (i.e. W_3 couples only to W^+W^- not to W_3W_3 or BB; hence the ρ_T^0 will not couple to them either).

We show in Figure 56 the mass spectrum of W^+W^- pairs produced in pp collisions at 20, 40 and 100 TeV, with and without the technirho enhancement. Both intermediate bosons are required to satisfy $|y| < 1.5$. The yields are slightly higher in the neighborhood of the ρ_T in $\bar{p}p$ collisions. This is a 25 percent effect at 40 TeV.

We show in Figure 57 the mass spectrum of $W^\pm Z^0$ pairs produced in pp collisions at 20, 40 and 100 TeV, with and without the $\rho_T^\pm$ enhancement. Again both intermediate bosons are required to satisfy $|y| < 1.5$.

The technirho enhancement amounts to nearly a doubling of the cross section in the resonance region for W^+W^- pair production and an even greater signal to background (S/B) ratio in the $W^\pm Z^0$ case. However, because the absolute rates are small, the convincing observation of this enhancement makes nontrivial demands on both collider and experiment.

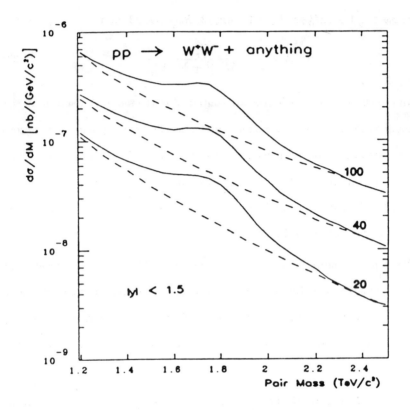

Figure 56: Mass spectrum of W^+W^- pairs produced in pp collisions, according to the parton distribution of Set 2 in EHLQ. The cross sections are shown with (solid lines) and without (dashed lines) the technirho enhancement of Eq. 5.19. $M_{\rho_T} = 1.77$ TeV/c^2 and $\Gamma_{\rho_T} = 325$ GeV.

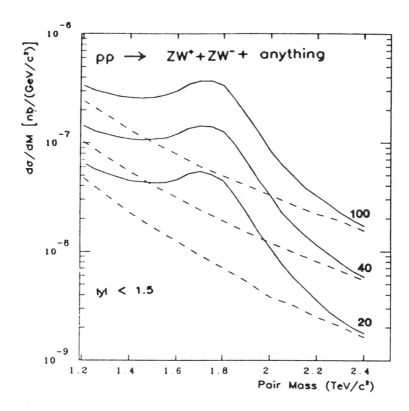

Figure 57: Mass spectrum of W^+Z^0 and W^-Z^0 pairs produced in pp collisions, according the the parton distributions of Set 2 in EHLQ. The cross sections are shown with (solid lines) and without (dashed lines) the technirho enhancement of Eq. 5.19. $M_{\rho_T} = 1.77$ TeV/c^2 and $\Gamma_{\rho_T} = 325$ GeV.

Table 7: Detecting the ρ_T of the Minimal Technicolor Model at a pp Supercollider. For an assumed integrated luminosity $\int dt \mathcal{L} = 10^{40}(cm)^{-2}$, the total signal/background rates (S/B) are given for the channels $W^+ W^-$ (column 2) and $W^\pm Z^0$ (column 4). Detecting 25 excess events with a 5σ S/B requires miminum detection effeciencies ϵ_W and ϵ_Z given in column 3 and 5 respectatively.

$\sqrt{s}$	$W^+ W^-$		$W^\pm Z^-$	
(TeV)	S/B	ϵ_W	S/B	$\sqrt{\epsilon_W \epsilon_Z}$
10	—	—	28/10	1
20	100/110	.52	152/50	.41
40	240/300	.36	420/130	.24

An estimate of the background of standard gauge boson pairs can be obtained be integrating that cross section over the resonance region

$$1.5 TeV/c^2 \leq \mathcal{M} \leq 2.1 TeV/c^2 \qquad (5.21)$$

The resulting signal and background events for a standard run with integrated luminosity of $10^{40} cm^{-2}$ are given Table 7. We require that the enhancement consist of at least 25 detected events, and that the signal represent a five standard deviation excess over the background. This criterion translates into minimum detection for the gauge bosons also listed in Table 7.

Since the leptonic branching ratio for the Z^0 is only 3 percent per charged lepton, we can conclude (from Table 7) that detection of the technirho at $\sqrt{s} = 40$ TeV requires observation of at least the Z^0 in its hadronic decay modes. Realistically it will also be necessary to detect the $W^\pm$ in their hadronic modes.

In these cases the two jet backgrounds to the $W^\pm$ or Z^0 must be separated. The severity of the 2 jet + W and 4 jet backgrounds is still an open question under intense study[89].

Whatever the conclusions of present studies, it is safe to say that discovering the technirho signature of the minimal technicolor model is one of the hardest challenges facing experimentalist at the future SSC.

5.2 Extended Technicolor

5.2.1 Generating Fermion Masses

The minimal model just presented illustrates the general strategy and some of the consequences of a technicolor implementation of dynamical electroweak symmetry breaking. However it does not provide a mechanism for generating masses for the ordinary quarks and leptons. Various methods of overcoming this problem have been proposed, in this section we consider the original proposal - extended technicolor[90,91] as a prototype.

The basic idea of extended technicolor (ETC) is to embed the technicolor group G_T into a larger extended technicolor group $G_{ETC} \supseteq G_T$ which couples quarks and leptons to technifermions. This extended gauge group is assumed to break down spontaneously $G_{ETC} \to G_T$ at an energy scale

$$\Lambda_{ETC} \sim 30 - 300 TeV \tag{5.22}$$

producing masses for the ETC gauge bosons of order

$$M_{ETC}^2 \approx g_{ETC}^2 \Lambda_{ETC}^2 . \tag{5.23}$$

Since the ETC bosons couple technifermions to ordinary fermions ETC boson exchange induces an effective four fermion interaction at energy scales below Λ_{ETC}:

$$\mathcal{L}_{ETC} = \frac{g_{ETC}^2}{M_{ETC}^2} \bar{q}_L \gamma^\mu Q_L \overline{Q}_R \gamma_\mu q_R + h.c \qquad (5.24)$$

where by Eq. 5.23:

$$g_{ETC}^2/M_{ETC}^2 = 1/\Lambda_{ETC}^2 \ . \qquad (5.25)$$

Now when technicolor becomes strong and the chiral symmetries of the technifermions are spontaneously broken at scale Λ_T, forming the condensate

$$<0|\Psi_L\overline{\Psi}_R|0> + h.c. \approx \Lambda_T^3 \ , \qquad (5.26)$$

the effective Lagrangian of Eq. 5.24 becomes

$$\mathcal{L}_{ETC} = \frac{\Lambda_T^3}{\Lambda_{ETC}^2}(\bar{q}_L q_R + h.c) \ . \qquad (5.27)$$

This is just a mass term for the ordinary fermion field q. Hence, by this mechanism the ETC interactions can generate a mass

$$m_q = \frac{\Lambda_T^3}{\Lambda_{ETC}^2} \qquad (5.28)$$

for the ordinary fermions.

5.2.2 The Farhi-Susskind Model[92]

In any of the more nearly realistic technicolor models produced so far, there are at least four flavors of technifermions. As a consequence, the chiral flavor group is larger than the $SU(2)_L \otimes SU(2)_R$ of the minimal technicolor model(Eq. 5.2), so that more than three massless technipions result from the spontaneous breakdown of chiral symmetry. These addition technipions remain as physical spinless particles. Of course, these cannot and do not remain massless, but acquire calculable masses considerably less than 1 TeV/c^2. These particles are therefore accessible to present and planned hadron colliders.

At present there is no completely realistic model that incorporates the ideas of ETC. In particular, the lack of an obvious analog of the Glashow-Iliopoulos-Maiani(GIM) mechanism[93] is precisely the feature of all known ETC models that makes them phenomenologically problematic[90,94,95]. Recently several attempts have been made to construct a GIM-like mechanism for ETC theories[96]. However, no proposal has yet been a complete success.

Here we consider a simple toy technicolor model due to Farhi and Susskind[92], which has quite a rich spectrum of technipions and technivectormesons. This model has been developed further by Dimopoulos[97], Peskin[98], Preskill[99], and Dimopoulos, Raby, and Kane[100]. Of course this model is not correct in detail, but many of the observable consequences should not be affected by these problems.

In the Farhi-Susskind model the technicolor group is $SU(4)$. The technifermions transform under $SU(3) \otimes SU(2)_L \otimes U(1)_Y$ as:

$$\begin{pmatrix} U \\ D \end{pmatrix}_L \quad 3 \quad 2 \quad \quad Y$$
$$U_R \quad \quad \quad \quad \quad Y+1$$
$$\quad \quad \quad \quad 3 \quad 1$$
$$D_R \quad \quad \quad \quad \quad Y-1$$
$$\begin{pmatrix} N \\ E \end{pmatrix}_L \quad 1 \quad 2 \quad \quad -3Y$$
$$N_R \quad \quad \quad \quad \quad -3Y+1$$
$$\quad \quad \quad \quad 1 \quad 1$$
$$E_R \quad \quad \quad \quad \quad -3Y-1$$

The choice $Y = 1/3$ gives the technifermions the same charges as the corresponding ordinary fermions.

The global flavor symmetries G_f of the massless technifermions in this model

are:
$$G_f = SU(8)_L \otimes SU(8)_R \otimes U(1)_V \qquad (5.29)$$

which are spontaneously broken by the strong technicolor interactions at the scale Λ_T to the nonchiral subgroup:

$$SU(8)_V \otimes U(1)_V \ . \qquad (5.30)$$

Associated with each spontaneously broken chiral symmetry is a massless Goldstone boson. There are $8^2 - 1 = 63$ such Goldstone bosons in this model. As before, there are three Goldstone bosons which are associated with the electroweak symmetries. When the electroweak gauge interactions are included these Goldstone bosons combine with the gauge fields to make the massive physical $W^{\pm}$ and Z^0 particles. The others Goldstone bosons will acquire masses when the ETC, $SU(3)$ color, and electroweak interactions are included. For this reason these remaining states are sometimes called Pseudo-Goldstone-Bosons (PGB's). More commonly, these additional states are called technipions in analog with corresponding states in QCD.

5.2.3 Masses for Technipions

The method of analysis used to determine the masses for technipions is a generalization of the Dashen's analysis for pion masses in QCD[101]. Let me briefly review this idea here.

Consider a Hamiltonian H_0 invariant under a set of symmetries with charges Q_a: i.e.

$$[Q_a, H_0] = 0 \ . \qquad (5.31)$$

Some of these symmetries may be spontaneously broken by the dynamics of the theory, as in the theory we are considering. We denote the spontaneously broken symmetries by Q_a^A; since they are axial global symmetries in the case at hand. While the remaining unbroken symmetries will be denoted by Q_a^V, as they are in fact vector symmetries here.

The vacuum state of the theory $|\Omega>$ will therefore annihilate the unbroken charges

$$Q_a^V|\Omega_0>= 0 \qquad (5.32)$$

while for spontaneously broken charges

$$e^{-iQ_a^A\Lambda_a}|\Omega_0>= |\Omega(\Lambda)>\neq 0 . \qquad (5.33)$$

That is, the spontaneously broken charges are not symmetries of the vacuum. They rotate the vacuum into other states which because the charges commute with H_0 are degenerate in energy with the vacuum. This is exactly what happened in the Higgs potential of the Weinberg-Salam model discussed in Sec.3; spontaneous symmetry breaking occurs when the Hamiltonian has a degenerate set of lowest energy states (i.e. associated with a rotational invariance under the charge Q_a^A). The physical theory must chose one vacuum (i.e. align along one direction), thus breaking the symmetry. The physical degree of freedom associated with rotation in the direction of the original degeneracy (ie. rotations generated by Q_a^A) is a Goldstone boson. The Goldstone bosons are massless because these rotations leave the energy of the system unchanged.

Now consider what happens when a small perturbation δH_I is added which explicitly breaks one of the symmetries that is spontaneously broken in the unperturbed theory described by H_0. The degeneracy of the vacuum states is broken by δH_I and there is now a unique lowest energy state. If we define an energy $E(\Lambda_a)$ by:

$$E(\Lambda_a) \equiv <\Omega_0|e^{-iQ_a^A\Lambda_a}\delta H_I e^{iQ_a^A\Lambda_a}|\Omega_0> \qquad (5.34)$$

then if the minimum of $E(\Lambda_a)$ occurs for $\Lambda_a = \Lambda_a^P$ the physical vacuum state will be

$$|\Omega_{\text{phy}}> = e^{iQ_a^A\Lambda_a^P}|\Omega_0> \qquad (5.35)$$

Reexpressing E in terms of the physical vacuum

$$E(\tilde{\Lambda}) = <\Omega_{\text{phy}}|e^{-iQ_a^A\tilde{\Lambda}_a}\delta H_I e^{iQ_a^A\tilde{\Lambda}_a}|\Omega_{\text{phy}}> \ . \qquad (5.36)$$

Now the minimun occurs at $\tilde{\Lambda}_a = 0$ for each a. Hence at $\tilde{\Lambda}_a = 0$

$$\frac{\partial E}{\partial \tilde{\Lambda}_a} = 0 \quad \Rightarrow <\Omega_{\text{phy}}|[Q_a, \delta H_I]|\Omega_{\text{phy}}> = 0 \qquad (5.37)$$

and

$$\frac{\partial^2 E}{\partial \tilde{\Lambda}_a \partial \tilde{\Lambda}_a} = M_{ab}^2 \qquad (5.38)$$

or equivalently

$$<\Omega_{\text{phy}}|[Q_a, [Q_b, \delta H_I]]|\Omega_{\text{phy}}> = M_{ab}^2 \ . \qquad (5.39)$$

The matrix M_{ab}^2 is simply related to the mass squared matrix for the pseudo Goldstone bosons associated with the the spontaneously broken symmetries of H_0. If the PGB decay constants are defined by

$$<0|Q_a^A|\Pi_b> = iF_\Pi \delta_{ab} \qquad (5.40)$$

then

$$m_{ab}^2 = \frac{1}{F_\pi^2} M_{ab}^2 \ . \qquad (5.41)$$

5.2.4 Colored Technipion Masses

One mechanism by which technipions get masses is via the explicit symmetry breaking resulting when the color and electroweak gauge interactions of the technifermions are considered. These radiative corrections have been considered in detail by Peskin and Chadha[102], Preskill[99], and Baluni[103].

The lowest order color gluon exchange leads to a explicit symmetry breaking interaction

$$\delta H_I = -g^2 \int d^4x D_{\mu\nu}(x) J_a^\mu J_b^\nu(0) \tag{5.42}$$

where $D_{\mu\nu}$ is the gluon propagator and

$$J_a^\mu(x) = \overline{\Psi}(x)\gamma^\mu T_a \Psi(x) \tag{5.43}$$

with T_a is a flavor matrix of the technifermions. Defining

$$Q_a^A \equiv \int d^3x \overline{\Psi}(x)\gamma^0\gamma_5 X_a \Psi(x) \tag{5.44}$$

then the mass matrix for the technipions is given from Eq. 5.39 and Eq. 5.41 by

$$m_{ab}^2 = \frac{1}{F_\Pi^2} \frac{\partial}{\partial \Lambda_a} \frac{\partial}{\partial \Lambda_b} \int d^4x g^2 D_{\mu\nu}(x) < \Omega_{\text{phy}}|T\tilde{J}_a^\mu(x)\tilde{J}_a^\nu(0)|\Omega_{\text{phy}} > |_{\Lambda=0} \tag{5.45}$$

where

$$\tilde{J}_a^\mu(x) \equiv e^{-iQ_a^A \Lambda_a} J_a^\nu(x) e^{iQ_a^A \Lambda_a} \tag{5.46}$$

Using Eq. 5.43, Eq. 5.45, and Eq. 5.46 and fact that the technicolor interactions are flavor blind the mass matrix for the colored technipions can be written as:

$$m_{ab}^2 = \frac{g^2}{4\pi} \text{Tr}([X_a, T_\alpha][X_b, T_\alpha])\{M_{\text{TC}}\}^2 \tag{5.47}$$

where all the technicolor strong dymanics is contained in the factor M_{TC}

$$\{M_{\text{TC}}\}^2 = \frac{4\pi}{F_\Pi^2} \int d^4x D_{\mu\nu}(x) < \Omega|T(J_V^\mu(x)J_V^\nu(0) - J_A^\mu(x)J_A^\nu(0))|\Omega > \}. \tag{5.48}$$

The magnitude the dynamic term in Eq. 5.48 can be estimated by analog with QCD. Dashen proved that

$$m_{\pi^+}^2 - m_{\pi^0}^2 = \alpha M_{\text{QCD}}^2 \tag{5.49}$$

where

$$M_{\text{QCD}}^2 = \frac{4\pi}{f_\pi^2} \int d^4x D_{\mu\nu}(x) < 0|T(J_V^\mu(x)J_V^\nu(0) - J_A^\mu(x)J_A^\nu(0))|0> . \tag{5.50}$$

with $D_{\mu\nu}$ the photon propagator. Experimentally the value of M_{QCD} is given by

$$M_{\text{QCD}}^2/m_\rho^2 = .3 \tag{5.51}$$

and the dependence on the gauge group $SU(N)$ can also be estimated using large N arguments[87]. The result is

$$\frac{\sqrt{N}M}{f_\pi} \approx 8 \tag{5.52}$$

Thus for $SU(4)$ Farhi-Susskind model, $F_\Pi = 126$ GeV and dynamic factor in Eq. 5.48 is

$$M_{\text{TC}} = 500 \; GeV/c^2 . \tag{5.53}$$

Turning explicitly to the technipions in the Farhi-Susskind model, we find 32 color octet technipions:

$$(P_8^+ \;\; P_8^0 \;\; P_8^-) \to \overline{Q}\gamma_5 \frac{T^a}{2}\frac{\lambda^\alpha}{2}Q \tag{5.54}$$

$$P_8^{0'} \to \overline{Q}\gamma_5 \frac{\lambda^\alpha}{2}Q \tag{5.55}$$

$$\tag{5.56}$$

all with mass

$$m(P_8) = (3\alpha_s)^{1/2} \; 500 \; GeV/c^2 \approx 240 \; GeV/c^2 \tag{5.57}$$

, and 24 color triplet technipions:

$$(P_3^+ \quad P_3^0 \quad P_3^-) \to \overline{L}\gamma_5 \frac{\tau^a}{2} Q_i \qquad (5.58)$$

$$P_3^{0'} \to \overline{L}\gamma_5 Q_i \qquad (5.59)$$

$$(\overline{P}_3^- \quad \overline{P}_3^0 \quad \overline{P}_3^+) \to \overline{Q}^i \gamma_5 \frac{\tau^a}{2} L \qquad (5.60)$$

$$\overline{P}_3^{0'} \to \overline{Q}^i \gamma_5 L \qquad (5.61)$$

$$(5.62)$$

with mass

$$m(P_8) = (\frac{4}{3}\alpha_s)^{1/2}\, 500\ GeV/c^2 \approx 160\ GeV/c^2 . \qquad (5.63)$$

5.2.5 Color Neutral Technipion Masses

The total number of technipions in the Farhi-Susskind model is 63. As we have shown in the last section, 56 of these are colored and receive masses from radiative corrections involoving color gluon exchange. The remaining 7 technipions are color neutral. Three of these are true Goldstone bosons remaining exactly massless:

$$(\Pi_T^+,\ \Pi_T^0,\ \Pi_T^-) \to \frac{1}{\sqrt{2}}(\overline{Q}\gamma_5\frac{\tau^a}{2}Q + \overline{L}\gamma_5 L) \qquad (5.64)$$

and become the longitudinal components of the W^+, Z^0, and W^- by the Higgs mechanism. So finally we are left with four additional color neutral technipions:

$$(P^+,\ P^0,\ P^-) \to \frac{1}{\sqrt{6}}(\overline{Q}\gamma_5\frac{\tau^a}{2}Q - 3\overline{L}\gamma_5\frac{\tau^a}{2}L)$$

$$P'^0 \to \frac{1}{\sqrt{6}}(\overline{Q}\gamma_5 Q - 3\overline{L}\gamma_5 L) \qquad (5.65)$$

$$(5.66)$$

The mechanism for mass generation is more complicated for these neutral technipions. It is discussed in detail by Peskin and Chadha[102] and Baluni[103]. The main points are:

- Before symmetry breaking effects are included the electroweak gauge interaction do not induce any masses for the technipions P^+, P^0, P^-, and P'^0.

- Including the symmetry breaking effects, in particular the nonzero mass for the Z^0, the charged P's acquire a mass[90,102,103]

$$m_{\text{EW}}(P^+) = m_{\text{EW}}(P^-) = \sqrt{\frac{3\alpha}{4\pi} log(\frac{\Lambda^2}{M_Z^2})} M_Z \approx 6 GeV/c^2 . \quad (5.67)$$

while the neutral states P^0 and P'^0 remain massless.

- The lightest neutral technipions can only acquire mass from the symmtry breaking effects of the ETC interactions.

The effects of ETC gauge boson exchanges induce masses of the order of[90,103]:

$$m_{\text{ETC}}^2 \approx \frac{1}{F_\Pi^2} \frac{\Lambda_{\text{TC}}^6}{\Lambda_{\text{ETC}}^2} \quad (5.68)$$

where the ETC scale Λ_{ETC} is related to the quark (and lepton) m_q by Eq. 5.28

$$m_{\text{ETC}}^2 \approx \frac{m_q \Lambda_{\text{TC}}^3}{F_\Pi^2} \quad (5.69)$$

However the scale of quark masses range from $m_u \approx 4 MeV/c^2$ to $m_t > 25 GeV/c^2$. Therefore which value to use for the ETC scale is very uncertain. A reasonable guess[90,104] for the total masses $m = \sqrt{m_{\text{EW}}^2 + m_{\text{ETC}}^2}$ of these lightest technipions are:

$$7 GeV/c^2 \leq m(P^\pm) \leq 45 GeV/c^2$$

$$2 GeV/c^2 \leq m(P^0) \text{ or } m(P'^0) \leq 45 GeV/c^2 \tag{5.70}$$

$$\tag{5.71}$$

5.2.6 Technipion Couplings

The coupling of technipions to ordinary quarks and leptons depend on the details of the ETC interactions in the particular model. However, in general, the couplings of these technipion are Higgs-like. Thus the naive expectation is that the technipion coupling to ordinary fermions pairs will be roughly proportional to the sum of the fermion masses. A discussion of various possiblities can be found in Lane[105].

In addition, there are couplings to two (or more) gauge bosons which arise from a triangle (anomaly) graph containing technifermions, analogous to the grahp responsible for the decay $\pi^0 \to \gamma\gamma$. The details of these couplings can be found elsewhere[1,105].

The major decay modes of the technipions are summarized in Table 8.

5.3 Detecting Technipions

The masses of the color neutral technipions are within the range of present experiments. Some constraints already exist on the possible masses and couplings of these technipions.

The strongest contraints on the charged technipions ($P^\pm$) come from limits on their production in e^+e^- colisions at PEP and PETRA[106]. A charged technipion decaying into $\tau\bar{\nu}_\tau$ or light quarks is ruled out for

$$m(P^\pm) < 17 GeV/c^2 \tag{5.72}$$

Table 8: Principal decay modes of technipions if Pf_1f_2 couplings are proportional to fermion mass.

Technipion	Principal decay modes
P^+	$t\bar{b}$, $c\bar{b}$, $c\bar{s}$, $\tau^+\nu_\tau$
P^0	$b\bar{b}$, $c\bar{c}$, $\tau^+\tau^-$
$P^{0\prime}$	$b\bar{b}$, $c\bar{c}$, $\tau^+\tau^-$, gg
P_3, P_3' (if unstable)	$t\tau^-$, $t\nu_\tau$, $b\tau^-$, ... or $\bar{t}t$, $\bar{t}b$, ...
$P_8{}^+$	$(t\bar{b})_8$
$P_8{}^0$	$(t\bar{t})_8$
$P_8{}^-$	$(t\bar{t})_8$, gg

; however decays into $b\bar{c}$ are not constrained by these experiments.

For the neutral technipions the constraints are indirect and generally rather weak. A detailed discussion all the existing limits is contained in Eichten, Hinchliffe, Lane, and Quigg[107].

Finally consider the detection prospects in hadron colliders for technipions. Mt discussion will draw heavily on the detailed analysis presented by Eichten, Hinchliffe, Lane, and Quigg[107] for present collider energies and EHLQ for supercollider energies.

The principal production mechanisms for the color-singlet technipions in $\bar{p}p$ collisions are:

- The production of $P^\pm$ in semiweak decays of heavy quarks.

- The production of the weak-isospin-singlet states P'^0 by the gluon fusion mechanism.

- pair production of $P^\pm P^0$ through the production of real or virtual $W^\pm$ bosons.

- Pair production of P^+P^- by the Drell-Yan mechanism, especially near the Z^0 pole.

each of these mechanisms will be discussed in turn.

If the the top quark is heavy enought for the decay:

$$t \to P^+ + (b \text{ or } s \text{ or } d) \tag{5.73}$$

to be kinematically allowed, then this decay will proceed at the semiweak rate[107]:

$$\Gamma(t \to P^+ q) \approx \frac{G_F\sqrt{2}}{\pi}|\mathcal{M}|^2(m_t^2 + m_q^2 - M_P^2)p \tag{5.74}$$

where
$$p = \frac{[m_t^2 - (m_q + M_{P+})^2]^{\frac{1}{2}}[m_t^2 - (m_q - M_{P+})^2]^{\frac{1}{2}}}{2m_t}. \quad (5.75)$$

With more or less conventional couplings of the $P^{\pm}$ to quarks and leptons, the coupling matrix $|\mathcal{M}| \approx 1$ and thus this decay mode of the top quark will swamp the normal weak decays. Hence the production of top quarks in hadron colliders will be a copious source of charged technipions if the decay is kinematically allowed. On the other hand, seeing the top quark though the normal weak decays will put strong constraints of the mass and couplings of any charged color neutral technipion.

The single production of the neutral isospin singlet technipion, P'^0, proceeds by the gluon fusion mechanism as for the usual Higgs scalar. The production rate is given by:

$$\frac{d\sigma}{dy}(ab \to P'^0 + \text{anything}) = \frac{\alpha_s^2}{96\pi F_\pi^2}\tau f_g^{(a)}(x_a, M_P^2) f_g^{(b)}(x_b, M_P^2) \quad (5.76)$$

where $\tau = M_P^2/s$. This differential cross section for P'^0 production at $y = 0$ is shown in Figure 58 as a function of technipion mass, M_P, for the $S\bar{p}pS$ and Tevatron colliders. The corresponding rates for Supercollider energies are shown in EHLQ (Fig. 181).

The principal decays of the P'^0 are: gg, $\bar{b}b$, and $\tau^+\tau^-$. The relative branching fractions are shown in Figure 59. Comparing the rates of P'^0 production with the background of QCD $\bar{b}b$ jet events (see for example Fig. 16), it becomes clear that detecting P'^0 in its hadronic decays is not possible. The background is more that two orders of magnitude that the signal. The only hope for detection is the leptonic final states - principally $\tau^+\tau^-$. For this channel the signal to background ratio is good; but this crucially depends on the reconstruction of

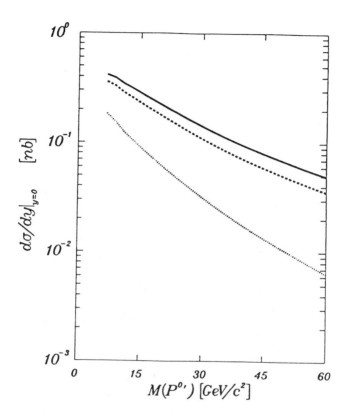

Figure 58: Differential cross section for the production of color singlet technipion P'^0 at $y = 0$ in $\bar{p}p$ collisions, for $\sqrt{s} = 2$ TeV (solid curve), 1.6 TeV (dashed curve), and 630 GeV (dotted curve). (From Ref. 107)

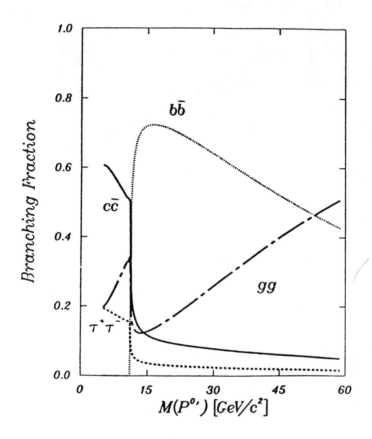

Figure 59: Approximate branching ratios for P'^0 decay. (From EHLQ)

the P'^0 invariant mass which is difficult since each of the τ decays contains a undetectable neutrino.

Finally there is the pair production of color singlet technipions through the chains:
$$\bar{p}p \;\rightarrow\; W^\pm + \text{anything} \atop \phantom{\bar{p}p \;\rightarrow\;}\hookrightarrow P^\pm + P^0 \tag{5.77}$$

and
$$\bar{p}p \;\rightarrow\; Z^0 + \text{anything} \atop \phantom{\bar{p}p \;\rightarrow\;}\hookrightarrow P^+ + P^- \tag{5.78}$$

where the intermediate bosons may be real or vitrual. The couplings of these technipion pairs to the $W^\pm$ and Z^0 are typically $1-2\%$. For more details see Ref. 107.

The cross section for $P^\pm P^0$ pairs produced in $\bar{p}p$ collisions at present collider energies is shown in Figure 60. Both $P^\pm$ and P^0 are required to have rapidities $|y_i| < 1.5$. The cross sections are appreciable only if $(M_{P^+} = M_{P_-}) < m_W/2$, for which the rate is determined by real $W^\pm$ decays.

Under the usual assumption that these lightest technipions couple to fermion pairs proportional to the fermion masses, the signal for these events would be four heavy quark jets, eg. $t\bar{b}b\bar{b}$. If heavy quark jets can be tagged with reasonable efficiency this signal should be observable. However, the couplings of $P^\pm$ and P^0 to fermion pairs are the result of the ETC model-dependent mixings and in general are more complicated than the simple mass proportional form usually assumed. Thus the search for scalar particles from $W^\pm$ decays should be as broad and thorough as practible.

Similarly, the cross section for production of P^+P^- pairs is only of experimental interest in present collider energies if $M_{P^\pm} < m_Z/2$. These cross sections

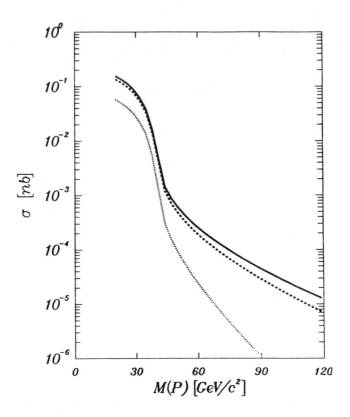

Figure 60: Cross section of the production of P^+P^0 and P^-P^0 (summed) in $\bar{p}p$ collisions as a function of the common (by assumption) mass of the technipions, for $\sqrt{s} = 2$ TeV (solid curve), 1.6 TeV (dashed curve), and 630 GeV (dotted curve). (From Ref. 107)

are shown in Figure 61. The rate of production of P^+P^- is low. It is not likely that this channel could be detected at a hadron collider in the near future. However this signal should be observable at the e^+e^- "Z^0 factories" at SLC and LEP.

5.3.1 Colored Technipions

The principal production mechanisms for the colored technipions in $\bar{p}p$ collisions are:

- Production of the weak isospin singlet state $P_8^{\prime 0}$ by gluon fusion.

- Production of $(P_3\overline{P}_3)$ or $(P_8\overline{P}_8)$ pairs in gg and $\bar{q}q$ fusion.

The gluon fusion mechanism for the single production of $P_8^{\prime 0}$ is the same as just discussed for $P^{\prime 0}$ production. The differential cross section is 10 times the cross section for $P^{\prime 0}$ production given in Eq. 5.76. The differential cross section (summed over the eight color indices) at $y = 0$ is shown as a function of the technipion mass in Figure 62 for present collider energies. The expected mass (Eq. 5.57) is approximately 240 GeV/c^2. The production for Supercollider energies is shown in EHLQ (see Fig. 184).

The principal decay modes are expected to be gg and $\bar{t}t$. The rates for the expected mass $M(P^{\prime 0}) = 240$ GeV/c^2 are too small for detection for $\sqrt{s}$ below 2 TeV. The best signals for detection are decays into top quark pairs.

Pairs of colored technipions are produced by the elementary subprocesses shown in Figure 63. The main contribution comes from the two gluon initial states (just as in the case of heavy quark pair production discussed in Lecture 2). Details about the production cross sections may be found in Ref. 107.

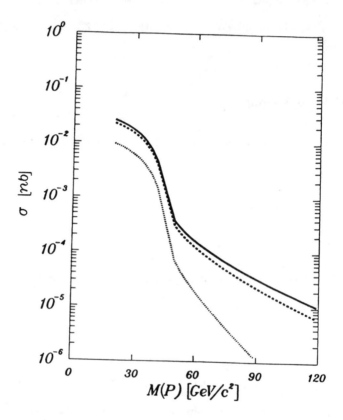

Figure 61: Cross section of the production of P^+P^- pairs in $\bar{p}p$ collisions as a function of the P^+ mass, for $\sqrt{s} = 2$ TeV (solid curve), 1.6 TeV (dashed curve), and 630 GeV (dotted curve). (From Ref. 107)

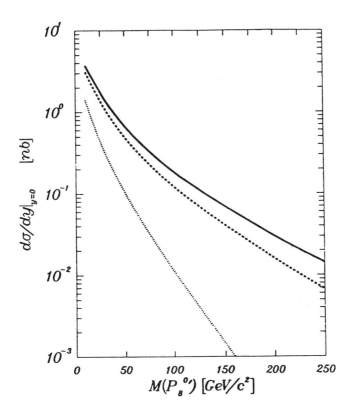

Figure 62: Differential cross section for the production of the color-octet technipion P_8^0 at $y = 0$ in $\bar{p}p$ collisions, for $\sqrt{s} = 2$ TeV (solid curve), 1.6 TeV (dashed curve), and 630 GeV (dotted curve). (From Ref. 107)

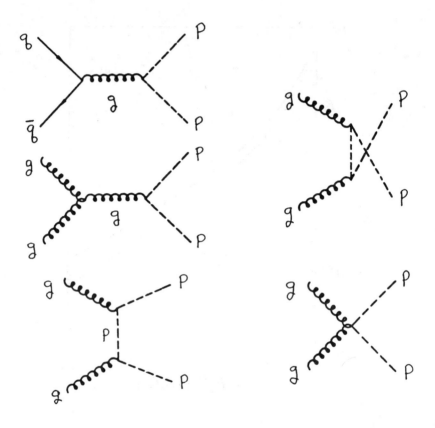

Figure 63: Feynman diagrams for the production of pairs of technipions. The curly lines are gluons, solid lines are quarks, and dashed lines are technipions. The graphs with s-channel gluons include the $\rho_8'^0$ enhancement.

The total cross sections for the process $\bar{p}p \to P_3\bar{P}_3$ are shown in Figure 64; and the total cross sections for the process $\bar{p}p \to P_8\bar{P}_8$ are shown in Figure 65. In both cases rapidity cuts $|y| < 1.5$ have been imposed. The expected mass (Eq. 5.63) for the P_3 technipion is approximately 160 GeV/c^2 and for the P_8 approximately 240 GeV/c^2 (Eq. 5.57). The corresponding cross sections for Supercollider energies are given in EHLQ (Figs. 187-190). The implications of these production rates for discovery of colored technipions are preesented in next section.

5.3.2 Discovery Limits

If the technicolor scenario correctly describes the breakdown of the electroweak gauge symmetry, there will be a number of spinless technipions, all with masses less than the technicolor scale of about 1 TeV. The simple but representative model of Farhi and Susskind[92] was considered here.

A rough appraisal of the minimum effective luminosities required for the observation of technipions of this model is given in Table 9 for present and future hadron colliders. The discovery criteria require that for a given charged state, the enhancement consists of at least 25 events, and that the signal represent a five standard deviation excess over background in the rapidity interval $|y| < 1.5$. The top quark mass was assumed to be 30 GeV.

We can conclude that a 40 TeV $p^{\pm}p$ collider with a luminosity of at least $10^{39} cm^{-2}$ will be able to confirm or rule out technicolor.

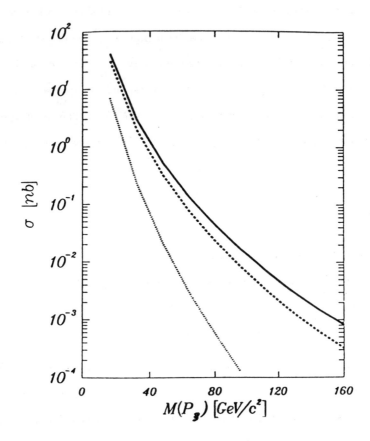

Figure 64: Cross sections for the production of $P_3\overline{P}_3$ pairs in $\bar{p}p$ collisions, for $\sqrt{s} = 2$ TeV (solid curve), 1.6 TeV (dashed curve), and 630 GeV (dotted curve). (From Ref. 107)

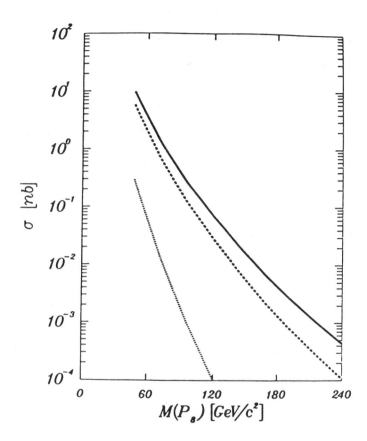

Figure 65: Cross sections for the production of $P_8\overline{P}_8$ pairs in $\overline{p}p$ collisions, for $\sqrt{s} = 2$ TeV (solid curve), 1.6 TeV (dashed curve), and 630 GeV (dotted curve). (From Ref. 107)

Table 9: Minimum *effective* integrated luminosities in cm^{-2} required to establish signs of extended technicolor (Farhi-Susskind Model) in various hadron colliders. To arrive at the required integrated luminosities, divide by the effeciencies ϵ_i to identify and adequately measure the products.

	Collider Energy			
	2 TeV	10 TeV	20 TeV	40 TeV
Channel	$\bar{p}p$	pp	pp	pp
$P^{0\prime} \to \tau^+\tau^-$	5×10^{36}	8×10^{35}	3×10^{35}	2×10^{35}
$P_8^{0\prime} \to t\bar{t}$ $(m(P_8^{0\prime}) = 240 GeV/c^2)$	2×10^{36}	7×10^{34}	3×10^{34}	10^{34}
$P_3\bar{P}_3$ $(m(P_3) = 160 GeV/c^2)$	2×10^{38}	2×10^{36}	4×10^{35}	2×10^{35}
$(m(P_3) = 400 GeV/c^2)$	—	10^{38}	2×10^{37}	4×10^{36}
$P_8\bar{P}_8$ $(m(P_8) = 240 GeV/c^2)$	10^{38}	2×10^{35}	5×10^{34}	2×10^{34}
$(m(P_8) = 400 GeV/c^2)$	—	2×10^{36}	4×10^{35}	10^{35}
$P_T^\pm \to W^\pm Z^0$	—	2×10^{39}	7×10^{38}	3×10^{38}

6 COMPOSITENESS ?

In the previous lectures, it was assumed that the quarks, leptons, and gauge bosons all are elementary particles. One extension of this standard picture, to which a considerable amount of attention has been given, is the possibility that the quarks and leptons are composite particles of more fundamental fields. However, the gauge bosons will still be assumed to be elementary excitations; so any masses for these gauge bosons are generated by spontaneous symmetry breakdown through the Higgs mechanism.

There is no experimental data to indicate any substructure for the quarks and leptons. Therefore all speculation about compositeness is theoretically motivated. Consequently a good fraction of this lecture is devoted to the theoretical aspects of composite model building. So far no obviously superior model has been proposed. Since the idea of quark and lepton compositeness is still in the early stages of development, the emphasis here is on the motivation for composite models and on the general theoretical constraints on composite models. After a general theoretical discussion we turn to the expected experimental consequences of compositeness. First the present limits on quark and lepton substructure will be reviewed. Then the signals for compositeness in the present generation of colliders as well as in supercolliders will be explained. Finally, the signals of crossing a compositeness threshold will be mentioned.

6.1 Theoretical Issues

6.1.1 Motavation

Several factors have contributed to speculation that the quarks and leptons are not elementary particles.

- The most obvious suggestion of compositeness is the proliferation of the number of quarks and leptons in a repeated pattern of left-handed doublets and right-handed singlets. This three generation spectra is suggestive of an excitation spectrum of more fundamental objects. Finding a repeated pattern has been a precursor to the discovery of substructure before; for example, the periodic table of elements in atomic physics.

- The complex pattern of the quark and lepton masses together with the mixing angles needed to describe the difference between the stong and electroweak flavor eigenstates suggests that these parameters are not fundamental.

- It is, moreover, very likely that at least the Higgs sector of the Weinberg-Salam model is not correct at energies above the electroweak scale. Therefore the scalar particles which implement the symmetry breakdown may be composites formed by a new strong interaction, such as technicolor. Although there is no compelling reason to associate a composite quark-lepton scale with these composite scalars, certainly it is an option which introduces a minimal amount of new physics.

For these reasons the idea of compositeness presently enjoys wide theoretical interest.

6.1.2 Consistency Conditions for Composite Models

To begin the theoretical discussion of composite models we will, following 't Hooft[108], consider a prototype composite theory of quarks and leptons consisting of a non-Abelian gauge interaction called metacolor which is described by a simple gauge group $\mathcal{G}$ with coupling constant g_M. Assuming that the gauge interaction is asymptotically free there will be some scale Λ_M at which the coupling becomes strong

$$\alpha_M \equiv \frac{g_M^2}{4\pi} = 1 \qquad (6.1)$$

This is the characteristic scale for the dynamics and hence for the masses of the physical states.

In addition this prototype theory has a set of massless fundamental spin 1/2 fermions, sometimes called preons, which carry metacolor. The massless fermions will be represented here by Weyl spinors. (The ordinary Dirac representation can be constructed whenever both a Weyl spinor and its complex conjugate representation appear.)

Metacolor dynamics is similar to QCD except that the gauge interaction will not in general be vectorlike. A theory is termed vectorlike if the fermion representation under the gauge group R is real; that is, every irreducible representation is accompanied by its complex conjugate representation hence $R^* = R$. The fermions will exhibit global symmetries described by a global chiral flavor symmetry group G_f. No global symmetry whose current conservation is spoiled by the presence of the metacolor gauge interactions will be included in G_f. Therefore the symmetry structure of the fermions consists of two relevant groups:

- the gauge group - $\mathcal{G}$, and

- the global flavor group - G_f.

The physical masses of the quarks and leptons are very small relative to the compositeness scale; this is one essential feature that any prototype model of composite quarks and leptons must explain. Therefore, with the assumption that the gauge interaction $\mathcal{G}$ is confining, there must exist a sensible limit of the theory in which the quarks and leptons are massless composite states. Thus the most relevant feature of any prototype composite model is its spectrum of massless excitations, of which the spin 1/2 particles are the candidates for quarks and leptons.

The spectrum of massless composites is directly related to the pattern global chiral symmetry breaking which occurs as metacolor becomes strong. In QCD the $SU_L(n) \otimes SU_R(n) \otimes U(1)$ flavor symmetry breaks down to the vector subgroup. In a metacolor theory one expects that the global symmetry group breaks down to a subgroup:

$$G_f \to S_f \tag{6.2}$$

at energy scale Λ_M. Associated with each spontaneously broken symmetry is a composite spin zero Goldstone bosons. Any massless composite fermions will form representations under the remaining unbroken subgroup S_f of the global symmetry group G_f.

A few simple examples of asymptotically free metacolor gauge groups $\mathcal{G}$, and fermion representations R; and the associated flavor groups G_f are presented below:

Gauge Group	Fermion Representation	Global Group
$SU(N)$	$m(\square \otimes \overline{\square})$	$SU(m) \otimes SU(m) \otimes U(1)$
$O(10)$	$m(\text{spinor})$	$SU(m)$
$SU(N)$	$\square\!\square \otimes (N-4)\, \overline{\square}$	$SU(N-4) \otimes U(1)$
$SU(3)$	$\square\!\square\!\square \otimes 2\, \square\!\square$	$SU(2) \otimes U(1)$
$SU(8)$	$\square\!\square\!\square\!\square \otimes 5\, \overline{\square}$	$SU(5) \otimes U(1)$

The first example shows how the standard $SU(N)$ vectorlike theory is denoted for m flavors of Dirac fermions in the fundamental representation. The flavor symmetries are just the usual $SU_L(m) \otimes SU_R(m) \otimes U(1)$. All the other examples are non-vector theories (i.e. the fermion representation is not real) and thus are prototypes for metacolor theories. The first such example is $O(10)$ with fermions in the lowest dimensional spinor representation, a **16**. The number m of spinor representations is limited by the requirement that the theory be asymptotically free. In order to have a sensible theory, the fermion representation must be such that any gauge anomalies must cancelled. $O(10)$ is anomaly safe; however, in the remaining examples, the anomalies are cancelled by judiciously choice of the fermion representations. The next example is a generalization of the Georgi-Glashow $SU(5)$ model[109] with fermions in the fundamental and antisymmetric tensor representations. If one wants to consider fermion representations of rank greater than two, then only an $SU(N)$ gauge group with a low N will maintain the asymptotic freedom of the gauge interactions.

Several general characteristics of the global symmetry breaking are relevant here:

- For real fermion representations, when the gauge interaction becomes strong the axial symmetries are broken and only the vector symmetries remain unbroken[86]. This case is uninteresting because the only massless particles are the Goldstone bosons associated with the broken axial symmetries. There are no massless composite fermions. Vectorlike gauge theories are not good candidates for a prototype theory of composite quark and/or leptons.

- General arguments guarantee that only spin 0 and spin 1/2 particles can couple to global conserved currents[110]. Hence only spin 0 and 1/2 massless states are relevant to the realization of the global symmetries in a metacolor theory.

The most powerful consistency condition on the massless spectrum of any proposed composite model is provided by 't Hooft[108]. These constraints provide a framework for studying the possible massless spectra of a metacolor model even though they do not imply a unique solution. To understand these constraints consider any global current $j^\mu(x)$ which is conserved at the Lagrangian level:

$$\partial_\mu j^\mu(x) = 0 \qquad (6.3)$$

This current involves preon fields and is associated with it a conserved charge

$$Q = \int d^3x\, j^0(x) \qquad (6.4)$$

When this current is coupled to a weak external gauge field, via a $j^\mu A_\mu$ interaction, the conservation may be destroyed by an anomaly, such as occurs for the axial $U(1)$ current in QCD. The divergence of the current in the presence of the external gauge field is proportional to $\mathcal{F}\tilde{\mathcal{F}}$

$$\partial_\mu j^\mu = \frac{T_f}{4\pi^2} \mathcal{F}^{\mu\nu} \tilde{\mathcal{F}}_{\mu\nu} \qquad (6.5)$$

where $T_f = Tr(Q_f^3)$ and $\mathcal{F}^{\mu\nu} = \partial^\mu A^\nu - \partial^\nu A^\mu$. Q_f is the charge matrix for the massless preon fields. Any current for which $T_f \neq 0$ is called anomalous. It is important to remember that this anomaly is in a global current and not in the metacolor interactions which are required to be anomaly free for the consistency of the theory. This global anomaly may also be seen in one fermion loop contribution to the three point correlation function.

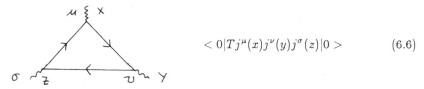

$$< 0|Tj^\mu(x)j^\nu(y)j^\sigma(z)|0 > \qquad (6.6)$$

At the preon level the anomalous contribution to this three point function is simple. The structure of the anomaly is given by bose symmetry in the three currents and current conservation while the coefficient is proportional to $T_f = Tr(Q_f^3)$. It is only necessary to consider a general diagonal charge of the global symmmetry group to determine the complete anomaly structure. All the off diagonal anomalies can be reconstructed from the coefficients of a general diagonal current.

We are now ready to state 't Hooft's condition explicitly. He states that the value of the anomaly calculated with the massless physical states of the theory must be the same as the value calculated using the fundamental preon fields of the underlying Lagrangian[108]. In the absence of the gauge interactions, these massless states are just the preons and therefore 't Hooft's condition can be restated that the gauge interactions do not modify the anomalies. It has been shown[111] that this constraint follows from behaviour is to be expected from general axioms of field theory. One important consequence of this condition is that if $T_f \neq 0$, then there must be massless physical states associated with the charge Q_f. This condition is the strongest constraint we have at present on composite model building.

To further elucidate 't Hooft's consistency condition consider adding some metacolor singlet fermions to the theory to cancel the anomalies in the global currents. Then including these spectator fermions the global symmetries are anomaly free and may themselves by gauged. Thus, at distances large relative to the metacolor interaction scale, there must still be no anomaly when all the massless physical states all included.

We will assume that the metacolor gauge interaction is confining. It should be remembered, however, that this is an ad hoc assumption. It is not presently possible to calculate (even by lattice methods) the behaviour of nonvectorlike theories.

The fundamental dynamical question for composite models is how 't Hooft's constraint is satisfied. There are two possiblities:

- If the global symmetry which has the anomaly is spontaneously broken when the metacolor interaction becomes strong, i.e. $Q_f \notin S_f$, then the massless physical state required by the anomaly consistency condition is just the Goldstone boson associated with the spontaneously broken symmetry. The strength of the anomaly T_f determines the coupling of the Goldstone boson to the other matter fields.

- If the anomalous symmetry remains unbroken when the metacolor interaction becomes strong, $Q_f \in S_f$, then there must be massless spin $1/2$ fermions in the physical spectrum which couple to the charge Q_f and produce the anomaly with the correct strength, T_f. Therefore, for unbroken symmetries, there must be a set of massless composite physical states for which the trace $Tr(Q^3_{\text{physical}})$ over the charges of the massless phyiscal fermions is equal the trace $Tr(Q_f^3)$ over the charges of the elementary preon fields.

Thus 't Hooft's consistency condition implies a relation between the symmetry breaking pattern $G_f \to S_f$ and the massless spectrum of fermions. However, it does not completely determine the massless fermion spectrum for a given Lagrangian. In his original paper[108] 't Hooft added two additional conditions. The first condition requires that if a mass term for a preon $(m\bar\psi\psi)$ is added to the Lagrangian, then, at least in the limit that the mass of this preon field becomes large, all composite fermions containing this preon acquire a mass and therefore no longer contribute to the anomaly. It is reasonable to expect this decoupling. The other condition is that the metacolor gauge interactions are independent of flavor except for mass terms. So that the solution to the anomaly constraints depend only trivially on the number of flavors in any given representation.

For vectorlike theories these two additional constraints allow definite conclusions about the massless spectrum of the theory. However in nonvector theories these additional conditions are generally not meaningful. For example, in our examples, a mass term cannot be introduced for any of the preon fields without explicitly breaking the metacolor gauge invariance. We will not consider these additional constraints further.

6.1.3 A Simple Example

It is instructive to give one explicit example which implements t'Hooft's condition and constrains the massless the physical spectrum. Unfortunately, this simple model (and in fact all other known models) is too naive to be phenomenologically relevant. Consider the model with metacolor gauge group $\mathcal{G} = SU(N)$ and preons in the antisymmetric tensor representation A_{ij} and $N-4$ fundamental representations ψ^i. The number of fundamental representations is fixed by the requirement that the gauge interaction has no anomalies.

The global symmetry group of this model is $G_f = SU(N-4) \otimes U(1)$. The origin of the U(1) symmetry can be seen as follows. For each type of representation a $U(1)$ symmetry can be defined; however only one combination of these two $U(1)$ symmetries is free of an anomaly associated with coupling of the current to two metacolor gauged currents (the generalization of the axial $U(1)$ anomaly in QCD). The coefficient of this coupling for each of the global $U(1)$ currents is:

Representation	$U(1)$ Anomaly Coefficient
A_{ij}	$(N-2)$
ψ^i	1

Hence the combination of global $U(1)$ charges which remains conserved in the presence of gauge interactions is:

$$Q = \int d^3x [(N-4)\overline{A}^{ij}\gamma^0 A_{ij} - (N-2) \sum_{\text{flavors}} \overline{\psi}_i \gamma^0 \psi^i]. \qquad (6.7)$$

Assuming confinement, any spin 1/2 massless physical state must be a singlet under the gauge group $\mathcal{G}$. One possible candidate for such a composite field is

$$F_{nm} = (A_{ij}\sigma_2\psi_n^i)(\sigma_2\psi_m^j) \qquad (6.8)$$

where n and m are flavor indices. In particular, we consider the symmetric tensor representation ($F_{nm} = F_{mn}$) under the global symmetry group $SU(N-4) \in G_f$.

The dimension of this representation is $(N-4)(N-3)/2$. The $U(1)$ charge of the F_{nm} fields is $-N$. Although it cannot be proven that the F_{nm} represents massless fields[112], it is consistent with 't Hooft's condition for these states to be massless. To show this we need to demonstrate that all the anomaly conditions are satisfied by these massless fields. Comparing the anomalies for the preons and these physical states gives:

Anomaly	Preons	Composites F_{nm}
$Tr[(SU(N-4))^3]$	N	$(N-4)+4=N$
$Tr[(SU(N-4))^2 Q]$	$-N(N-2)$	$-N[(N-4)+2]$
		$=-N(N-2)$
$Tr[Q^3]$	$(N-4)^3(N-1)N/2$	$-N^3(N-4)(N-3)/2$
	$-(N-2)^3(N-4)N$	
	$=-N^3(N-4)(N-3)/2$	

The anomalies match exactly between the elementary and the composite particles. Therefore this model provides a non-trivial example in which massless composite fermions can be introduced in such a way that 't Hooft's consistency condition is satisfied with the global symmetry group G_f completely unbroken. It should be remembered that the anomaly matching does not guarantee that the states F_{nm} are in fact massless composites in this theory or that the maximal flavor group remains unbroken only that it is a consistent possibility. It could also happen that only a subgroup of G_f is remains unbroken and then there will be massless Goldstone bosons and some of these states F_{nm} may acquire masses. In any case, the existence of the solution above for the case G_f is completely unbroken ensures that for any subgroup $S_f \in G_f$ the subset of the fermions which remain massless together with the Goldstone bosons associated with the broken symmetries satisfy 't Hooft's consistency condition.

The consistency condition of 't Hooft provides some guideline to which massless composite fermions could be produced by an strong metacolor dynamics. It is also possible to envision mechanisms which would provide the small explicit symmetry breaking required to generated small masses for initially massless composite quarks and leptons. However, it is very difficult to understand the generation structure of quarks and leptons as an excitation spectra of the metacolor interactions. Excited states would be expected to have a mass scale determined by the strong gauge interactions; but all of the generations of observed quarks and leptons have very small masses on the energy scale Λ_M of the composite binding forces. Hence all masses and mixings would be required to originate from explicit symmetry breaking not directly associated with the metacolor strong interactions.

In this brief introduction to the theoretical issues of composite model building it is clear that many of the original advantages of composite models remain unattained.

6.2 Phenomenological Implications of Compositeness

If the quarks and leptons are in fact composite, what are the phenomenological consequences of this substructure? At energies low compared to the compositeness scale the interactions between bound states is characterized by the finite size of the bound states, indicated by a radius R. Since the interactions between the composite states are strong only within this confinement radius, the cross section for scattering of such particles at low energies should be essentially geometric, that is, approximately $4\pi R^2$. The compositeness scale can also be

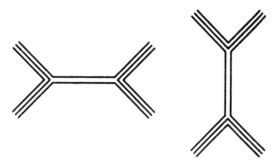

Figure 66: Elastic scattering between composite states at energies much below the compositeness scale. The dominant term is simply the exchange of the lowest-lying massive composite boson.

characterized by a energy scale $\Lambda^* \sim 1/R$.

Another naive view of the scattering process would replace constitutent exchange by an exchange of a composite massive boson as shown in Figure 66. This approximation is analagous to the one particle exchange approximation for the usual strong interactions at low energies; for example, ρ exchange in $\overline{N}N$ collisions. The strength of the coupling $g_M^2/4\pi$ may be estimated by taking this analog one step further. The coupling $g_{\rho\overline{N}N}/4\pi \approx 2$ suggests that the coupling $g_M^2/4\pi \approx 1$ is not unreasonable.

The interaction at low energies is given by an effective four fermion interaction, or contact term, of the general form:

$$\frac{g_M^2}{M_V^2}(\overline{f}_1\gamma^\mu f_2)(\overline{f}_3\gamma_\mu f_4) \qquad (6.9)$$

Using $g_M^2/4\pi = 1$ and identifying M_V with Λ^* the effective interaction is of the expected geometric form.

6.2.1 Limits From Rare Processes

The possible contact terms in the effective low energy Lagrangian are of the general form:

$$\frac{4\pi}{\Lambda^{*d}} \mathcal{O} \qquad (6.10)$$

where $\mathcal{O}$ is a local operator of dimension $4 + d$ constructed of the usual quark, lepton, and gauge fields. Ignoring quark and lepton masses, the contribution of these contact terms of the effective Lagrangian to the amplitude of some physical process involving quarks, leptons, or gauge fields must be proportional to the energy scale $\sqrt{s}$ of the process considered raised to a power determined by the dimension of the operator. High dimension operators are suppressed by high powers of $\sqrt{s}/\Lambda^*$; and hence are highly suppressed at ordinary energies. Some possible operators which would contribute to rare processes at low energies are given in Table 10. The present limits on rare processes involving ordinary quarks and leptons provide severe restrictions on the scale Λ^* for the associated operator as shown in Table 10. For example, if the $K_L^0 - K_S^0$ mass difference has a contribution from a contact term as shown in Table 10, then the scale of that interaction $\Lambda^* > 6,100$ TeV. Therefore these flavor changing contact terms can not be present in any composite model intended to describe dynamics at the TeV energy scale. Thus, in addition to the theoretical constraints imposed by 't Hooft, rare processes such as those listed in Table 10 provide strong phenomenological constraints on composite model building.

Table 10: Limits of contact terms from rare processes. The interaction type assumed for each rare process is shown along with the resulting limit of the compositeness scale $\Lambda^\star$.

Process	Contact Interaction	Limit on $\Lambda^\star$ (TeV)
$(g-2)_e$	$\dfrac{m_e}{\Lambda_\star^2}\,\bar{e}\sigma^{\alpha\beta}e\,F_{\alpha\beta}$	.03
$(g-2)_\mu$	$\dfrac{m_\mu}{\Lambda_\star^2}\,\bar{\mu}\sigma^{\alpha\beta}\mu\,F_{\alpha\beta}$	.86
$\mu \to e\,\gamma$	$\dfrac{4\pi}{\Lambda_\star^2}\,\bar{\mu}\gamma^\alpha \tfrac{1}{2}(1-\gamma_5)e\,\bar{e}\gamma_\alpha \tfrac{1}{2}(1-\gamma_5)e + (\mu \leftrightarrow e)$	60
$\mu \to 3e$	$\dfrac{4\pi}{\Lambda_\star^2}\,\bar{\mu}\gamma^\alpha \tfrac{1}{2}(1-\gamma_5)e\,\bar{e}\gamma_\alpha \tfrac{1}{2}(1-\gamma_5)e$	400
$\mu\,N \to e\,N$	$\dfrac{4\pi}{\Lambda_\star^2}\,\bar{\mu}\gamma^\alpha \tfrac{1}{2}(1-\gamma_5)e\,\bar{d}\gamma_\alpha \tfrac{1}{2}(1-\gamma_5)d$	460
$K_L \to e^\pm\,\mu^\mp$	$\dfrac{4\pi}{\Lambda_\star^2}\,\bar{s}\gamma^\alpha \tfrac{1}{2}(1-\gamma_5)d\,\bar{e}\gamma_\alpha \tfrac{1}{2}(1-\gamma_5)\mu$	140
$K^+ \to \pi^+\,e^-\,\mu^+$	$\dfrac{4\pi}{\Lambda_\star^2}\,\bar{s}\gamma^\alpha \tfrac{1}{2}(1-\gamma_5)u\,\bar{e}\gamma_\alpha \tfrac{1}{2}(1-\gamma_5)\mu$	210
$\Delta M(K_L - K_S)$	$\dfrac{4\pi}{\Lambda_\star^2}\,\bar{s}\gamma^\alpha \tfrac{1}{2}(1-\gamma_5)d\,\bar{s}\gamma_\alpha \tfrac{1}{2}(1-\gamma_5)d$	6,100

6.2.2 Limits On Lepton Compositeness

The correct strategy for composite model building has not yet emerged. All that is known is that the mass scale Λ^* which characterizes the preon binding interactions and the mass scale of the composite states is ≥ 1 TeV. Very little is known in a model-independent way about the composite models except for the experimental and theoretical restrictions discussed above. For example, it is also entirely possible that some of the quarks and leptons are elementary while others are composite. Therefore a conservative approach is to consider only those four fermion interactions which in addition to conserving $SU(3) \otimes SU(2) \otimes U(1)$ gauge symmetries are also completely diagonal in flavor. These interactions must be present in any composite model. For example, if the elctron is a bound state; then, in addition to the usual Bhabha scattering, there must be electron-positron scattering in which there is constituent interchange between the electron's and positron's preonic components. These diagonal contact terms test the compositeness hypothesis in a direct and model independent way. The effective Lagrangian for electron weak doublet compositeness is[113]

$$\mathcal{L}_{\text{eff}} = \frac{4\pi}{\Lambda^{*2}}[\frac{1}{2}\eta_{\text{LL}}(\bar{l}\gamma^\mu l)(\bar{l}\gamma_\mu l) + \eta_{\text{RL}}(\bar{l}\gamma^\mu l)(\bar{e}_R\gamma_\mu e_R)$$
$$+ \frac{1}{2}\eta_{\text{RR}}(\bar{e}_R\gamma^\mu e_R)(\bar{e}_R\gamma_\mu e_R)] \qquad (6.11)$$

where l is the left-handed (ν, e) doublet. All of the terms in Eq. 6.11 are helicity conserving in for $m_i \ll \sqrt{s} \ll \Lambda^*$. The coefficients η are left arbitrary here since they are model dependent.

For the left-handed components, a composite electron implies a composite neutrino since they are in the same electroweak doublet, but no such relation exists for the right-handed components. The interactions in Eq. 6.11 imply that there will be new terms in addition to the Bhabha scattering and Z^0 exchange graphs in the cross section for electron-positron scattering which in lowest order is given by:

$$\frac{d\sigma}{d\Omega}(e^+e^- \to e^+e^-) = \frac{\pi\alpha^2}{4s^2}[4A_0 + A_+(1+\cos\theta)^2 + A_-(1-\cos\theta)^2] \qquad (6.12)$$

where

$$A_0 = (\frac{s}{t})^2|1 + \frac{L_e R_e t}{\sin^2 2\theta_w t_z} + \frac{\eta_{\text{RL}} t}{\alpha\Lambda^{*2}}|^2$$

$$A_+ = \frac{1}{2}|1 + \frac{s}{t} + \frac{R_e^2}{\sin^2 2\theta_w}(\frac{s}{s_z} + \frac{s}{t_z}) + \frac{2\eta_{\text{RR}}}{\alpha\Lambda^{*2}}s|^2$$

$$\frac{1}{2}|1 + \frac{s}{t} + \frac{L_e^2}{\sin^2 2\theta_w}(\frac{s}{s_z} + \frac{s}{t_z}) + \frac{2\eta_{\text{LL}}}{\alpha\Lambda^{*2}}s|^2$$

$$A_- = \frac{1}{2}|1 + \frac{R_e L_e s}{s_z} + \eta_{\text{RL}}\frac{s}{\alpha\Lambda^{*2}}|^2 \qquad (6.13)$$

and

$$s_z = s - m_Z^2 + im_Z\Gamma_Z \qquad t_z = t - m_Z^2 + im_Z\Gamma_Z$$

$$L_e = -\cos 2\theta_w \qquad R_e = 2\sin^2\theta_w \qquad (6.14)$$

This formula is valid only for energies much below the compositeness scale Λ^*. The presence of a compositeness term can be tested by comparing the cross section of Eq. 6.12 with the experimental data to give limits on the contact terms for various interaction types η, whose explicit values depend on the particular composite model.

In Figure 67 the deviation:

$$\Delta_{ee} = \left(\frac{d\sigma/d\Omega|_{\text{measured}}}{d\sigma/d\Omega|_{\text{standard model}}} - 1\right) \qquad (6.15)$$

is plotted for e^+e^- collisions at $\sqrt{s} = 3$ GeV. At $\sqrt{s} = 35$ GeV the maximun deviation is approximately 4% for the left-left ($\eta_{\text{LL}} = \pm 1$, all other η's= 0) or

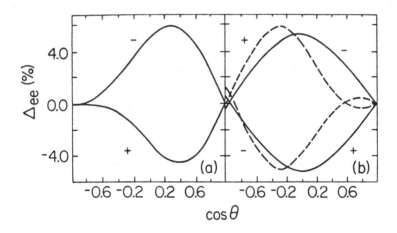

Figure 67: $\Delta_{ee}(\cos\theta)$, in percent, at $\sqrt{s} = 35$ GeV. (a) The LL and RR models with $\Lambda^* = 750$ GeV. (b) The VV model (solid lines) with $\Lambda^* = 1700$ GeV and the AA model (dashed lines) with $\Lambda^* = 1400$ GeV. The $\pm$ signs refer to the overall sign of the contact term in each case.

right-right ($\eta_{RR} = \pm 1$, all other η's = 0) couplings with $\Lambda^* = .75$ TeV and for the vector-vector coupling ($\eta_{LL} = \eta_{RR} = \eta_{RL} = \pm 1$) with $\Lambda^* = 1.7$ TeV or for the axial-axial coupling ($\eta_{LL} = \eta_{RR} = -\eta_{RL} = \pm 1$) with $\Lambda^* = 1.4$ TeV. The present limits obtained from various experiments at PEP and PETRA are shown in Table 11. It is clear from these experimental limits that the electron is still a structureless particle on the scale of one TeV.

At LEP energies ($\sqrt{s} = 100$ GeV) a deviation of about 6 % occurs for left-left or right-right couplings with $\Lambda^* = 2$ TeV, or vector-vector or axial-axial couplings with $\Lambda^* = 5$ TeV.

6.3 Signals for Compositeness in Hadron Collisions

Searches for compositeness in hadron collisions will naturally concentrate on looking for internal structure of the quarks. As in the case of a composite electron, if the quark is composite there will be an additional interaction between quarks which can be represented by a contact term at energy scales well below the compositeness scale. However, the reference cross section for elastic scattering of pointlike quarks, the QCD version of Bhabha scattering, has both nonperturbative and perturbative corrections and is therefore not as accurately known as Bhabha scattering in QED. Futhermore the extraction of the elementary subprocesses in the enviroment of hadron-hadron collisions involves knowledge of the quark and gluon distribution functions. Therefore larger deviations from QCD expectations will be required before a signal for compositeness can be trusted.

The most general contact interactions which:

Table 11: Present limits on electron compositeness for e^+e^- colliders. The four Fermi couplings considered are all left-left (LL), right-right (RR), vector (VV), and axial (AA). Both constructive (−) and destructive (+) interference between the contact term and the standard terms are displayed. The experimental limits are from the MAC[114], PLUTO[115], MARK-J[116], JADE[117], TASSO[118], and HRS[119] Collaborations and are in TeV.

Type	Sign	MAC	PLUTO	MARK-J	JADE	TASSO	HRS
LL	+	1.2	1.1	0.92	0.82	0.7	0.64
LL	−		.76	0.95	1.45	1.94	0.51
RR	+	1.2	1.1	0.92	0.81	0.7	0.64
RR	−		.76	0.95	1.44	1.94	0.51
VV	+	2.5	2.2	1.71	2.38	1.86	1.42
VV	−		1.9	2.35	2.92	2.91	1.38
AA	+	1.3	2.0	2.25	2.22	1.95	0.81
AA	−		1.6	0.94	2.69	2.28	1.06

- preserve $SU(3) \otimes SU(2) \otimes U(1)$,

- involve only the up and down quarks, and

- are helicity conserving

involve 10 independent terms.

$$\mathcal{L}_{\text{eff}} = \frac{4\pi}{\Lambda^{*2}} [\frac{\eta_1}{2} \bar{q}_L \gamma^\mu q_L \bar{q}_L \gamma_\mu q_L + \eta_2 \bar{q}_L \gamma^\mu q_L \bar{u}_R \gamma_\mu u_R$$
$$+ \eta_3 \bar{q}_L \gamma^\mu q_L \bar{d}_R \gamma_\mu d_R + \eta_4 \bar{q}_L \gamma^\mu \frac{\lambda_a}{2} q_L \bar{u}_R \gamma_\mu \frac{\lambda_a}{2} u_R$$
$$+ \eta_5 \bar{q}_L \gamma^\mu \frac{\lambda_a}{2} q_L \bar{d}_R \gamma_\mu \frac{\lambda_a}{2} d_R + \frac{\eta_6}{2} \bar{u}_R \gamma^\mu u_R \bar{u}_R \gamma_\mu u_R$$
$$+ \eta_7 \bar{u}_R \gamma^\mu \frac{\lambda_a}{2} u_R \bar{d}_R \gamma_\mu \frac{\lambda_a}{2} d_R + \eta_8 \bar{u}_R \gamma^\mu u_R \bar{d}_R \gamma_\mu d_R$$
$$+ \frac{\eta_9}{2} \bar{d}_R \gamma^\mu d_R \bar{d}_R \gamma_\mu d_R + \frac{\eta_{10}}{2} \bar{q}_L \gamma^\mu \frac{\tau_\alpha}{2} q_L \bar{q}_L \gamma_\mu \frac{\tau_\alpha}{2} q_L] \quad (6.16)$$

This complicated form for the contact terms will not be considered in full generality here[113]. To understand the nature of the bounds on quark substructure which can be seen in hadron collisions it is sufficient to take the simple example where only one of the 10 possible contact terms is considered. For this purpose only the left-left coupling contact terms will be considered:

$$\mathcal{L}_{\text{eff}} = \frac{\eta_{LL}}{2} \frac{4\pi}{\Lambda^{*2}} \bar{q}_L \gamma^\mu q_L \bar{q}_L \gamma_\mu q_L \quad (6.17)$$

for both signs $\eta_{LL} = \pm 1$ of the interaction.

A typical quark-antiquark elementary subprocess including a contact term due to quark compositeness is shown in Figure 68.

Analytically the differential cross section (anti)quark-(anti)quark scattering is given by:

$$\frac{d\sigma}{d\hat{t}}(i\ j \to i'\ j') = \alpha_s^2 \frac{\pi}{\hat{s}^2} |A_{2\to 2}(i\ j \to i'\ j')|^2 \quad (6.18)$$

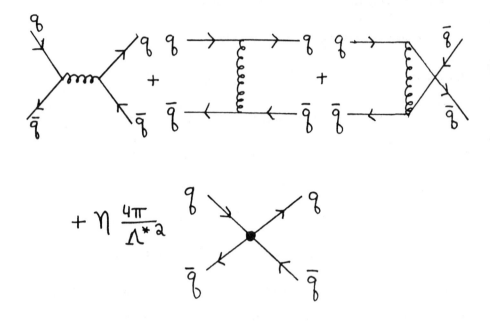

Figure 68: The Feynman diagrams contributing to the amplitude for the sub-process $\bar{q}q \to \bar{q}q$ in the presence of a contact interaction associated with quark compositeness. The first three diagrams are simply the order α_s contribution from QCD and the last diagram represents the contribution from the contact interaction of Eq. 6.18.

where

$$|A(u\bar{u} \to u\bar{u})|^2 = |A(d\bar{d} \to d\bar{d})|^2$$
$$= \frac{4}{9}[\frac{\hat{u}^2 + \hat{s}^2}{\hat{t}^2} + \frac{\hat{u}^2 + \hat{t}^2}{\hat{s}^2} - \frac{2\hat{u}^2}{\hat{t}\hat{s}}]$$
$$+\frac{8}{9}\frac{\eta_{LL}}{\alpha_s \Lambda^{*2}}[\frac{\hat{u}^2}{\hat{t}} + \frac{\hat{u}^2}{\hat{s}}] + \frac{8}{3}[\frac{\eta_{LL}\hat{u}}{\alpha_s \Lambda^{*2}}]^2$$

$$|A(uu \to uu)|^2 = |A(dd \to dd)|^2 = |A(\bar{u}\bar{u} \to \bar{u}\bar{u})|^2 = |A(\bar{d}\bar{d} \to \bar{d}\bar{d})|^2$$
$$= \frac{4}{9}[\frac{\hat{u}^2 + \hat{s}^2}{\hat{t}^2} + \frac{\hat{s}^2 + \hat{t}^2}{\hat{u}^2} - \frac{2\hat{s}^2}{\hat{t}\hat{u}}]$$
$$+\frac{8}{9}\frac{\eta_{LL}}{\alpha_s \Lambda^{*2}}[\frac{\hat{s}^2}{\hat{t}} + \frac{\hat{s}^2}{\hat{u}}] + \frac{8}{3}[\frac{\eta_{LL}}{\alpha_s \Lambda^{*2}}]^2(\hat{u}^2 + \hat{t}^2 + \frac{2}{3}\hat{s}^2)$$

$$|A(u\bar{u} \to d\bar{d})|^2 = |A(d\bar{d} \to u\bar{u})|^2$$
$$= \frac{4}{9}[\frac{\hat{u}^2 + \hat{t}^2}{\hat{s}^2}] + [\frac{\eta_{LL}\hat{u}}{\alpha_s \Lambda^{*2}}]^2$$

$$|A(ud \to ud)|^2 = |A(u\bar{d} \to u\bar{d})|^2 = |A(\bar{u}d \to \bar{u}d)|^2 = |A(\bar{u}\bar{d} \to \bar{u}\bar{d})|^2$$
$$= \frac{4}{9}[\frac{\hat{u}^2 + \hat{s}^2}{\hat{t}^2}] + [\frac{\eta_{LL}\hat{u}}{\alpha_s \Lambda^{*2}}]^2 \quad (6.19)$$

Note that the effects of the contact term grow linearly with $\hat{s}$ relative the the QCD terms in the amplitude for elastic scattering. There is no effect in lowest order on (anti)quark-gluon or gluon-gluon scattering. The inclusive jet production in $\bar{p}p$ collisions at $\sqrt{s} = 1.8$ TeV including the effect of this contact term in the (anti)quark-(anti)quark scattering amplitude is shown in Figure 69.

The present measurements of inclusive single jet production at the $S\bar{p}pS$ collider bounds the possible value of Λ^* associated with light quark compositeness. For the left-left coupling with $\eta_{LL} = -1$ the effects of a contact term for various values of Λ^* are shown with the UA2 data[39] in Figure 70. The analysis of the UA2 Collaboration[39] shows that $\Lambda^* \geq 370$ GeV is required to be consistent with their results. This limit is the best bound on light quark compositeness which

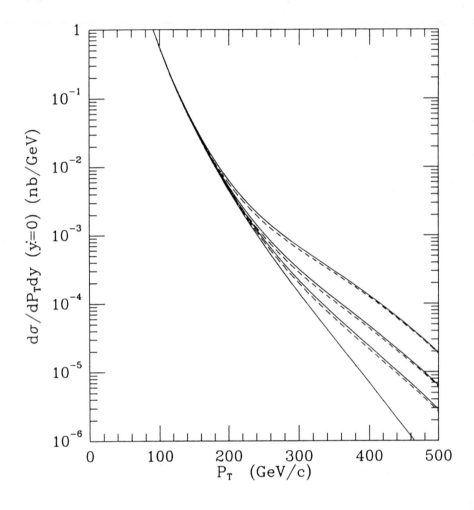

Figure 69: The inclusive jet production cross section $d^2\sigma/dp_\perp dy|_{y=0}$ in $\bar{p}p$ collisions at $\sqrt{s} = 1.8$ TeV including the effects of a contact interaction. The contact term was the LL type with $\eta_{LL} = -1$ (solid line) and $\eta_{LL} = +1$ (dashed line). The values of Λ^* are .75 (top pair of lines), 1.0 (middle pair of lines), and 1.25 (bottom pair of lines) TeV. The standard QCD prediction using the distributions of Set 2 is denoted by a the bottom single solid line.

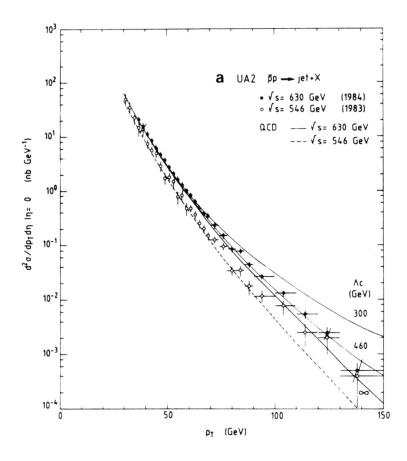

Figure 70: Inclusive jet production cross sections from the UA2 Collaboration (as shown in Fig. 22) with the effects of a composite interaction shown for $\sqrt{s} = 630$ GeV. The three solid lines (from top to bottom) represent the prediction for $d^2\sigma/dp_\perp dy|_{y=0}$ for the left-left contact term with $\Lambda^* = 300$ GeV, 460 GeV, and infinity respectively.

presently exists. Hence the light quarks do not have any structure below a scale of 370 GeV. Since the contact term in the cross section grows linearly with $\hat{s}$ while the standard terms fall off with increasing energy like $1/\hat{s}$ the contact will eventually dominate the cross section. This occurs when

$$\hat{s} \approx \alpha_s \Lambda^{*2} \;. \tag{6.20}$$

Therefore the contact term dominates at an energy scale well below the compositeness scale Λ^* itself.

6.3.1 Quark-Lepton Contact Term

In generalized Drell-Yan processes, a quark-antiquark initial state annihilates into a lepton pair via an imtermediate virtual γ or Z^0. Therefore composite effects can contribute only if both the quark and lepton are composite and they have some constituent in common. Whether these conditions are meet is so is more dependent on the particular composite model.

A contact term associated with compositeness of the first generation which can contribute to Drell-Yan processes is of the general form:

$$\begin{aligned}\mathcal{L}_{\text{eff}} = \;& \frac{4\pi}{\Lambda^{*2}}[\eta_{\text{LL}}\bar{q}_L\gamma^\mu q_L \bar{l}_L\gamma_\mu l_L + \eta_{\text{LR}}\bar{q}_L\gamma^\mu q_L \bar{e}_R\gamma_\mu e_R \\ & + \eta_{\text{RU}}\bar{u}_R\gamma^\mu u_R \bar{l}_L\gamma_\mu l_L + \eta_{\text{RD}}\bar{d}_R\gamma^\mu d_R \bar{l}_L\gamma_\mu l_L \\ & + \eta_{\text{RRU}}\bar{u}_R\gamma^\mu u_R \bar{e}_R\gamma_\mu e_R + \eta_{\text{RRD}}\bar{d}_R\gamma^\mu d_R \bar{e}_R\gamma_\mu e_R \\ & + \eta_V \bar{q}_L\gamma^\mu \frac{T_a}{2} q_L \bar{l}_L\gamma_\mu \frac{T_a}{2} l_L]\end{aligned} \tag{6.21}$$

where $q_L = (u_L, d_L)$ and $l_L = (\nu_L, e_L)$. Again the nature of the bounds are illustrated by a simple case of a left-left coupling ($\eta_{\text{LL}} = \pm 1$ and all other η's= 0).

When the contact term is added to the standard γ and Z^0 contributions to the Drell-Yan process we obtain[113]:

$$\sigma(\bar{q}q \to \bar{e}e) = \frac{\pi\alpha^2}{27\hat{s}}[A(\hat{s}) + B(\hat{s})] \qquad (6.22)$$

where

$$A(\hat{s}) = 3[(\frac{1}{3} + \frac{L_q L_e}{sin^2 2\theta_w}\frac{\hat{s}}{\hat{s}_z} + \eta_{LL}\frac{\hat{s}}{\alpha\Lambda^{*2}})^2 + (\frac{1}{3} + \frac{R_q R_e}{sin^2 2\theta_w}\frac{\hat{s}}{\hat{s}_z})^2]$$

$$B(\hat{s}) = 3[(\frac{1}{3} - \frac{R_q L_e}{sin^2 2\theta_w})^2 + (\frac{1}{3} - \frac{L_q R_e}{sin^2 2\theta_w})^2] \qquad (6.23)$$

and L_e, R_e, and $\hat{s}_z$ are given by Eq. 6.16. Of course the cross section would be similarly modified if the μ or τ is composite and shares constituents with the light quarks.

The effect on electron pair production in $\bar{p}p$ collisions at $\sqrt{s} = 1.8$ TeV is shown in Figure 71 for various compositeness scales Λ^*. The effect of the contact term is quite dramatic. Whereas the standard Drell-Yan process drops very rapidly with increasing lepton pair mass above the Z^0 pole, the contact term causes the cross section to essentially flatted out at a rate dependend on the the value of Λ^*. This is do to the combination of the elementary cross section which grows linearly with pair mass and the rapidly dropping lunimosity of quark-antiquark pairs as the subprocess energy increases. Hence the probablity of observing a lepton pair with invariant mass significantly greater than the Z^0 mass becomes substantial. By this method contact scales up to approximately 1.0 TeV can be probed with an integrated hadron luminosity of $10^{37} cm^{-2}$ at this energy.

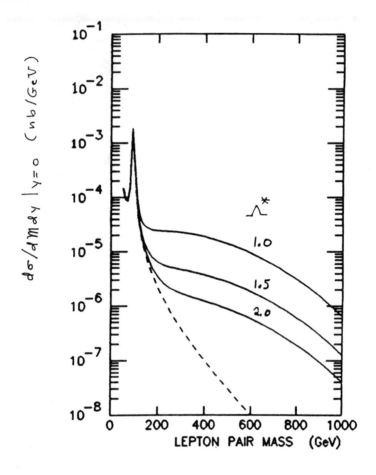

Figure 71: Cross section $d\sigma/dMdy|_{y=0}$ for dilepton production in $\bar{p}p$ collisions at $\sqrt{s} = 1.8$ TeV, according to the parton distributions of Set 2. The curves are labeled by the contact interaction scale Λ^* (in TeV) for a LL interaction type with $\eta_{LL} = -1$ (solid lines). (The curves for $\eta_{LL} = +1$ are very similar to the corresponding $\eta_{LL} = -1$ curve and therefore are not separately displayed.) The standard model prediction for the Drell-Yan cross section is denoted by a dashed line.

6.3.2 Compositeness at the SSC

The discover range for compositeness is greatly extended at a supercollider. For example in pp collisions at $\sqrt{s} = 40$ TeV the effects of a left-left contact term in the inclusive jet production is shown in Figure 72 for compositeness scales of $\Lambda^* = 10$, 15, and 20 TeV. In pp collisions the effects of the interference between the usual QCD processes and the composite interaction are significantly larger than for $\bar{p}p$ collisions as can be seen by comparing Fig. 72 and Fig. 68.

The effects of a left-left contact term contributing to the Drell-Yan processes for pp collisions at $\sqrt{s} = 40$ TeV is shown in Figure 73.

6.4 Summary of Discovery Limits

The discovery limits from contact terms associated with quark and/or lepton substructure is given in Table 12. The same discovery criteria were used for present hadron colliders as for the supercollider which are detailed in EHLQ. The discovery criteria LEP and HERA are found in Ref.120. Compositeness scales (for the first generation of quarks and leptons) as high as 20-25 TeV can be probed at an SSC.

6.5 Crossing the Compositeness Threshold

Finally it is interesting to consider what signals will be seen in hadron colliders as the compositeness scale Λ^* is crossed. As the subprocess energy becomes comparable to the compositeness scale not only the lowest mass composite states (the usual quarks and lepton) can be produced but also excited quarks and leptons. These excited quarks could be in color representations other that the

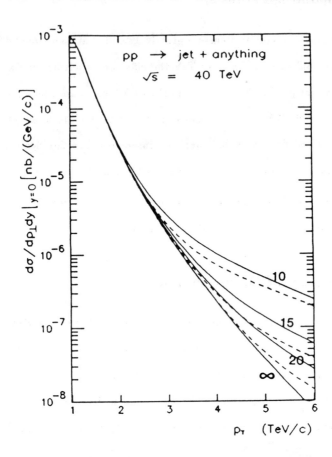

Figure 72: Cross section $d\sigma/dp_\perp dy|_{y=0}$ for jet production in pp collisions at $\sqrt{s} = 40$ TeV, according to the parton distributions of Set 2. The curves are labeled by the compositeness scale Λ^* (in TeV) for a LL interaction type and $\eta_{LL} = -1$ (solid line) and $\eta_{LL} = +1$ (dashed line). The QCD prediction for the cross section is denoted by the bottom solid line. (From EHLQ).

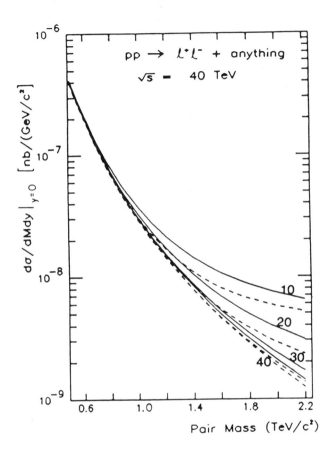

Figure 73: Cross section $d\sigma/d\mathcal{M}dy|_{y=0}$ for dilepton production in pp collisions at $\sqrt{s} = 40$ TeV, according to the parton distributions of Set 2. The curves are labeled by the contact interaction scale Λ^* (in TeV) for a LL interaction type with $\eta_{LL} = -1$ (solid lines) and $\eta_{LL} = +1$ (dashed lines). (From EHLQ)

Table 12: Compositeness scale Λ^* probed at various planned colliders. The left-left interaction type is assumed. The discovery limit is in TeV.

Collider	$\sqrt{s}$ (TeV)	$\int dt\mathcal{L}$ $(cm)^{-2}$	Subprocess tested		
			$e^+e^- \to e^+e^-$ Λ^*	$\bar{q}q \to \bar{q}q$ Λ^*	$\bar{q}q \to e^+e^-$ Λ^*
HERA (ep)	.314	10^{39}	—	—	3
LEP I (or SLC) (e^+e^-)	.10	10^{39}	3.5	—	—
LEP II (e^+e^-)	.20	10^{39}	7	—	—
$S\bar{p}pS$ $(\bar{p}p)$	.63	3×10^{37}	—	.60	1.1
$TEVI$ $(\bar{p}p)$	2.0	10^{38}	—	1.5	2.5
SSC (pp)	40	10^{40}	—	17	25

Table 13: Expected discovery limits for fermions in exotic color representations at present and planned colliders. It is assumed that 100 produced events are sufficient for discovery.

Collider		$\sqrt{s}$	$\int dt \mathcal{L}$	Mass limit (Gev/c^2) Color Representation		
		(TeV)	(cm)$^{-2}$	3*	6	8
$S\bar{p}pS$	$\bar{p}p$	.63	3×10^{36}	65	85	88
upgrade			3×10^{37}	90	110	115
$TEVI$	$\bar{p}p$	1.8	10^{37}	150	200	205
upgrade		2	10^{38}	220	285	290
SSC	pp	40	10^{38}	1,350	2,000	2,050
			10^{39}	2,000	2,750	2,800
			10^{40}	2,750	3,700	3,750

standard triplets. The masses of the lightest excited quarks would naively be expected to be of the same order as Λ^*. It is of course possible that some might be considerably lighter. In this hope the cross sections for pair production of excited quarks in $\bar{p}p$ at $\sqrt{s} = 1.8$ TeV are shown in Figure 74 for color representations 3*, 6, and 8.

The discover limits for fermions in exotic color representations at various colliders are given in Table 13.

What happens to the $\bar{q}q$ total cross section? The behaviour of this total cross section has been studied recently by Bars and Hinchliffe[121]. At energies at and

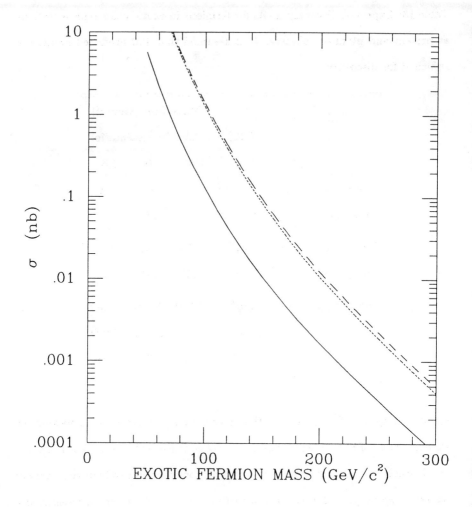

Figure 74: Total cross sections for production of excited quarks in $\bar{p}p$ collisions at $\sqrt{s} = 1.8$ TeV as a function of their mass. Color representations 3*, 8, and 6 are denoted by solid, dashed, and dotted lines respectively. The parton distributions of Set 2 was used.

above the compositeness scale this cross section would have the same general behaviour as the $\bar{p}p$ total cross section at and above 1 GeV. Using this rough analog, we would expect a resonance dominated region at energy scales a few times the compositeness scale and then at much higher energies the total cross section should rise slowly. However most of this cross section is within an angle of approximately $\arcsin(2\Lambda^*/\sqrt{\hat{s}})$ to the beam directions. At energies $\sqrt{\hat{s}} \gg \Lambda^*$, the large angle scattering will again exhibit the $1/\hat{s}$ behaviour expected for preon scattering via single metacolor gluon exchange.

The behaviour of the $\bar{q}q$ subprocess has to be combined with the appropriate parton distribution functions to obtain the physical cross sections in hadron-hadron collisions. The resulting inclusive jet cross section for pp collisions at $\sqrt{s} = 40$ TeV is shown in Figure 75 for a particular model of Bars and Hinchliffe[121] with various compositeness scales. These models exhibit the general behaviour discussed above. Quark-antiquark scattering is mainly inelastic at subprocess energies above the compositeness scale. Thus the two jet final state will be supplanted as the dominate final state by multijets events (possibly with accompanying lepton pairs). This will provide unmistakable evidence that the composite threshold has been crossed.

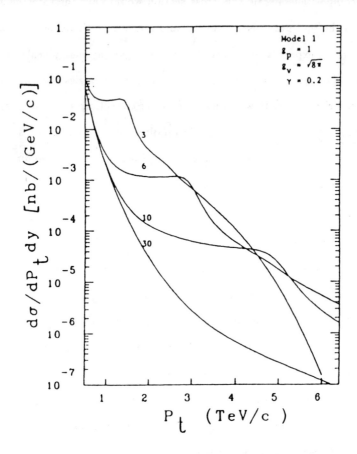

Figure 75: The differential cross section $d\sigma/dp_\perp dy|_{y=0}$ in pp collisions at $\sqrt{s} = 40$ TeV for a model of composite interactions at and above the scale of compositeness. In this model proposed by Bars and Hinchliffe[121] there is a resonance in quark-quark scattering due to the composite interactions. The expected cross section is shown for various values of the resonance's mass: $M_V = 3, 6, 10$, and 30 TeV. For other details on the model and the parameter values used in these curves see Ref.121 (Fig.7)

7 SUPERSYMMETRY ?

One set of symmetries normally encountered in elementary particle physics are the space-time symmetries of the Poincare group:

- P^μ – the momentum operator – the generator of translations.
- $M^{\mu\nu}$ – the Lorentz operators – the generators of rotations and boosts.

These symmetries classify the elementary particles by mass and spin.

The other symmetries usually encountered are internal symmetries such as color, electric charge, isospin, etc. For each non-Abelian internal symmetry there is a set of charges $\{Q_a\}$ which form a Lie Algebra:

$$-i[Q_a, Q_b] = f_{abc} Q_c \qquad (7.1)$$

under which the Hamiltonian is invariant:

$$-i[Q_a, H] = 0 \qquad (7.2)$$

If these symmetries are not spontaneously broken the physical states form representations under the associated Lie group, $\mathcal{G}$.

Because the charges are associated with internal symmetries they commute with the generators of space-time symmetries

$$\begin{aligned} -i[Q_a, P^\mu] &= 0 \\ -i[Q_a, M^{\mu\nu}] &= 0 \, . \end{aligned} \qquad (7.3)$$

We have already seen that such symmetries play a central role in the physics of the standard model. The internal symmetries $SU(3) \otimes SU(2)_L \otimes U(1)_Y$ determine all the basic gauge interactions. Global symmetries such as fermion number and flavor symmetries also play an important role.

Supersymmmetry is a generalization of the usual internal or space-time symmetry sharing aspects of both. Formally the concept of a Lie algebra is generalized to a structure called a graded Lie algebra[122] which is defined by both commutators and anticommutators. A systematic development of the formal aspects of supersymmetry is outside the scope of these lectures but can be found in Wess and Bagger[123].

The simpliest example of a global supersymmetry is $N = 1$ supersymmetry which has a single generator $\mathcal{Q}_\alpha$ which transforms as spin $\frac{1}{2}$ under the Lorentz group:

$$-i[\mathcal{Q}_\alpha, P^\mu] = 0$$
$$-i[\mathcal{Q}_\alpha, M^{\mu\nu}] = (\sigma^{\mu\nu}\mathcal{Q})_\alpha \tag{7.4}$$

where $\sigma^{\mu\nu}$ are the Pauli matrices. Finally the generator $\mathcal{Q}$ and the Hermitian conjugate generator $\overline{\mathcal{Q}}$ must have the following anticommutation relations:

$$\{\mathcal{Q}_\alpha, \mathcal{Q}_\beta\} = 0$$
$$\{\overline{\mathcal{Q}}_\alpha, \overline{\mathcal{Q}}_\beta\} = 0$$
$$\{\mathcal{Q}_\alpha, \overline{\mathcal{Q}}_\beta\} = -2(\gamma_\mu)_{\alpha\beta}P^\mu \tag{7.5}$$

These are the relations for $N = 1$ global supersymmetry. The generator $\mathcal{Q}$ is a spin $\frac{1}{2}$ fermionic charge. If this is a symmetry of the Hamiltonian, then

$$-i[\mathcal{Q}_\alpha, H] = 0 \tag{7.6}$$

and assuming this symmetry is realized algebraically the physical states of the system can be classified by these charges. Since the supercharge has spin $\frac{1}{2}$, states differing by one-half unit of spin will belong to the same multiplet. This fermion-boson connection will allows a solution to the naturalness problem of the standard model (discussed in Section 4).

7.1 Minimal $N = 1$ Supersymmetric Model

The minimal supersymmmetric generalization of the standard model is to extend the standard model to include a $N = 1$ supersymmetry. The supercharge $\mathcal{Q}$ acts on an ordinary particle state to generate its superpartner. For a massless particle with helicity h (i.e. transforming as the $(0, h)$ representation of the Lorentz group) the action of the charge $\mathcal{Q}$ produces a superpartner degenerate in mass with helicity $h - \frac{1}{2}$ (i.e. transforming as $(0, h - \frac{1}{2})$). Applying the supercharge again vanishes sinc the anticommutator of the supercharge with itself is zero (Eq. 7.5). Hence the supermultiplets are doublets with the two particles differing by one-half unit of spin. The number of fermion states (counted as degrees of freedom) is identical with the number of boson states. For massless spin 1 gauge bosons these superpartners are massless spin $\frac{1}{2}$ particles called gauginos (gluino, wino, zino, and photino for the gluon, W, Z, and photon respectively). For spin $\frac{1}{2}$ fermions these superpartners are spin 0. If the fermion is massive the superpartner will be a scalar particle with the same mass as the associated fermion. The superpartners of quarks and leptons are denoted scalar quarks (squarks) and scalar leptons (sleptons). The superpartners of the Higgs scalars of the standard model are spin $\frac{1}{2}$ fermions called Higginos.

Since the supercharge commutes with every ordinary internal symmetry Q_a

$$-i[\mathcal{Q}_\alpha, Q_a] = 0 \ . \tag{7.7}$$

all the usual internal quantum numbers of the superparticle will be identical to those of its ordinary particle partner. In nearly all supersymmetric theories, the superpartners carry a new fermionic quantum number R which is exactly conserved[124]. All the ordinary particles and their superpartners are shown in Table 14.

No superpartner of the ordinary particles has yet been observed, thus supersymmetry must be broken. This scale of supersymmetry breaking is denoted:

$$\Lambda_{ss} . \tag{7.8}$$

Even in the presence of supersymmetry breaking it is normally possible to retain a R quantum number for superpartners which is absolutely conserved[125]. This means that the lightest superpartner will be absolutely stable. If the supersymmetry is spontaneously broken there is an additional massless fermion the Goldstino $\mathcal{G}$, which is the analogy of the Goldstone boson in the case of spontaneous breaking of an internal symmetry. In more complete models with local supersymmetry, such as supergravity, there is a superHiggs mechanism in which the Goldstino becomes the longitudinal component of a massive spin $\frac{3}{2}$ particle - the gravitino[126]. Hence the existence of the Goldstino as a massless physical is dependent of the way global $N = 1$ supersymmetry is incorporated into a more complete theory and the mechanism of supersymmetry breaking.

The gauge interactions of the ordinary particles and the invariance of the action under supersymmetric transformations completely determine the interactions of fermions, gauge bosons, squarks, sleptons, and gauginos among themselves. The details of the Lagrangian can be found in, for example, Dawson, Eichten, and Quigg[127] (hereafter denoted DEQ).

On the other hand, the masses of the superpartners associated with super-

Table 14: Fundamental Fields of the Minimal Supersymetric Extension of the Standard Model

Particle		Spin	Color	Charge
gluon	g	1	8	0
gluino	$\tilde{g}$	1/2	8	0
photon	γ	1	0	0
photino	$\tilde{\gamma}$	1/2	0	0
intermediate bosons	$W^{\pm}, Z^0$	1	0	$\pm 1, 0$
wino, zino	$\tilde{W}^{\pm}, \tilde{Z}^0$	1/2	0	$\pm 1, 0$
quark	q	1/2	3	2/3, -1/3
squark	$\tilde{q}$	0	3	2/3, -1/3
electron	e	1/2	0	-1
selectron	$\tilde{e}$	0	0	-1
neutrino	ν	1/2	0	0
sneutrino	$\tilde{\nu}$	0	0	0
Higgs bosons	$H^+ \quad H'^0$ $H^0 \quad H'^-$	0	0	$\pm 1, 0$
Higginos	$\tilde{H}^+ \quad \tilde{H}'^0$ $\tilde{H}^0 \quad \tilde{H}'^-$	1/2	0	$\pm 1, 0$

symmetry breaking and the interactions of the Higgs scalars and Higginos are not similarly specified.

The Higgs sector of the minimum supersymmetric extension of the standard model requires two scalar doublets:

$$\begin{pmatrix} H^+ \\ H^0 \end{pmatrix} \quad \begin{pmatrix} H'^0 \\ H^- \end{pmatrix} \tag{7.9}$$

and their Higgino superpartners:

$$\begin{pmatrix} \tilde{H}^+ \\ \tilde{H}^0 \end{pmatrix} \quad \begin{pmatrix} \tilde{H}'^0 \\ \tilde{H}^- \end{pmatrix} \tag{7.10}$$

Two Higgs doublets are required because the Higginos associated with the usual Higgs doublet have nonzero weak hypercharge Q_Y and therefore contribute to the $U(1)_Y$ and $(U(1)_Y)^3$ anomalies; to recover a consistent gauge theory another fermion doublet must be introduced with the opposite Q_Y charge.

One complication introduced when supersymmetry breaking is included is that color neutral gauginos and Higginos can be mix. So in general the true mass eigenstates will be linear combinations of the original states. For the charged sector the wino $(\tilde{w}^\pm)$ and charged Higgino $(\tilde{H}^\pm)$ can mix. For the neutral sector the zino $(\tilde{z}^0)$, photino $(\tilde{\gamma})$, and the two neutral Higgsinos $(\tilde{H}^0, \tilde{H}'^0)$ can mix. The effects of these mixings will not be discussed further here[128].

The usual Yukawa couplings between Higgs scalars and quarks or leptons generalize in the supersymmetric theory to include Higgs-squark and Higgs-slepton couplings, as well as Higgsino-quark-squark and Higgsino- lepton-slepton transitions. Just as there is a Kobayshi-Maskawa matrix which mixes quark flavors and introduces a CP-violating phase, so too, will there be mixing matrices in the quark-squark and squark-squark interactions. There may also be mixing in the lepton-slepton and slepton-slepton interactions. These mixings have some constraints which arise from the experimental restrictions on flavor-changing neutral currents. For a possible Super-GIM mechanism to avoid these constraints see Baulieu, Kaplan, and Fayet[129].

The actual masses and mixings are extremely model dependent. Again for simplicity it will be assumed in the phenomenological analysis presented here that:

- There is no mixing outside the quark-quark sector

- The masses of the superpartners will be treated as free parameters.

It is straightforward to see that the supersymmetric extension of the standard model can satisfy 't Hooft's naturalness condition. The mass of each Higgs scalar is equal by supersymmetry to the mass of the associated Higgsino for which a small mass can be associated with an approximate chiral symmetry. Defining the parameter ξ to be the mass of the Higgs scalar over the energy scale of the effective Lagrangian, the limit $\xi \to 0$ is associated with a chiral symmetry if supersymmetry is unbroken. Hence the scale of supersymmetry breaking Λ_{ss} must be not be much greater than the electroweak scale if supersymmetry is to solve the naturalness problem of the standard model. Therefore the masses of superpartners should be accessible to the present or planned hadron collider.

Since the masses of superpartners are not tightly constrained by theory we begin by investigating the experimental constraints on their masses.

7.2 Present Bounds on Superpartners

The present bounds on superpartners are discussed in DEC and in the review by Haber and Kane[130]. I will give a short summary of the situation. Limits on superpartner masses arise from a large variety of sources including:

- Searches for direct production in hadron and lepton colliders as well as in fixed target experiments.

- Limits are rare processes such as flavor changing neutral currents induced by the effects of virtual superpartners.

- Effects of virtual superpartners on the parameters of the standard model.

- Cosmological bounds on the abundance of superpartners.

Before beginning to discuss some of these limits one point must be stressed. In the absence of a specific model all the superpartner masses and even the scale of supersymmetry breaking must be taking as free parameters. This greatly complicates the analysis of limits and weakens the results. In general each limit depends not only on the mass of the superpartner in question but also on:

- The rate for the reaction involved; and therefore the masses of other superpartners which are enter as virtual states in the process and the scale of supersymmetry breaking.

- The decay chain of the superpartner. Which decays are kinematically allowed again depends on the masses of other superpartners.

This interdependence of the mass limits makes it difficult to reduce the results to a single mass limit for each superpartner.

7.2.1 Photino Limits

The simpliest models of supersymmetry breaking have a color and charge neutral fermion (the photino) as the lightest superpartner. Three cases can be distinguished:

- The photino is the lightest superpartner and $m_{\tilde{\gamma}} < 1$ MeV/c^2.

- The photino is the lightest superpartner and $m_{\tilde{\gamma}} > 1$ MeV/c^2.

- The photino decays into a photon and a Goldstino.

In the first case the photinos are stable spin $\frac{1}{2}$ fermions. An upper bound on the photino mass arises by demanding the the density of photinos in the present universe is less than the closure density[131]:

$$\rho_{\tilde{\gamma}} = 109 cm^{-3} m_{\tilde{\gamma}} < \rho_{\text{critical}} = (3.2 - 10.3) \times 10^3 eV/c^2 cm^{-3} \qquad (7.11)$$

which implies that

$$m_{\tilde{\gamma}} < 100 (eV/c^2) \qquad (7.12)$$

If the photino is the lightest superpartner and heavier than 1 MeV/c^2 Goldberg has pointed out that photino pairs can decay into ordinary fermion pairs by the virtual exchange of the associated sfermion. The annihilation rate is dependent on the photino and sfermion masses. By integrating the rate equation numerically over the history of the universe, the present photino number density can be estimated[132]. This leads to a sfermion mass dependent upper bound on the photino mass.

The resulting limits on a the mass of a stable photino are summarized in Figure76.

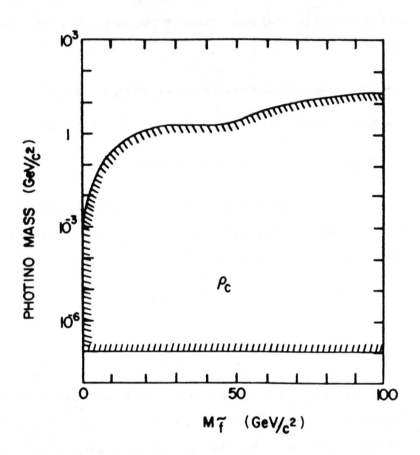

Figure 76: Cosmological limits of the allowed photino mass as a function of the mass of the lightest scalar partner of a fermion. This figure assumes that the photino is stable and is the lightest superrsymmetric particle. (From DEQ)

The photino may decay by:

$$\tilde{\gamma} \to \gamma + \tilde{\mathcal{G}} \qquad (7.13)$$

if a massless Goldstino $\tilde{\mathcal{G}}$ exists. One constraint in this case is that the photons produced in photino decays must have thermalized with the cosmic microwave background[133]. This requires that the photino lifetime ($\tau_{\tilde{\gamma}}$) is less than 1000 seconds. Since

$$\tau_{\tilde{\gamma}} = 8\pi \frac{\lambda^4}{m_{\tilde{\gamma}}^5} \qquad (7.14)$$

the limit on photino mass becomes

$$m_{\tilde{\gamma}} > 1.75 MeV/c^2 (\frac{\Lambda_{ss}}{1Tev/c^2})^{4/5} \qquad (7.15)$$

The constraints from laboratory experiments on the photino mass are obtained from:

- The axion searches[134]:

$$\Psi \to \gamma + \text{unobserved neutrals} \qquad (7.16)$$

can be reinterpreted as photino searches.

- Limits on $\Psi \to$ unobserved neutrals imply that the scale of supersymmetry breaking $\Lambda_{ss} \geq 10$ GeV[135]. A stronger limit[136], $\Lambda_{ss} \geq 50$ GeV, can be inferred from constraints on emission of photinos from white dwarf or red giant stars if the Goldstino or gravitino mass is less than 10 keV/c^2.

- Limits on e^+e^- production of photons plus missing energy from CELLO[137] imply limits on the processes:

$$\begin{aligned} e^+e^- &\to \tilde{\gamma}+\tilde{\gamma} \to \gamma+\gamma+\tilde{\mathcal{G}}+\tilde{\mathcal{G}} \\ e^+e^- &\to \gamma\tilde{\gamma}\tilde{\gamma} \end{aligned} \qquad (7.17)$$

The resulting limits on the mass of an unstable photino are given in Figure 77.

7.2.2 Gluino Limits

The gluino is the spin $\frac{1}{2}$ partner of the gluon. It is a color octet and charge zero particle. Again for the gluino there are three decay alternatives:

- The gluino is stable or long-lived.
- The gluino decays into photino and a quark-antiquark pair.
- The gluino decays into a gluon and a Goldstino.

If the gluino is long-lived ($\tau_{\tilde{g}} \geq 10^{-3}$ sec) then it would be bound into a long-lived R-hadron (so called because of the R quantum number of gluinos). Thus stable particle searches can be used to put limits on the mass of such R-hadrons. For charged hadrons these limits are[138]:

$$1.5 GeV/c^2 \leq m_R \leq 9 GeV/c^2 \qquad (7.18)$$

if $\tau_{\tilde{g}} \geq 10^{-8}$ sec. While for neutral hadrons the limits are[139]:

$$2 GeV/c^2 \leq m_R \leq 9 GeV/c^2 \qquad (7.19)$$

if $\tau_{\tilde{g}} \geq 10^{-7}$ sec. It seems that gluinos with $m_{\tilde{g}} \leq 1.5 GeV/c^2$ and $\tau_{\tilde{g}} \geq 10^{-8}$ sec could have escaped detection.

In the second decay scenario the decay chain is:

$$\begin{array}{l} \tilde{g} \rightarrow \bar{q}\tilde{q} \\ \phantom{\tilde{g} \rightarrow \bar{q}} \hookrightarrow q + \tilde{\gamma} \end{array} \qquad (7.20)$$

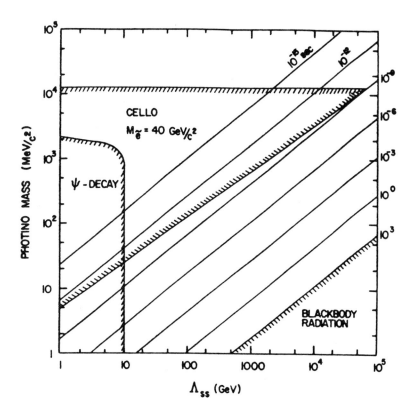

Figure 77: Limits on the allowed photino mass as a funcion of the supersymmetry breaking scale Λ_{ss}. This figure assumes that the photino decays to a photon and a massless Goldstino. The various limits from Ψ decay, the search for the process $e^+e^- \to \tilde{\gamma}\tilde{\gamma} \to \gamma\gamma\tilde{\mathcal{G}}\tilde{\mathcal{G}}$ by the CELLO Collaboration, and blackbody radiation are discussed in the text. (From DEQ)

; and therefore the decay rate is sensitive to the squark mass. For $m_{\tilde{\gamma}} = 0$ the lifetime is:

$$\tau(\tilde{g} \to \bar{q}q\tilde{\gamma}) = \frac{48\pi m_{\tilde{q}}^4}{\alpha_s \alpha_{EM} e_q^2 m_{\tilde{g}}^5} . \qquad (7.21)$$

There are stringent bounds on the mass and lifetime of the gluino from beam dump experiments both the E-613 experiment at Fermilab[140] and the CHARM Collaboration at CERN[141]. The limits on gluino mass as a function of lifetime (or alternatively squark mass) are summarized in Figure 78 for the assumption that the resulting photino is stable. Note that for squark masses in the range 200-1,000 GeV/c^2 there is no limit on gluino mass for this gluino scenario. The possibility that the photino is unstable to decay into photon and Goldstino requires a somewhat more complicated analysis. In that case the limit from E-613 beam dump experiments constrain the relation between the gluino mass, the supersymmetry breaking scale, and the photino mass. Details of these constraints can be found in DEQ.

The final possibility is that the gluino can decay into a gluon and a Goldstino. The lifetime of the gluino is given by

$$\tau(\tilde{g} \to g + \tilde{\mathcal{G}}) = \frac{8\pi \Lambda_{ss}^4}{m_{\tilde{g}}^5} = 1.65 \times 10^{-23} sec (\frac{\Lambda_{ss}}{1 GeV/c^2})^4 (\frac{1 GeV/c^2}{m_{\tilde{g}}})^5 \qquad (7.22)$$

Again beam dump experiments constrain the relationship between $m_{\tilde{g}}$ and Λ_{ss}. The resulting limits are shown in Figure 79.

In *all* scenarios for gluino decay it is possible to find ranges of parameters for which light (a few GeV/c^2) gluinos are allowed by experiment. This corresponds to a gap in experimental technique for lifetimes between 10^{-10} and 10^{-12} sec in hadron initiated experiments.

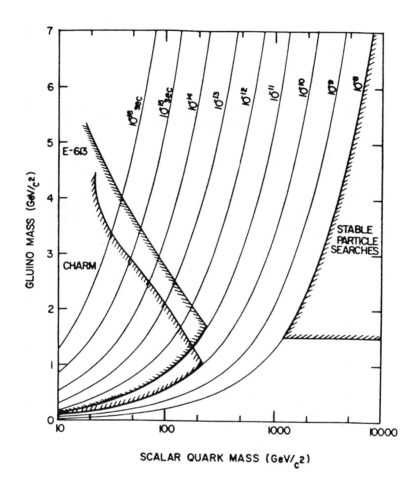

Figure 78: Limits on the gluino mass as a function of the lightest squark mass. The gluino is assumed to decay to a $\bar{q}q$ pair and a massless photino. The limits are from beam-dump experiments and stable particle searches as discussed in the text. The corresponding gluino lifetimes are also shown. (From DEQ)

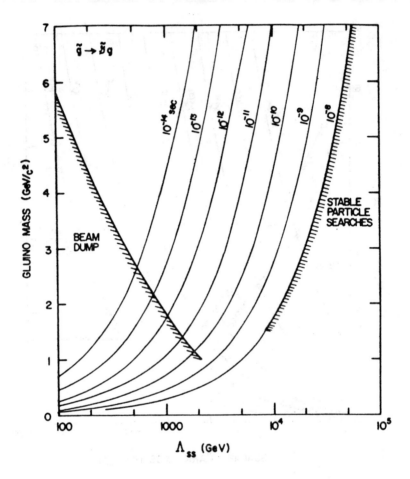

Figure 79: Limits on the gluino mass as a function of the supersymmetry breaking scale Λ_{ss}. The limits are from the Fermilab beam-dump experiment[140] and the stable particle searches[138,139] and assume that the gluino decays to a gluon and a massless Goldstino. The corresponding gluino lifetimes are also shown. (From DEQ)

7.2.3 Squark Limits

A squark is a spin zero color triplet particle with the flavor and charge of the associated quark. There are four sources of limits on squark masses.

- Free quark searches. The MAC Collaboration at PEP[142] find a limit for e^+e^- production of fractionally charged long-lived ($\tau \gg 10^{-8}$ sec) particles which corresponds to a lower bound on the mass of any squark of 14 GeV/c^2.

- Stable hadron searches. Stable hadron searches in hadron initiated reactions exclude a charged squark bearing hadron with mass in the range:

$$1.5 GeV/c^2 \leq m_{\tilde{q}} \leq 7 GeV/c^2 \qquad (7.23)$$

for lifetimes $\tau \geq 5 \times 10^{-8}$ seconds[138]. The JADE Collaboration at PETRA looked for

$$e^+e^- \to \tilde{q}\tilde{q}^* \qquad (7.24)$$

in both charged and neutral final state hadrons. Their exclude long-lived squarks in the range[139]:

$$2.5 GeV/c^2 \leq m_{\tilde{q}} \leq 15.0 GeV/c^2 \quad \text{for} \quad |e_{\tilde{q}}| = 2/3$$

$$2.5 GeV/c^2 \leq m_{\tilde{q}} \leq 13.5 GeV/c^2 \quad \text{for} \quad |e_{\tilde{q}}| = 1/3 \qquad (7.25)$$

- Narrow resonance searches in e^+e^- collisions. Squark-antisquark bound states could be produced as narrow resonances in e^+e^- collisions. The production rates have been estimated by Nappi[144] who concludes that $|e_{\tilde{q}}| = 2/3$ squarks with masses below 3 GeV/c^2 can be ruled out. No limits exist from this process for $|e_{\tilde{q}}| = 1/3$ squarks.

- Heavy Lepton searches. If a squark decays to a quark and a (assumed massless) photino the decay signature in e^+e^- collisions is similar to that for a heavy lepton decay - two acoplanar jets and missing energy. The JADE Collaboration[145] has excluded squarks with this decay pattern for:

$$3.1 GeV/c^2 \leq m_{\tilde{q}} \leq 17.8 GeV/c^2 \quad \text{for} \quad |e_{\tilde{q}}| = 2/3$$

$$7.4 GeV/c^2 \leq m_{\tilde{q}} \leq 16.0 GeV/c^2 \quad \text{for} \quad |e_{\tilde{q}}| = 1/3$$

Summarizing these limits:

1. Stable squarks must have masses exceeding ≈ 14 GeV/c^2.

2. If the photino is nearly massless, unstable $|e_{\tilde{q}}| = 2/3$ squarks are ruled out for masses ≤ 17.8 GeV/c^2; while for $|e_{\tilde{q}}| = 1/3$ squarks a window exists for masses below 7.4GeV/c^2, otherwise their mass must exceed 16 GeV/c^2.

3. If the photino is massive all that can be said is that $m_{\tilde{q}} \geq 3$ GeV/c^2 if the lifetime is less than 5×10^{-8} sec and $|e_{\tilde{q}}| = 2/3$.

7.2.4 Limits on Other Superpartners

The limits on the wino, zino, and sleptons come from limits on production in e^+e^- collisions. The wino is a spin 1/2 color singlet particle with unit charge. If the photino is light the wino can decay via:

$$\tilde{w} \rightarrow W\tilde{\gamma} \qquad (7.26)$$
$$\phantom{\tilde{w} \rightarrow W}\hookrightarrow \bar{q}q$$

and hence the heavy lepton searches will be sensitive to a wino as well. The Mark J Collaboration at PETRA have set the limit[146]:

$$m_{\tilde{w}} \geq 25 GeV/c^2 \qquad (7.27)$$

For the zino, the JADE collaboration obtains the bound[147]

$$m_{\tilde{z}} \geq 41 GeV/c^2 \qquad (7.28)$$

assuming a massless photino and $m_{\tilde{e}} = 22$ GeV/c^2.

For the charged sleptons the limits are[148]:

$$m_{\tilde{e}} \geq 51 GeV/c^2 \qquad (7.29)$$

assuming $m_{\tilde{\gamma}} = 0$; and[149]:

$$m_{\tilde{\mu}} \geq 16.9 GeV/c^2$$

$$m_{\tilde{\tau}} \geq 15.3 GeV/c^2 \ . \qquad (7.30)$$

7.3 Discovering Supersymmetry In Hadron Colliders

All the lowest order (Born diagrams) cross sections $d\tilde{\sigma}/dt$ and $\tilde{\sigma}$ have been calculated in DEQ for

$$(\tilde{q} \quad \tilde{l} \quad \tilde{\nu} \quad \tilde{g} \quad \tilde{\gamma} \quad \tilde{z}^0 \quad \tilde{H}^0 \quad \tilde{H}'^0 \quad \tilde{w}^\pm \quad \tilde{H}^\pm)^2 \qquad (7.31)$$

final states in parton-parton collisions; including the mixing in the neutral $(\tilde{\gamma}, \tilde{z}^0, \tilde{H}^0, \tilde{H}'^0)$ and charged $(\tilde{w}^\pm, \tilde{H}^\pm)$ fermion sectors. Many of these processes have also been studied by others as well: see DEQ for complete references.

The overall production rate for pair production of superpartners is determined by the strength of the basic process. These relative rates for the various final states are:

Final State	Production Mechanism	Strength
$(\tilde{q},\tilde{g})^2$	QCD	α_s^2
$(\tilde{q},\tilde{g}) \times (\tilde{\gamma},\tilde{z},\tilde{H}^0,\tilde{H}^{\prime 0})$	QCD-Electroweak	$\alpha_s \alpha_{EW}$
$\tilde{l}\tilde{\nu}, \tilde{l}\tilde{l}^*, \tilde{\nu}\tilde{\nu}^*$	decays of real (or virtual) $W^\pm$ and Z^0	α_{EW} ($\alpha_{EW}{}^2$)
$(\tilde{\gamma},\tilde{z}^0,\tilde{H}^0,\tilde{H}^{\prime 0},\tilde{w}^\pm,\tilde{H}^\pm)^2$	Electroweak	$\alpha_{EW}{}^2$

We will consider each of these processes in turn beginning with the largest rates: squark and gluino production.

The lowest order processes for gluino and squark production are shown in Figure 80. The underlined graphs in Figure 80 depend only on the masses of the produced superpartner and are therefore independent of all other supersymmetry breaking parameters. Hence hardon colliders allow clean limits (or discovery) on the masses of gluinos and squarks. The cross sections for gluino production are large, since gluinos are produced by the strong interactions. The total cross section for gluino pairs in $\bar{p}p$ collisions is shown in Figure 81 as a function of gluino mass at $\sqrt{s}$ = 630, 1.8, and 2 TeV. The squark masses were all taken to be 1 TeV/c^2 so there would be not significant contribution from diagrams involving squark intermediate states. The typical effects of the diagrams with squark intermediate states is also illustrated in Fig. 81 by including the cross section for gluino pair production for $\sqrt{s}$ = 630 GeV with $m_{\tilde{q}} = m_{\tilde{g}}$. Because of the dominance of gluon initial states, the dependence of the gluino pair production cross section on the squark mass is small except at the highest $\sqrt{\hat{s}/s}$. In any case, the cross section excluding the contribution from intermediate squarks gives lower bound on the gluino production for a given mass gluino $(m_{\tilde{g}})$.

Typically the supersymmetry breaking leads to gluinos not much heavier than the lightest squark. In the case that the up squark mass (assuming $m_{\tilde{u}} =$

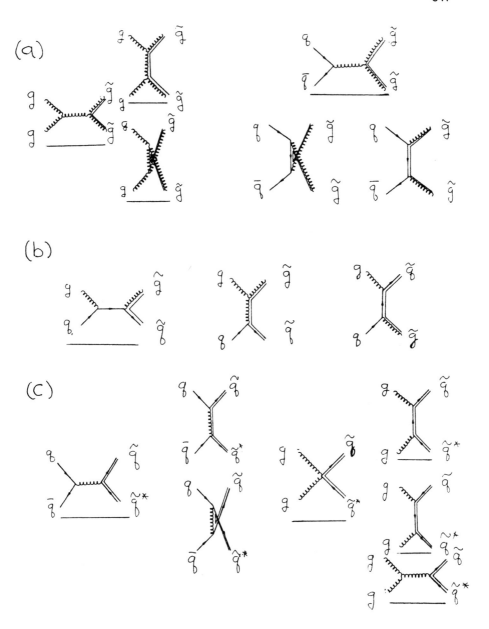

Figure 80: Feynman diagrams for the lowest order production of (a) gluino pairs, (b) gluino in association with a up squark, and (c) up squark-antisquark pair.

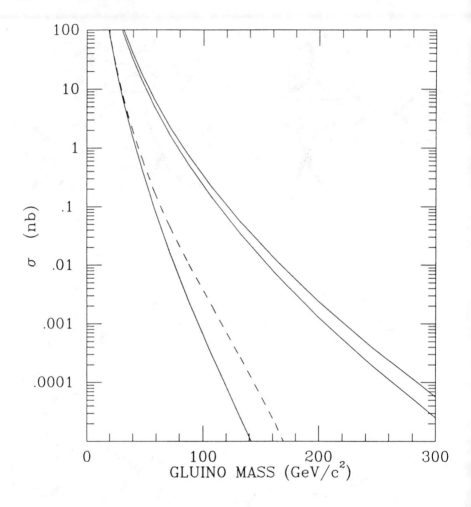

Figure 81: Total cross section for gluino pair production in $\bar{p}p$ collisions as a function of gluino mass. The rates for squark mass $m_{\tilde{q}} = 1$ TeV/c^2 are shown for $\sqrt{s} = 630$ GeV (lower solid line), 1.8 TeV (middle solid line), and 2.0 TeV (upper solid line); as well as for $m_{\tilde{q}} = m_{\tilde{g}}$ at $\sqrt{s} = 630$ GeV (dashed line). The rapidity of each of the gluinos is restricted to $|y_i| \leq 1.5$.

$m_{\tilde{q}}$) equals the gluino mass the total cross section for the reaction

$$p\bar{p} \to \tilde{u} + \tilde{q}^* + \text{anything} \qquad (7.32)$$

where $\tilde{q}^* = \tilde{u}, \tilde{d}, \tilde{u}^*$, or $\tilde{d}^*$ is shown in Figure 82 as a function of the up squark mass for $\sqrt{s}=$ 630, 1.8, and 2.0 TeV. This can be compared for $\sqrt{s} = 630$ GeV to the cross section for up squark production with $m_{\tilde{g}} = 1$ TeV as shown in Fig. 82. Clearly for squark production the total cross sections depend more strongly on other superpartner's (specifically the gluino's) mass.

For gluino and squark masses approximately equal there is also a comparable contribution from squark-gluino associated production. For example, for $m_{\tilde{u}} = m_{\tilde{g}} = 50$ GeV/c^2 the cross section for associated production is approximate 7 nanobarns at $\sqrt{s} = 2$ TeV.

The detection signatures for gluino and squarks are similar but model and mass dependent. Here I will consider only the usual scenario in which the lightest superparticle is the photino. Other possiblities exist, for example if the Goldstino is massless then the photino can decay:

$$\tilde{\gamma} \to \tilde{g} + \gamma \, . \qquad (7.33)$$

In another possible model the lightest superpartner is the the sneutrino. For a discussion of these alternatives see for example Haber and Kane[130] and Dawson[150]. The basic signature of squark or gluino production is some number of jets accompanied by sizable missing energy. The decay chains for the squark and gluino are:

$$\tilde{q} \to \tilde{g} + q$$
$$\tilde{g} \to q + \bar{q} + \tilde{\gamma} \qquad (7.34)$$

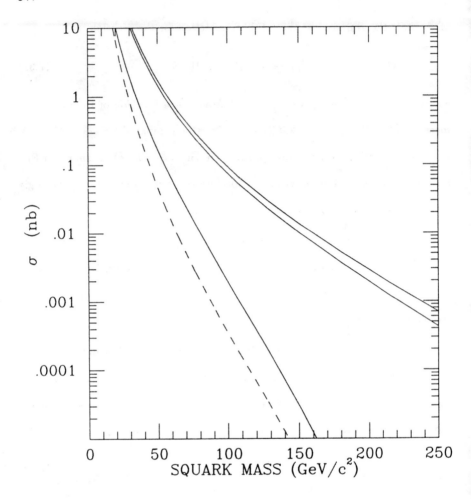

Figure 82: Total cross section for up squark production in $\bar{p}p$ collisions as a function of up squark mass. The rates for gluino mass equal up squark mass $m_{\tilde{g}} = m_{\tilde{u}}$ are shown for $\sqrt{s} = 630$ GeV (bottom solid line), 1.8 TeV (middle solid line), and 2.0 TeV (top solid line); as well as for $m_{\tilde{g}} = 1$ TeV/c^2 at $\sqrt{s} = 630$ GeV (dashed line). The rapidity of the up squark (and the associated squark) is restricted to $|y_i| \leq 1.5$.

if $m_{\tilde{g}} < m_{\tilde{q}}$ and :

$$\tilde{g} \rightarrow \tilde{q} + \bar{q}$$

$$\tilde{q} \rightarrow q + \tilde{\gamma} \qquad (7.35)$$

if $m_{\tilde{q}} < m_{\tilde{g}}$. The number of jets which are experimentally distinguishable depends on the masses of the superpartners and the energy of the hadron collisions in a complicated experiment dependent way[151,152,153]. Clearly there are backgrounds from ordinary QCD jets which can have missing transverse energy for a variety of reasons (weak decays of a heavy quark in the jet, energy measurement inefficiencies, dead spots in the detector, etc.). Even though each decay chain above leads to an event with at least two final state quarks or gluons, the experimental requirements for a jet imply that a number of these events will appear to have only one jet - a monojet event[151].

The backgrounds for detection of squarks and gluinos in the present colliders are:

- One monojet background is the production of $W^{\pm}$ which then decays by the chain:

$$W^{\pm} \rightarrow \bar{\nu}\tau$$
$$\phantom{W^{\pm} \rightarrow \bar{\nu}\tau}\hookrightarrow \text{hadrons} + \nu \qquad (7.36)$$

There are of course distinguishing features of these background events. The missing transverse energy E_T of the background events will be ≤ 30 GeV since the primary $W^{\pm}$ is produced nearly at rest while for squark or gluino production the missing energy is not bounded in the same way. Also the multiplicity of charged tracks from the τ decay will be low (usually only 1 or 3) while from squark or gluino production the multiplicity should be comparable to a ordinary QCD jet of similar energy. These differences are helpful in the analysis of the monojet events.

- Another monojet background is the associated production:

$$\bar{p}p \rightarrow g(\text{ or } q)Z^0 \atop \phantom{\bar{p}p \rightarrow g(\text{ or } q)}\hookrightarrow \nu\bar{\nu} \tag{7.37}$$

The rate of these background events are reasonably low when minimum missing E_T cuts are imposed[151]. Also because the final state in squark or gluino production has more than one quark or gluon, monojet events arising from supersymmetric particle production typically will not be as "clean" (no significant addition energy deposition) as the monojet events from associated Z^0 production events. If the charged lepton is undetected or misidentified, associated $W^{\pm}$ production and leptonic decay can also mimic monojet events.

- The main background to multijet events with missing E_T is heavy quark weak decays inside jets. For example the decay of a b quark in one jet:

$$b \rightarrow c + \bar{\nu} + l \tag{7.38}$$

can produce large missing E_T in a two jet event. This background can be reduced if methods are found to identify charged leptons in a jet[153].

There has been a great deal of recent work on reducing these backgrounds to the detection of superpartners[151,152,153]. My best guess is that 1000 produced events will be required to obtain a clear signal for either a gluino or squark in the collider enviroment.

It also seems likely that experiments at the $S\bar{p}pS$ and TeV I Colliders can be designed to close any gaps in present limits for light gluinos ($m_{\tilde{g}} = 1 - 3$

GeV/c^2) and charge $-1/3$ squarks ($m_{\tilde{q}} \leq 8$ GeV/c^2), but careful study of this possibility will be required.

The other superpartners can be produced in hadron collisions in the following ways:

- The photino, wino, and zino can be produced in association with a squark or gluino.

- The photino, wino, zino, slepton, and sneutrino can be produced in the decays of $W^{\pm}$ or Z^0 bosons if kinematically allowed.

For present collider energies no other production mechanisms are significant.

The photino is generally assumed to be the lightest superpartner. The major mechanism for producing photinos in hadron-hadron collisions is the associated production processes:

$$p\bar{p} \to \tilde{g} + \tilde{\gamma} + \text{anything} \tag{7.39}$$

and

$$p\bar{p} \to \tilde{q} + \tilde{\gamma} + \text{anything} \tag{7.40}$$

These production mechanisms are shown in Figure 83.

The total cross section for production of $\tilde{q} + \tilde{\gamma}$ in $\bar{p}p$ collisions as a function of the squark mass where the photino mass is assumed to be zero is given in Figure 84 for $\sqrt{s} = .63$, 1.8, and 2 TeV. These production cross sections are smaller than the squark pair production cross sections in Fig. 82 by roughly $\alpha_{\text{EW}}/\alpha_s$ but this reaction produces a clear signature: a jet (if $m_{\tilde{q}} < m_{\tilde{g}}$) or three jets (if $m_{\tilde{q}} > m_{\tilde{g}}$) on one side of the detector and no jet on the other; hence the missing transverse energy will be large. For the one jet case there is the Z^0 plus

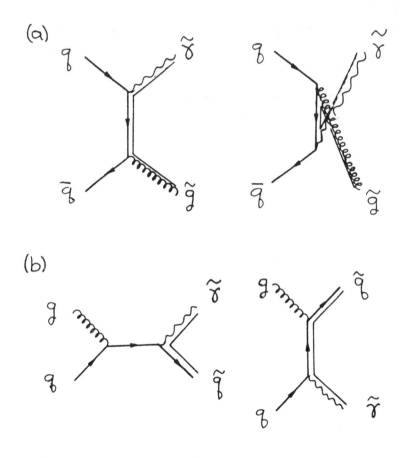

Figure 83: Lowest order diagrams for associated production of photino and (a) gluino or (b) squark.

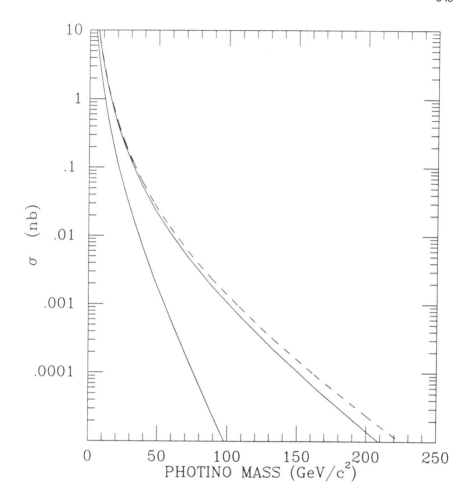

Figure 84: Total cross section for associated production in $\bar{p}p$ collisions of a photino and light squark (up or down) as a function of the photino mass (assuming $m_{\tilde{\gamma}} = m_{\tilde{u}} = m_{\tilde{d}}$). The rates are shown for $\sqrt{s} = 630$ GeV (lower solid line), 1.8 TeV (upper solid line), and 2 TeV (dashed line). The rapidity of both the photino and the squark is restricted to $|y_i| \leq 1.5$. The parton distributions of Set 2 were used.

jet background discussed previously, but the rate and characteristic of these events are well understood theoretically and hence relatively small deviations from expectations would be significant. For the production of $\tilde{g} + \tilde{\gamma}$ the same comments apply. Because of the striking signature of these events 100 produced events should be sufficient for discovery of the photino (and associated gluino or squark) through this mechanism.

The bounds on the wino and zino masses are not model independent but these gauginos are likely heavier than 40 GeV/c^2. The total cross sections for associated production of a massive wino or zino with a squark or gluino are quite small. For $m_{\tilde{w}} = m_{\tilde{z}} = m_{\tilde{q}} = m_{\tilde{g}} = 50$ GeV/c^2:

Process	Total Cross Section (nb) $\sqrt{s} = 630$ GeV	Total Cross Section (nb) $\sqrt{s} = 2$ TeV
$\tilde{w}^{\pm} + \tilde{g}$	5×10^{-3}	5×10^{-2}
$\tilde{z}^0 + \tilde{g}$	3×10^{-3}	2×10^{-2}

It should be remembered that these electroweak gauginos are in general mixed with the Higgsinos. The physical mass eigenstates are linear combinations of the gauginos and associated Higgsinos. This mixing also effects the production cross sections. For example, for some mixing parameters, the total cross section for production of $\tilde{w}^{\pm} + \tilde{g}$ is $\bar{p}p$ collisions at $\sqrt{s} = 2$ TeV is 1.5×10^{-1} approximately three times larger than the unmixed case above[127].

Assuming a light photino, the wino and zino decay into quark-antiquark photino or lepton pair and photino. Since the decays into quark final states have the same charactistics as gluino decay with 1/100 the signal, observation of winos or zinos in their hadronic decays is hopeless. For the leptonic decays, the leptons will be hard to detect as their energies will typically be rather low and the background rates high from heavy quark decays. Therefore it is likely that at least 1000 produced events will be required to observe either the wino or zino.

The discovery limits for gauginos produced in associated production are summarized in Table 15 for present collider energies.

The other mechanism for superpartner production at present colliders is via the decay of real $W^\pm$ and Z^0 bosons. If $m_{\tilde{w}} + m_{\tilde{\gamma}} < m_W$ or $m_{\tilde{w}} + m_{\tilde{z}} < m_W$ or $2m_{\tilde{w}} < m_Z$ winos will be a product of W or Z decays. Ignoring any phase space suppression the branching ratio for $W \to \tilde{w} + \tilde{\gamma}$ is a few percent. At $\sqrt{s} = 2$ TeV a one percent branching ratio corresponds to a total cross section of .22 (nb) or equivalently to 2×10^4 events for an integrated luminosity of $10^{38} cm^{-2}$. Comparing these rates to the discovery limits for the wino given in Table 15 for the associated production mechanism, it is clear that real decays of $W^\pm$ and Z^0 bosons is the main production mechanism for the masses accessible in present generation colliders. The decays of $W^\pm$ and Z^0 bosons are also a possible source of sleptons and sneutrinos if $m_{\tilde{l}} + m_{\tilde{\nu}} < m_W$, $2m_{\tilde{l}} < m_Z$, or $2m_{\tilde{\nu}} < m_Z$.

7.3.1 Supersymmetry at the SSC

At SSC energies the discovery limits for superpartners are greatly extended. For example the total cross section for gluino pair production in pp collisions as a function of the gluino mass is shown in Figure 85 for various supercollider energies. Even with the very conservative assumption that 10,000 produced events are required for detection, the discovery limit is 1.6 TeV/c^2 at $\sqrt{s} = 40$ TeV for integrated luminosity of $10^{40} cm^{-2}$.

At supercollider energies there are additional production mechanisms for

Table 15: Expected discovery limits for superpartners at $S\bar{p}pS$ and Tevatron Colliders, based on associated production of scalar quarks and gauginos. All superpartner masses are set equal.

	Mass limit (Gev/c^2)					
Superpartner	$\sqrt{s} = 630$ GeV			$\sqrt{s} = 2$ TeV		
	$\int dt \mathcal{L}$ (cm)$^{-2}$					
	10^{36}	10^{37}	10^{38}	10^{36}	10^{37}	10^{38}
Gluino or squark (1000 events)	45	60	75	85	130	165
Photino (100 events)	35	60	90	45	90	160
Zino (1000 events)	17	30	50	22	50	95
Wino (1000 events)	20	35	55	32	60	110

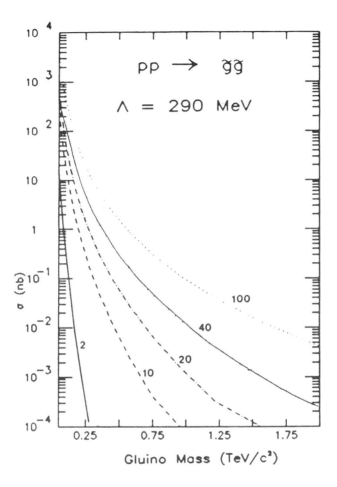

Figure 85: Cross sections for the reaction $pp \to \tilde{g}\tilde{g}$ + anything as a function of the gluino mass, according to the parton distributions of Set 2. Rates shown for collider energies $\sqrt{s} = 2$, 10, 20, 40, and 100 TeV. Both gluinos are restricted to the interval $|y_i| < 1.5$. The squark mass is set equal to the gluino mass. (From EHLQ)

superpartners including:

- Pair production of the electroweak gauginos from quark-antiquark initial states.

- Production of electroweak gauginos, sleptons, sneutrinos, and even Higgsinos via the generalized Drell-Yan mechanism (i.e. virtual $W^\pm$, Z^0, and γ).

The details about the production and detection of superpartners at SSC energies may be found in EHLQ and Ref.153. The discovery limits for all the superpartners at $\sqrt{s} = 40$ TeV are summarized in Table 16.

If supersymmetry plays a role in resolving the naturalness problem of the standard model, the scale of supersymmetry breaking can not be much higher than the electroweak scale; and therefore the masses of the superpartners should also be in this mass range. It is clear from Table 16 that in this case superpartners will be discovered at or below SSC energies.

Table 16: Expected discovery limits for superpartners at the SSC for various integrated luminosities. Associated production of gauginos and squarks is assumed. All superpartner masses are set equal.

Superpartner	p p collisions $\sqrt{s}=40$ TeV Mass limit (Gev/c^2) $\int dt \mathcal{L}$ (cm)$^{-2}$		
	10^{38}	10^{39}	10^{40}
Gluino (1000 events)	900	1,600	2,500
Squark (up and down) (1000 events)	800	1,450	2,300
Photino (100 events)	350	750	1,350
Zino (1000 events)	250	500	825
Wino (1000 events)	300	550	1,000
pair production			
Top squark (1000 events)	500	850	1,350
Slepton (100 events)	100	200	400

8 CONCLUDING REMARK

Hadron-hadron colliders will be one of the main testing grounds for both the standard model and possible new physics. Specific applications have been detailed in these seven lectures. However I would like to conclude these lectures with a general remark. The advances of the last decade have brought us to a deep understanding of the fundamental constituents of matter and their interactions. Progress toward a fuller synthesis will require both theoretical and experimental breakthroughs. The present generation of hadron (and also lepton) colliders are bound to provide much additional information. But the full exploration of the physics of the TeV scale will require the next generation of hadron colliders - the supercollider - as well.

ACKNOWLEDGMENTS

It is a pleasure to thank the organizers Tom Appelquist, Mark Bowick, and Feza Gursey for their hospitality. I would also like to thank my scientific secretaries David Pfeffer, David Lancaster, and Chris Burges who were of assistance in the preparation of the written version of these lectures. Particular credit should go to David Lancaster since I relied heavily on his draft of lecture 4 in my final version. I would also like to thank my collaborators on EHLQ and DEQ: Sally Dawson, Ian Hinchliffe, Ken Lane, and Chris Quigg, on whose hard work much of the material in these lectures is based.

REFERENCES AND FOOTNOTES

1. E. Eichten, I. Hinchliffe, K. Lane, and C. Quigg, Rev. Mod. Phys. **56**, 579 (1984).

2. See for example: L.B. Okun, *Leptons and Quarks* (North Holland, Amsterdam, 1981); D.H. Perkins, *Introduction to High Energy Physics* , 2nd ed. (Addison-Wesley, Reading, Massachusetts, 1982); or C. Quigg, *Gauge Theories of the Strong, Weak, and Electromagnetic Interactions* (Benjamin/Cummings, Reading, Massachusetts, 1983).

3. N. Minard, in *International Symposium on Physics of Proton -Antiproton Collisions*, University of Tsukuba, KEK, March 13-15 (1985).

4. G. 't Hooft, in *Recent Developments in Gauge Theories, Proceedings of the 1979 NATO Advanced Study Institute*, Cargese, edited by G. 't Hooft et al. (Plenum, New York), p.101.

5. M. Kobayashi and T. Maskawa, Prog. Theor. Phys. **49K**, 652, (1973).

6. For a general reference to the physics potential of the present and future colliders see: *Proceedings of the 1982 DPF Summer Study on Elementary Particle Physics and Future Facilities*, edited by R. Donaldson, R. Gustafson, and F. Paige, (Fermilab, Batavia, Illinois, 1982).

7. For a review of the physics potential of LEP see, for example: CERN 76-16, *Physics with 100 GeV Colliding Beams*, CERN (1976); CERN 79-01, *Proceedings of the LEP Summer Study*, CERN (1979). For a review of the physics potential of the SLC at SLAC see, for example: *SLAC Linear Collider Conceptual Design Report*, SLAC Report-229, (1980).

8. For a review of the physics potential of HERA see: *Proceedings of the Discussion Meeting on HERA Experiments*, Genoa, Italy, October 1-3 1984. DESY-HERA preprint 85/01 (1985).

9. In July 1983, the New Facilities Subpanel of the High Energy Physics Advisory Panel (HEPAP) recommended to HEPAP that a 20 TeV proton-proton collider be the next high energy facility (to be completed in the 1990's). That recomemdation was accepted by HEPAP and forwarded to the DOE. Many details of the proposed SSC are contained in: *Proceedings of the 1984 DPF Summer Study on the Design and Utilization of the Superconducting Super Collider*, edited by R. Donaldson, and J. Morfin (Fermilab, Batavia, Illinois, 1984), p.i .

10. Proceedings of the CERN/ICFA Workshop:*Large Hadron Collider in the LEP Tunnel*, CERN 84-10 (1984).

11. For a general discussion of QCD see any of the textbooks in Ref. 2.

12. D. J. Gross and F. Wilczek Phys. Rev. Lett. **30**, 1343 (1973) and Phys. Rev. D**8**, 3633 (1973); H. D. Politzer Phys. Rev. Lett. **30**, 1346 (1973).

13. For a sampling of the models see: F. Paige and S. Protopopescu, Brookhaven Report BNL-29777 (1980); Andersson et. al.,Phys. Rep.**97**, 31 (1983) ; R. D. Field and S. Wolfram, Nucl. Phys. **B213**, 65 (1983) ; T. Gottschalk, Nucl. Phys. **B239**, 349 (1984); and G. Marchesini and B. R. Webber, Nucl. Phys. **B238**, 1 (1984) .

14. This was first shown by : G. Sterman and S. Weinberg, Phys. Rev. Lett. **39**, 1436 (1977) . For a general review of the status of jets in QCD see S. Ellis lectures in the *Proceedings of the 11th SLAC Summer Institute*

on Particle Physics, Edited by P. McDonough, SLAC-Report No. 267 (Stanford, California, 1984).

15. S. Orito (JADE Collaboration) presentation at *Proceedings of the 1979 International Symposium on Lepton and Photon Interactions At Very High Energies*, edited by T. B. W.Kirk and H. D. I. Abarbanel, p.52, (Fermilab, Batavia, Illinois, 1979).

16. G. Hanson, et.al., Phys. Rev. Lett. **35**, 1609 (1975) .

17. JADE Collaboration event published in: P. Soding and G. Wolf, Ann. Rev. Nucl. Part. Sci. **31**, 231 (1981).

18. See for example: H. J. Behrend et. al., (CELLO Collaboration), DESYPreprint, DESY 82-061 (1982). Also see Ref. 19.

19. J. Dorfan, in *Proceedings of the 1983 International Symposium on Lepton and Photon Interactions At High Energies*, edited by D. Cassel and D. Kreinick, p.686, (Cornell University, Ithaca, New York, 1983).

20. M. Block and R. Cahn Rev. Mod. Phys. **57**, 563 (1985) .

21. D. Froldevaux (UA2 Collaboration), presented at *International Symposium on Physics of Proton -Antiproton Collisions*, University of Tsukuba, KEK, March 13-15 (1985).

22. G. Altarelli and G. Parisi, Nucl. Phys. **B126**, 298 (1977) .

23. H. Abramowicz, et. al. (CDHS Collaboration), Z. Phys. **C13**, 199 (1982); and Z. Phys. **C17**, 283 (1983) .

24. See EHLQ (Ref. 1) p.588 for details of the experimental measurements and theoretical expectations.

25. F. Eisele in the *Proceedings of the 21th International Conference on High Energy Phyiscs*, edited by P. Petiau and M. Porneuf, J. Phys. (Paris) **43**, Supp. 12, C3-337 (1982).

26. F. Bergsma, et.al. (CHARM Collaboration), Phys. Lett. **123**, 269 (1983)

27. D. B. MacFarlane, et. al. (CCFRR Collaboration), Fermilab-Pub-83/108-EXP (1983).

28. The situation is summarized by F. Sciulli in *1985 International Symposium on Lepton and Photon Interactions At High Energies*, Kyoto, Japan, August 19-24 (1985)..

29. A. Mueller, these proceedings.

30. M. Gluck, E. Hoffmann, and E. Reya, Z. Phys. **C13**, 119 (1982) .

31. A simple derivation of these results is given, for example, in: C. Quigg, *Gauge Theories of the Strong, Weak, and Electromagnetic Interactions* (Benjamin/Cummings, Reading, Massachusetts, 1983), p.241.

32. S. Dawson in *Proceedings of the 1984 DPF Summer Study on the Design and Utilization of the Superconducting Super Collider*, edited by R. Donaldson, and J. Morfin (Fermilab, Batavia, Illinois, 1984), p. 96

33. For a discussion of the small x behaviour see A. Mueller's lectures in these proceedings. The $ln(1-x)$ behaviour can actually be resummed but in any case is no problem because the distributions are neglilsmall near x=1 for large Q^2 as can be seen from Figs 8, 9, 11, and 12.

34. The elementary cross sections have been calculated by manny authors, and Phys. Rev. D**18**, 1501 (1978).

35. G. Arnison et.al.(UA1 Collaboration), Phys. Lett. **123**, 115 (1983).

36. G. Arnison et.al.(UA1 Collaboration), CERN Preprint, CERN-EP/83-118 (1983).

37. P. Bagnaia et.al. (UA2 Collaboration), Phys. Lett. **129**, 130 (1983).

38. P. Bagnaia et.al. (UA2 Collaboration), CERN Preprint, CERN-EP/84-12 (1984).

39. J. A. Appel et.al. (UA2 Collaboration) Phys. Lett. **160**, 349 (1985).

40. W. Scott (UA1 Collaboration), presented at *International Symposium on Physics of Proton -Antiproton Collisions*, University of Tsukuba, KEK, March 13-15 (1985).

41. F. Pastore (UA2 Collaboration), presented at *Topical Conference on Proton-Antiproton Collider Physics*, Saint-Vincent, Aosta Valley, Italy Feb. 25-Mar.1 (1985).

42. These calculations have now been completed by: S. J. Parke and T. Taylor, Phys. Lett. **157**, 81 (1985) and Fermilab Preprint PUB-85/118-T (85); J. Gunion and Z. Kunszt, Phys. Lett. **159**, 167 (1985); and Z. Kunszt, CERN Preprint TH-4319 (1985).

43. F. A. Berends, et al., Phys. Lett. **103**, 124 (1981).

44. R. K. Ellis and J. Owens, *Proceedings of the 1984 DPF Summer Study on the Design and Utilization of the Superconducting Super Collider*, edited by

R. Donaldson, and J. Morfin (Fermilab, Batavia, Illinois, 1984), p. 207.

45. Z. Kunszt et.al., Phys. Rev. D**21**, 733 (1980) .

46. F. Halzen and P. Hoyer, Phys. Lett. **154**, 324 (1985).

47. For methods of separating light quarks from gluons see: T. Sjostrand, *Proceedings of the 1984 DPF Summer Study on the Design and Utilization of the Superconducting Super Collider*, edited by R. Donaldson, and J. Morfin (Fermilab, Batavia, Illinois, 1984), p.84.

48. For methods of identifying heavy quarks see: K. Lane, *Proceedings of the 1984 DPF Summer Study on the Design and Utilization of the Superconducting Super Collider*, edited by R. Donaldson, and J. Morfin (Fermilab, Batavia, Illinois, 1984), p.729.

49. S. Coleman, "Secret Symmetry", in *Laws of Hadronic Matter*, 1973 Erice School, Academic Press, New York (1975), p. 139.

50. G. Arnison et al. (UA1 Collaboration), CERN Preprint CERN-EP/85-185 (1985).

51. J. A. Appel, et al. (UA2 Collaboration), CERN Preprint CERN-EP/85-166 (1985).

52. For a full discussion of how the W signal is extracted experimental see Ref. 51 and the references contained therein.

53. W, Marciano and A. Sirlin, Phys. Rev. D**29**, 945 (1984) .

54. G. Altarelli, R. K. Ellis, M. Greco, and G. Martinelli, Nucl. Phys. **B246**, 12 (1984) .

55. G. Altarelli, R. K. Ellis, and G. Martinelli, Z. Phys. **C27**, 617 (1985).

56. The feasiblity of studying rare $W^\pm$ decays at the SSC has not yet been studied in detail. Some of the issues of $W^\pm$ identification are presented in: I. Hinchliffe, *Proceedings of the 1984 DPF Summer Study on the Design and Utilization of the Superconducting Super Collider*, edited by R. Donaldson, and J. Morfin (Fermilab, Batavia, Illinois, 1984), p.1 and the references cited therein.

57. A few of the monojet events seen by UA1 are consistent with the Z^0 plus jet with the Z^0 decaying into $\bar{\nu}\nu$. This can be determined from a measurement of the same process but where the Z^0 decays into e^+e^-. See Ref. 50. For the measurement of jet plus $Z^0 \to e^+e^-$ by the UA2 Collaboration see Ref. 51.

58. For a calculation of all the pair cross sections including a possible anomolous magnetic moment for the $W^\pm$ see: R. W. Brown and K. O. Mikaelian, Phys. Rev. **D19**, 922 (1979).

59. S. J. Brodsky and R. W. Brown, Phys. Rev. Lett. **49**, 966 (1982).

60. The requirement of anomaly cancellation in a renormalizable gauge theory was emphasized by: D. Gross and R. Jackiw, Phys. Rev. **D6**, 477 (1972).

61. S. Drell and T. M. Yan, Phys. Rev. Lett. **25**, 316 (1970); and Ann. Phys. **66**, 578, 1971.

62. J. C. Pati and A. Salam, Phys. Rev. **D10**, 275 (1974); R. N. Mohapatra and J. C. Pati, Phys. Rev. **D11**, 566 (1975); R. N. Mohapatra and G. Senjanovic, Phys. Rev. **D23**, 165 (1981).

63. For a general review of unified models: H. Georgi and S. Weinberg, Phys. Rev. D**17**, 275 (1978). Prospects for detecting additional Z^0's from E_6 models see, for example: J. L. Rosner, Univ. of Chicago preprint, EFI85-34, 1985.

64. A. D. Linde, Pis'ma Zh. Eksp. Teor. Fiz. **23**, 73 (1976). S. Weinberg, Phys. Rev. Lett. **36**, 294 (1973).

65. Spontaneous breaking of a continuous symmetry in the context of a scalar doublet model was first discussed by: J. Goldstone, Nuovo Cim. **19**, 154 (1961).

66. E. Gildener and S. Weinberg, Phys. Rev. D**13**, 3333 (1976). A general analysis of the effect on symmetry breaking of the one loop corrections to the effective potential was given by: S. Coleman and E. Weinberg, Phys. Rev. D**7**, 1888 (1973); also, see: S. Weinberg Phys. Rev. D**7**, 2887 (1973).

67. The case where $\mu^2 = 0$ was also considered in the first two papers in Ref. 66.

68. The presentation I give here follows closely the analysis given by: B. W. Lee, C. Quigg, and H. B. Thacker, Phys. Rev. D**16**, 1519 (1977).

69. The proof is given in the Appendix of Ref.68.

70. J. M. Cornwall, D. N. Levin, and G. Tiktopoulos, Phys. Rev. D**10**, 1145 (1974).

71. G. 't Hooft, Nucl. Phys. **B35**, 167 (1971).

72. M. Chanowitz and M. K. Gaillard, Nucl. Phys. **B261**, 379 (1985).

73. The definition of the ρ parameter is given in Eq. 3.17.

74. M. Chanowitz, M. A. Furman, and I. Hinchliffe, Nucl. Phys. **B153**, 402 (1979).

75. M. Veltman, Nucl. Phys. **B123**, 89 (1979). An upper bound on the Higgs boson mass can also be obtained by similar considerations: M. Veltman, Acta. Phys. Polon. **8**, 475 (1977).

76. W. J. Marciano and A. Sirlin, Phys. Rev. **D29**, 75 (1984).

77. H. Georgi, S. L. Glashow, M. E. Machacek, and D. V. Nanopoulos, Phys. Rev. Lett. **40**, 692 (1978).

78. J. F. Gunion, P. Kalyniak, M. Soldate, and P. Galison, SLAC-PUB-3604 (1985); J. F. Gunion, Z. Kunszt, and M. Soldate, SLAC -PUB -3709 (1985).

79. This mechanism was proposed by: R. N. Cahn and S. Dawson, Phys. Lett. **136**, 196 (1984).

80. G. 'tHooft, in *Recent Developments in Gauge Theories, Proceedings of the 1979 NATO Advanced Study Institute, Cargese*, edited by G. 't Hooft *et al.* (Plenum, New York), p. 135

81. One other (weaker) definition of naturalness is that the form of the Lagrangian be preserved by radiative corrections. A zero (or very small) mass parameter in the Higgs potential is not natural in this sense as was shown in the second paper of Ref. 66.

82. See, for example: S. Coleman, in *1971 International School of Subnuclear Physics*, Erice, Italy, edited by A. Zichichi (Editrice Compositori, Bologna, 1973).

83. M. K. Gaillard (in these proceedings).

84. S. Weinberg, Phys. Rev. **D19**, 1277 (1979).

85. L. Susskind, Phys. Rev. **D20**, 2619 (1979).

86. It has been shown that the global symmetry breaking must have this pattern if the gauge interaction is vectorlike (i.e. like QCD) and confining: C. Vafa and E. Witten, Nucl. Phys. **B234**, 173 (1984).

87. For a general discuusion of large N arguments see: E. Witten, Anns. Phys. **128**, 363 (1980). Also see Ref. 100.

88. This signature for technicolor was proposed in the original Kxpaper of Susskind (Ref. 85).

89. Some recent studies of these backgrounds to W pair production are contained in Ref. 78 as well as *Proceedings of the 1984 DPF Summer Study on the Design and Utilization of the Superconducting Super Collider*, edited by R. Donaldson, and J. Morfin (Fermilab, Batavia, Illinois, 1984).

90. E. Eichten and K. D. Lane, Phys. Lett. **90**, 125 (1980).

91. S. Dimopoulos and L. Susskind, Nucl. Phys. **B155**, 237 (1979).

92. E. Farhi and L. Susskind, Phys. Rev. **D20**, 3404 (1979).

93. S. L. Glashow, J. Iliopoulos, and L. Maiani, Phys. Rev. **D2**, 1285 (1980).

94. S. Dimopoulos and J. Ellis, Nucl. Phys. **B182**, 505 (1981).

95. S. C. Chao and K. D. Lane, Phys. Lett. **159**, 135 (1985).

96. A. M. Hadeed and B. Holdom, Phys. Lett. **159**, 379 (1985); S. Dimopoulos and H. Georgi, Phys. Lett. **140**, 148 (1985); Also Ref.95.

97. S. Dimopoulos, Nucl. Phys. **B166**, 69 (1980).

98. M. Peskin, Nucl. Phys. **B175**, 197 (1980).

99. J. Preskill, Nucl. Phys. **B177**, 21 (1981).

100. S. Dimopoulos, S. Raby, and G. L. Kane, Nucl. Phys. **B182**, 77 (1981).

101. R. Dashen, Phys. Rev., **183**, 1245 (1969).

102. S. Chadha and M. Peskin, Nucl. Phys. **B185**, 61 (1981); **B187**, 541 (1981).

103. V. Baluni, Phys. Rev. **D28**, 2223 (1983).

104. This crude estimate follows from the current algebra relation $F_\pi^2 M(P^0)^2 = m_q < Q\overline{Q} >$, with the identifications $< Q\overline{Q} > = F_\pi^3 \approx (250 GeV)^3$ and $m_q = (m_u + m_d)/2 \approx 7 GeV/c^2$. For an explicit model in which this behaviour is realized, see Ref. 100.

105. K. D. Lane, in *Proceedings of the 1982 DPF Summer Study on Elementary Particle Physics and Future Facilities*, edited by R. Donaldson, R. Gustafson, and F. Paige, (Fermilab, Batavia, Illinois, 1982).

106. For a review of experimental comstraints see: S. Komamiya, *1985 International Symposium on Lepton and Photon Interactions At High Energies*, Kyoto, Japan, August 19-24 (1985).

107. E. Eichten, I. Hinchliffe, K. D. Lane, and C. Quigg, Fermilab Pub-85/145-T (1985).

108. 't Hooft, in Lecture 3 of Ref.80.

109. H. Georgi and S. L. Glashow, Phys. Rev. Lett. **32**, 438 (1974).

110. S. Weinberg and E. Witten, Phys. Lett. **96**, 491 (1981).

111. S. Coleman and B. Grossman, Nucl. Phys. **B203**, 205 (1982).

112. The dynamics of chiral gauge theories has been studied in the large N limit by: E. Eichten, R. Peccei, J. Preskill, and D. Zeppenfeld, Nucl. Phys. **B268**, 161 (1986).

113. E. Eichten, K. D. Lane, and M. E. Peskin, Phys. Rev. Lett. **50**, 811 (1983).

114. G. B. Chadwick, MAC Collaboration, *Proceedings of the Europhysics Study Conference on Electroweak Effects at High Energies*, Erice, Italy (1983).

115. Ch. Berger et al., PLUTO Collaboration, Z. Phys. **C27**, 341 (1985).

116. MARK-J Collaboration, reported in S. Komamiya, *1985 International Symposium on Lepton and Photon Interactions At High Energies*, Kyoto, Japan, August 19-24 (1985)..

117. W. Bartel et al., JADE Collaboration, Z. Phys. **C19**, 197 (1983).

118. M. Althoff et al., TASSO Collaboration, Z. Phys. **C22**, 13 (1984).

119. M. Derrick, HRS Collaboration, ANL-HEP-CP-85-44 (1985).

120. M. Abolins et al., *Proceedings of the 1982 DPF Summer Study on Elementary Particle Physics and Future Facilities*, edited by R. Donaldson, R. Gustafson, and F. Paige, (Fermilab, Batavia, Illinois, 1982). Also see Ref 113.

121. I. Bars and I. Hinchliffe, Phys. Rev. D**33**, 704 (1986).

122. The only graded Lie Algebras consistent with relativistic quantum field theory are the supersymmetric algebras. This was proven by: R. Haag, J. Lopuszanski, and M. Sohnius, Nucl. Phys. **B88**, 257 (1975).

123. J. Wess and J. Bagger, *Supersymmetry and Supergravity*, Princeton University Press, Princeton, New Jersey (1983).

124. P. Fayet, Nucl. Phys. **B90**, 104 (1975); A. Salam and J. Strathdee, Nucl. Phys. **B87**, 85 (1975).

125. The possiblities for a residual R symmetry are analysed by: G. Farrar and S. Weinberg, Phys. Rev. D**27**, 2732 (1983).

126. L. Hall, J. Lykken, and S. Weinberg, Phys. Rev. D**27**, 2359 (1983); L. Alvarez-Gaume, J. Polchinski, and M. Wise, Nucl. Phys. **B221**, 495 (1983).

127. S. Dawson, E. Eichten, and C. Quigg, Phys. Rev. D**31**, 1581 (1985).

128. The mixing is treated in general for the minimal supersymmetric model in Ref. 127. Also see Ref. 130.

129. L. Baulieu, J. Kaplan, and P. Fayet, Phys. Lett. **141**, 198 (1984).

130. H. Haber and G. Kane, Phys. Rep. **117**, 75 (1985).

131. K. Freese and D. Schramm, Nucl. Phys. **B233**, 167 (1984).

132. H. Goldberg, Phys. Rev. Lett. **50**, 1419 (1983).

133. N. Cabibbo, G. Farrar, and L. Maiani, Phys. Lett. **105**, 155 (1981).

134. G. Snow (private communication).

135. P. Fayet, Phys. Lett. **84**, 421 (1979).

136. M. Fukugita and N. Sakai, Phys. Lett. **114**, 23 (1982); A. Bouquet and C. E. Vayonakis, Phys. Lett. **116**, 219 (1982); and J. Ellis and K. Olive, Nucl. Phys. **B223**, 252 (1983).

137. H. Behrend et al., CELLO Collaboration, Phys. Lett. **123**, 127 (1983).

138. B. Alper et al., Phys. Lett. **46**, 265 (1973); and D. Cutts et al., Phys. Rev. Lett. **41**, 363 (1978).

139. R. Gustafson et al., Phys. Rev. Lett. **37**, 474 (1976).

140. R. C. Ball et al., Phys. Rev. Lett. **53**, 1314 (1984).

141. F. Bergsma et al., CHARM Collaboration, Phys. Lett. **121**, 429 (1983).

142. E. Fernandez et al., MAC Collaboration, Phys. Rev. Lett. **54**, 1118 (1985).

143. W. Bartel el al., Z. Phys. **C6**, 295 (1980).

144. C. Nappi, Phys. Rev. **D25**, 84 (1982).

145. W. Bartel et al., JADE Collaboration, DESY-Preprint 84-112 (1984).

146. B. Adeva et al., Mark-J Collaboration, Phys. Rev. Lett. **53**, 1806 (1984).

147. W. Bartel et al., JADE Collaboration , Phys. Lett. **146**, 126 (1984).

148. G. Bartha et al., ASP Collaboration, Phys. Rev. Lett. **56**, 685 (1986).

149. See Ref. 127 for details on these limits.

150. S. Dawson, Nucl. Phys. **B261**, 297 (1985).

151. J. Ellis and H. Kowalski, Nucl. Phys. **B259**, 109 (1985).

152. R. M. Barnett, H. E. Haber, and G. Kane, Berkeley preprint, LBL-20102 (1986).

153. S. Dawson and A. Savoy-Navarro, *Proceedings of the 1984 DPF Summer Study on the Design and Utilization of the Superconducting Super Collider*, edited by R. Donaldson, and J. Morfin (Fermilab, Batavia, Illinois, 1984), p.263.

PARTICIPANTS OF THE THEORETICAL ADVANCED STUDY INSTITUTE

Yale University
June 9 - July 5, 1985

Accetta, Frank S. - Astsronomy and Astrophysics Center - Chicago
Atick, Joseph J. - Stanford University
Bennett, David P. - Stanford University
Bernstein, Marc D. - University of Washington
Burges, Christopher J. - Massachusetts Institute of Technology
Carter, Mark R. - Stanford University
Cobb, Wesley K. - Brandeis University
Cohn, Joanne D. - Enrico Fermi Institute
Crnkovic, Cedomir - Princeton University
Dannenberg, Alexander H. - Harvard University
Deng, Yao-Bing - University of Connecticut
Düsedau, Dieter W. - Massachusetts Institute of Technology
Font, Anamaria - University of Texas
Gaffney, John B. - Columbia University
Geiger, Angela O. - Massachusetts Institute of Technology
Goff, William E. - University of Illinois
Gozzi, Ennio - Max-Planck Institute
Griest, Kim E. - University of California - Santa Cruz
Hadeed, Anthony M. - University of Toronto
Hlousek, Zvonimir Z. - Brown University
Ingermanson, Randall S. - Lawrence Berkeley Laboratory
Jo, Sang G. - Massachusetts Institute of Technology
Jolicoeur, Thierry T.J. - CEN-Saclay

Kao, Yeong-Chuan - Lawrence Berkeley Laboratory
Kastor, David A. - University of Chicago
Keil, Werner H. - University of British Columbia
Klebanov, Igor - Princeton University
Korpa, Caba L. - Massachusetts Institute of Technology
Kumar, Alok - University of Maryland
Lancaster, David J. - University of Sussex - U.K.
Lee, Kimyeong - Columbia University
Liang, Yigao - Virginia Polytechnic Institute and State University
Lin, Jane Mei-Jech - National Taiwan University
Lin, Yeu-Chung - University of Massachusetts
Liu, Ying - Carnegie-Mellon University
Lizzi, Fedele - Syracuse University
London, David M. - Enrico Fermi Institute
MacKenzie, Richard B. - University of California - Santa Barbara
Mahajan, Shobhit - Lawrence Berkeley Laboratory
Mueller-Hoissen, Folkert - Max Planck Institute
Myers, Eric - Brookhaven National Laboratory
Naculich, Stephen G. - Princeton University
Olness, Fredrick I. - University of Wisconsin
Park, Q-Han - Brandeis University
Periwal, Vipul - Princeton University
Pfeffer, David M. - University of Florida
Phelps, Richard A. - University of Michigan
Quevedo, Fernando - University of Texas
Randall, Lisa J. - Harvard University
Rhie, Sun Hong - Stanford University
Roberge, Andre - University of British Columbia
Rodgers, Vincent G. - Syracuse University
Rosenfeld, Rogerio - Universidade de Sao Paulo
Ryzak, Zbigniew - Massachusetts Institute of Technology
Sadun, Lorenzo A. - University of California - Berkeley
Sienko, Tanya C. - University of Illinois
Sloan, John H. - Princeton University

Velasco, Eduardo S. - State University of New York at Stony Brook
Ware, James D. - Northeastern University
Wu, Ke - The City College of New York
Zoller, David J. - Enrico Fermi Institute